AF769901

Nonlinear Filters
2nd Edition

Springer
Berlin
Heidelberg
New York
Barcelona
Budapest
Hong Kong
London
Milan
Paris
Santa Clara
Singapore
Tokyo

Hisashi Tanizaki

Nonlinear Filters

Estimation and Applications

Second Revised and Enlarged Edition

With 18 Figures
and 45 Tables

Springer

WITHDRAWN

J. ROBERT VAN PELT LIBRARY
MICHIGAN TECHNOLOGICAL UNIVERSITY
HOUGHTON, MICHIGAN

Associate Professor
Hisashi Tanizaki
Kobe University
Faculty of Economics
Rokkodai, Nadaku
Kobe 657, Japan

Cataloging-in-Publication Data applied for

Die Deutsche Bibliothek - CIP-Einheitsaufnahme

Tanizaki, Hisashi:
Nonlinear filters : estimation and applications / Hisashi Tanizaki. - 2, rev. and enl. ed. - Berlin ; Heidelberg ; New York ; Barcelona ; Budapest ; Hong Kong ; London ; Milan ; Paris ; Santa Clara ; Singapore ; Tokyo : Springer, 1996
ISBN 3-540-61326-9

ISBN 3-540-61326-9 Springer-Verlag Berlin Heidelberg New York

ISBN 3-540-56772-0 Springer-Verlag New York Heidelberg Berlin

This work is subject to copyright. All rights are reserved, whether the whole or part of the material is concerned, specifically the rights of translation, reprinting, reuse of illustrations, recitation, broadcasting, reproduction on microfilms or in other ways, and storage in data banks. Duplication of this publication or parts thereof is only permitted under the provisions of the German Copyright Law of September 9, 1965, in its version of June 24, 1985, and a copyright fee must always be paid. Violations fall under the prosecution act of the German Copyright Law.

© Springer-Verlag Berlin · Heidelberg 1993, 1996
Printed in Germany

The use of registered names, trademarks, etc. in this publication does not imply, even in the absence of a specific statement, that such names are exempt from the relevant protective laws and regulations and therefore free for general use.

SPIN 10540183 42/2202-5 4 3 2 1 0 - Printed on acid-free paper

QA
297
.T36
1996

To My Wife Miyuki

SEP 02 1998

Preface

Preface to Second Edition

This book is a revision of *Nonlinear Filters: Estimation and Applications*, (Lecture Notes in Economics and Mathematical Systems, No.400), which was published from Springer-Verlag in 1993.

Compared with the first edition, I have made a substantial revision in the second edition.

First, titles in the following chapters, sections, terms and so on are changed as follows.

The First Edition		The Second Edition
Chapter 3 Nonlinear Filters based on Taylor Series Expansion	$\Longrightarrow$	Chapter 3 Traditional Nonlinear Filters
Chapter 4 Nonlinear Filters based on Density Approximation	$\Longrightarrow$	Chapter 4 Density-Based Nonlinear Filters
Section 4.3 Numerical Density Approximation by Piecewise Linear Functions: Modified Kitagawa Estimator	$\Longrightarrow$	Section 4.3 Numerical Integration Filter
Section 4.4 Simulation-based Density Estimator	$\Longrightarrow$	Section 4.4 Importance Sampling Filter
Chapter 5 Comparison of Nonlinear Filters: Monte-Carlo Experiments	$\Longrightarrow$	Chapter 5 Monte-Carlo Experiments
Chapter 6 An Application of Nonlinear Filters: Estimation of Permanent Consumption	$\Longrightarrow$	Chapter 6 Application of Nonlinear Filters
Monograph	$\Longrightarrow$	Book

Thus, the most appropriate title is taken or the title is made short.

Second, new contents are briefly summarized as follows.

- Section 2.3.3 Minimum Mean Square Linear Estimator
  As the third derivation method of the Kalman filter, I put this section. This section is not utilized for the proceeding chapters but it is added as a survey of the standard Kalman filter.
- Appendix A2.2 Conditional Normal Distribution
  The derivation method under normality assumption is discussed in Section 2.3.1. Lemmas and Proofs used in Section 2.3.1 are summarized in this appendix.
- Section 4.5 Density-based Monte-Carlo Filter
  This nonlinear filter is one of the recent new nonlinear filters.
- Section 4.6 Rejection Sampling Filter
  The rejection sampling filter is also one of the recent topics, where a recursive algorithm of random draws from filtering densities is derived.
- Appendix A4.5 Density-Based Monte-Carlo Filter
  For the filtering estimates by the density-based Monte-Carlo filter, the asymptotic properties are discussed.
- Appendix A4.6 Rejection Sampling
  Random number generation by rejection sampling is discussed in a general form.
- Appendix A5.1 On Initial Value of State-Variable
  In the Kalman filter algorithm, the filtering estimates are recursively obtained given the initial value. We analyze how the filtering estimates are sensitive to the initial value.
- Appendix A5.3 On Random Draws by Importance Sampling
  Monte-Carlo integration with importance sampling is utilized to the nonlinear filter in Section 4.4, where random draws have to be generated from the importance density. We discuss about the random draws generated from the importance density.
- Appendix A5.4 Rejection Sampling
  Using rejection sampling, random draws are generated by a computer. Precision of random draws are examined.
- Chapter 7 Prediction and Smoothing
  The density-based filtering algorithms discussed in Chapter 4 are extended to prediction and smoothing.

Moreover, the following chapters are substantially revised.

- Chapter 5 Monte-Carlo Experiments
  I changed a functional form of the nonlinear measurement and transition equations in some simulation studies.

- Chapter 6 Application of Nonlinear Filters
  In the first edition, only the U.S. data are used. In the second edition, the same type of state-space model is estimated for Japan, U.S., U.K., France, Spain, Italy, Canada and Germany. I focus on estimation of the unknown parameters and estimation of a ratio of per capita permanent consumption relative to per capita total consumption for the above countries.

Thus, the second edition is substantially changed compared with the first edition.

Finally, I am grateful to the editor Dr. Werner A. Mueller, who gave me a chance to revise the monograph.

March, 1996

Author: Hisashi Tanizaki
Associate Professor
Faculty of Economics, Kobe University
Rokkodai, Nadaku, Kobe 657, Japan

E-mail: tanizaki@kobe-u.ac.jp

Preface to First Edition

Acknowledgements

Originally, this monograph is a revision of my Ph.D. dissertation (" Nonlinear Filters: Estimation and Applications, " December, 1991) at the University of Pennsylvania. In writing the dissertation, many people took care of me in different ways.

In particular, Professor Roberto S. Mariano was very patient as my main thesis advisor, and he gave me many useful suggestions and comments. I wish to thank him for recommending me this line of research and giving me the suggestions and comments. Also, I would like to acknowledge the support of NSF grant SES 9011917 in Professor Roberto S. Mariano's supervision of the thesis.

Professor Marc Nerlove supported me financially during my staying in the U.S. in spite of my trouble with English. I would like to acknowledge the support of NSF grant SES 5-25056. Professor Francis X. Diebold has stimulated me through his studies. Although the dissertation was not directly related to their research, their work significantly affected me and will influence my future research. I am grateful to them for that.

Moreover, I do not forget to thank the classmates in the Graduate Program in Economics who had been encouraging me since I started studying at the University. Especially, I got many " invisible things " from Michael Blackman, Celia Chen, Ilaria Fornari and Til Schuermann (in alphabetical order) and asked them to correct my English.

Also, I am grateful to Jun Sato and Shinichi Suda (the same year Japanese Ph.D. students in Economics) and Suminori Tokunaga (Japanese Ph.D. student in Regional Science) for helping me. I could not continue to study there without them. When I was depressed just after coming to the U.S., they encouraged me.

Furthermore, I wish to thank Professors Mitsuo Saito, Kazuo Ogawa, Toshihisa Toyoda and Kazuhiro Ohtani. When I was in Japan before going to the U.S., I learned a lot of things under them at Kobe University. Especially, Professor Mistuo Saito, who was my main advisor in Japan, recommended that I continue to study econometrics at the University of Pennsylvania. And I am grateful to Shinichi Kitasaka, who was one of my classmates at Kobe University, for his suggestions and comments. Moreover, I have been stimulated by Shigeyuki Hamori's passion for research. He was one of my classmates at Kobe University, and he studied economics at Duke University at the same time that I was at the University of Pennsylvania. I am grateful to him for his valuable suggestions and advice.

Also, I wish to say " Thanks " to my parents. They have not understood what I have done, but they supported me with regard to everything. Thus, I believe that I could finish the dissertation thanks to all the professors and all the friends who know me.

Finally, I am grateful to Professor Marc Nerlove, again, who suggested me to revise the dissertation for possible publication, and Professor Wilhem Krelle, who was the referee of Springer-Varlag and gave me valuable suggestions and comments.

March, 1993

Preface

The purpose of this monograph is to develop nonlinear filters and demonstrate their applications. There are two approaches to nonlinear filters. One is approximating nonlinear measurement and transition equations by the Taylor series expansion and applying the approximated nonlinear functions directly to the standard linear recursive Kalman filter algorithm. Another is approximating the underlying density functions of the state vector by a Gaussian sum, numerical integration or Monte-Carlo integration.

For the nonlinear filters based on the Taylor series expansion, first, it is shown that we need to impose some approximations on the disturbances. Next, I propose a nonlinear filter which combines the extended Kalman filter with Monte- Carlo stochastic simulations, where each expectation in the algorithm is evaluated by generating random numbers. Also, for the single-stage iteration filter, a re-interpretation is given, which is different from the conventional one.

It is, however, known that applying the linearized nonlinear measurement and transition equations to the conventional linear algorithm leads to biased filtering estimates. Therefore, it is essential to approximate the underlying conditional density functions rather than the nonlinear measurement and transition equations. A small extension is given to density approximation by numerical integration, where the nodes are taken as random numbers. Furthermore, the nonlinear filter with importance sampling is proposed, where the importance sampling theory developed by Geweke is applied to the nonlinear filtering problem.

Monte-Carlo experiments are performed to examine the nonlinear filters. It is found that the nonlinear filters based on the density approximation are better estimators than those based on the Taylor series expansion by the criteria of BIAS (bias) and RMSE (root mean squared error).

Finally, as an application to the nonlinear filters, I consider estimating permanent and transitory consumption separately, where the findings are as follows; (i) according to the likelihood ratio test, the hypothesis of no transitory consumption is rejected, (ii) we do not have the excess sensitivity problems on consumption, taking into account nonlinearity of the Euler equation, variable gross rate of return on savings and transitory consumption, and (iii) consumption consists of three parts: permanent consumption, transitory consumption and the other factor which depends on income or trend.

Table of Contents

List of Tables

List of Figures

1. Introduction

1.1 Objectives of the Book

There is a large amount of literature on applications of the Kalman filter model, which is used for estimation of unobservable variables. As for applications, we can consider a time-varying parameter model, an estimation of seasonal components, an estimation of autoregressive-moving average (ARMA) model, prediction of final data and so on. Thus, the Kalman filter is particularly powerful and useful in the model that includes unobservable components.

The equation to be estimated is usually nonlinear. For example, the Euler equation derived from a utility maximization problem is nonlinear, depending on the underlying utility function of the representative agent. Therefore, we need to consider a nonlinear filter, rather than a linear filter, where measurement and transition equations are nonlinear in both the state-variables and the error terms.

Moreover, as Meinhold and Singpurwalla (1989) pointed out, Kalman filter models based on normality assumption are known not to be robust, which implies that the posterior density becomes unrealistic when there is a large difference between the prior density and the observed data. Therefore, we have to consider nonnormal error terms, rather than normal ones, in the state-space model. From such reasons, several nonlinear and/or nonnormal filters are introduced and developed in this book.

In the case of linear measurement and transition equations, we can obtain the conventional linear recursive Kalman filter algorithm explicitly under normality assumption for the error terms. For the nonlinear measurement and transition equations and/or the nonnormal error terms, however, it is difficult to obtain the explicit filtering algorithm in closed form. Therefore some approximations have to be made for estimation.

Thus, the purpose of this book is to introduce and develop nonlinear and nonnormal filters and demonstrate their economic applications. There are two approaches to nonlinear filtering algorithms. One is approximating the nonlinear measurement and transition equations by the Taylor series expansions. The filtering algorithms are obtained by applying the linearized nonlinear functions directly to the conventional linear recursive algorithm.

Another approach is approximating the underlying density functions of the state-vector. A recursive algorithm on the densities is derived from Bayes' formula. According to this density-based algorithm, the distribution functions of the error terms and the functional forms of the measurement and transition equations determine the conditional density functions of the state-vector. Therefore, the algorithms based on density functions are applicable in all the following cases:

- Nonlinear measurement and transition equations and Nonnormal error terms,
- Linear measurement and transition equations and Nonnormal error terms,
- Nonlinear measurement and transition equations and Normal error terms, and
- Linear measurement and transition equations and Normal error terms.

Thus, we can handle all the cases with the density-based approach. There, rather than the nonlinear functions, the underlying probability density functions are approximated.

In the first part of the book, we discuss what we have to approximate when applying the Taylor series expansions to the nonlinear functions. There we encounter some problems with the approximated error terms which are referred to as the residuals. The problems are shown as follows.

(i) The error terms (i.e., residuals) in the state-space model do not necessarily have zero-means.

(ii) In the measurement equation, the state-vector is correlated with the error term (i.e., residual), and similarly, in the transition equation, the lagged state-vector is correlated with the error term (i.e., residual).

(iii) The error term (i.e., the residual) in the measurement equation is correlated with the error term (i.e., residual) in the transition equation.

(iv) The error terms (i.e., residuals) are not normally distributed.

Thus, the approximated error terms (i.e., residuals) are not well-behaved. If we approximate the error terms (i.e., residuals) to be well-behaved and apply the linearized functions to the standard linear recursive Kalman filter algorithm, we have the filtering algorithms based on the Taylor series expansions. Clearly, however, ignoring the above problems (i) – (iv) on the error terms (i.e., residuals) implies that the filtering estimates of the state-vector are biased.

Next, in order to improve the biased filtering estimates, the nonlinear and nonnormal filters based on the underlying density functions are examined. There, We have to consider approximating the probability density functions of the state-vector, not the nonlinear measurement and transition equations. The functional forms of the measurement and transition equations and the distributions of the error terms determine the conditional density functions of the state-variable. Each density is approximated by a sum of Gaussian distributions, numerical integration, Monte-Carlo integration with importance sampling and so no.

Furthermore, by Monte-Carlo experiments, it is shown that the density approximation approach has a superior performance over the Taylor series expansion approach. Finally, an example is illustrated as an application to the nonlinear filters introduced in the book.

1.2 Review of Related Studies

For a nonlinear filtering problem, the most heuristic and easiest approximation is to use the Taylor series expansions and apply the expanded nonlinear measurement and transition functions directly to the standard linear recursive Kalman filter algorithm. The traditional nonlinear filters include the extended Kalman filter (EKF), the second-order nonlinear filter (SNF) and the single-stage iteration filter (SIF), which are described in Wishner, Tabaczynski and Athans (1969) and Gelb (1974), where it was concluded that the single-stage iteration filter (SIF) has superior mean square performance, followed by the second-order nonlinear filter (SNF). As for the nonlinear filters that utilize the Taylor series expansions, four nonlinear filtering algorithms (i.e., the extended Kalman filter (EKF), the second-order nonlinear filter (SNF), the Monte-Carlo simulation filter (MSF) and the single-stage iteration filter (SIF)) are introduced in Chapter 3 of this book. Of the four nonlinear filters, the extended Kalman filter (EKF), the second-order nonlinear filter (SNF) and the Monte-Carlo simulation filter (MSF) can be derived from the same theoretical basis. Since we have the following inequality: $\mathrm{E}\big(g(x)\big) \neq g\big(\mathrm{E}(x)\big)$ for a nonlinear function $g(\cdot)$ and a random variable x, the state-vector estimated by the extended Kalman filter (EKF), i.e., the first-order nonlinear filter, is clearly biased. The second-order nonlinear filter (SNF) might be a less biased estimator than the extended Kalman filter (EKF), because we can regard the second-derivative terms as bias correction terms. To reduce bias of the filtering estimates and evaluate expectations of the nonlinear functions more correctly, the Monte-Carlo stochastic simulations suggested by Brown and Mariano (1984, 1989), Mariano and Brown (1983, 1989) and Tanizaki (1995a) are applied to the nonlinear filtering problem, which is called the Monte-Carlo simulation filter (MSF) in this book and originally proposed in Tanizaki (1991) and Tanizaki and Mariano (1996).

The single-stage iteration filter (SIF) is different from the other three nonlinear filters in its interpretation. It can be derived by the mixed estimation, so-called Goldberger-Theil estimation (see Harvey (1981) and Diderrich (1985)). Wishner, Tabaczynski and Athans (1969) stated that the extended Kalman filter (EKF), the second-order nonlinear filter (SNF) and the single-stage iteration filter (SIF) can be derived from the same theoretical basis.

In this book (Chapter 3), however, it is shown that the single-stage iteration filter (SIF) is distinguished from the extended Kalman filter (EKF) and the second-order nonlinear filter (SNF) in its interpretation.

As another approach which avoids the biased filtering estimates, we consider the filtering algorithms based on the underlying distribution functions. The standard Kalman filter algorithm is represented by the first- and the second-moments of the underlying density functions (i.e., mean and variance), provided that the measurement and transition equations are linear and normality is assumed for the error terms (see, for example, Harvey (1989)). Unless the distributions of the error terms in the state-space model are normal and/or the measurement and transition equations are linear, we cannot derive an explicit expression for the filtering algorithm. The filtering estimates obtained are unrealistic if we approximate the nonlinear measurement and transition equations by the Taylor series expansions and apply the linearized nonlinear equations directly to the linear recursive Kalman filter algorithm. This is because applying the linearized nonlinear functions to the conventional Kalman filter algorithm implies that the nonnormal error terms with non-zero means are approximated as the normal error terms with zero means.

It is known that when the system is linear and normal the Kalman filter estimate is optimal in the sense that it minimizes the mean square error. When the normality assumption is dropped, there is no longer any guarantee that the Kalman filter gives the conditional mean of the state-vector. However, it is still an optimal estimator in the sense that it minimizes the mean square error within the class of all linear estimators, which implies that the Kalman filter under the normality assumption is the minimum mean square estimator and that the Kalman filter without the normality assumption is known as the minimum mean square linear estimator (see Harvey (1989)).

As Meinhold and Singpurwalla (1989) pointed out, Kalman filter models based on normality assumption are known to be non-robust, which implies that when there is a large difference between the prior density and the observed data, the posterior density becomes unrealistic. Therefore, approximation of the underlying densities, rather than that of the nonlinear functions, is essential to the nonlinear and/or nonnormal filtering problem.

It is known that a recursive filtering algorithm can be obtained from the conditional density functions of the state-variable, which is obtained from Bayes' formula. The nonlinear filters based on the underlying density functions were developed by Sorenson and Alspach (1971), Alspach and Sorenson (1972), Kitagawa (1987), Kramer and Sorenson (1988), Tanizaki (1991, 1995b), Carlin, Polson and Stoffer (1992), Tanizaki and Mariano (1994, 1995a, 1995b, 1996) and Mariano and Tanizaki (1995, 1996). Alspach and Sorenson (1972) and Sorenson and Alspach (1971) approximated the densities by a sum of Gaussian distributions, called the Gaussian sum filter (GSF). Carlin, Polson and Stoffer (1992) suggested applying an adaptive Monte-Carlo integration technique known as the Gibbs sampler to the density approximation.

Kitagawa (1987) and Kramer and Sorenson (1988) proposed an approach of representing the densities numerically by a piecewise linear function, where each density is represented as number of segments, location of nodes and the value at each node, and it is evaluated through numerical integration. According to the numerical integration approach, however, computational burden increases more than proportionally as the dimension of the state-vector increases. Programming is also extremely tedious if the dimension is high. We call this nonlinear and nonnormal filter the numerical integration filter (NIF) in this book.

Improving the above problems which are included in the numerical integration filter (NIF), Tanizaki (1991), Tanizaki and Mariano (1994) and Mariano and Tanizaki (1995) suggested a nonlinear filter by importance sampling through random numbers (i.e., Monte-Carlo integration with importance sampling), in which a recursive algorithm of the weight functions represented by the ratio of two densities is derived through Monte-Carlo stochastic simulations. Geweke (1988, 1989a, 1989b) and Shao (1989) developed an approximation of prior density in Bayesian framework, so-called importance sampling theory. Here, this approach is applied to the nonlinear and nonnormal filter. From the point of programming and computational time, the nonlinear and nonnormal filter based on the importance sampling theory can be easily extended to the higher-dimensional cases in practice, comparing with the numerical integration filter. This nonlinear filter is called the importance sampling filter (ISF) in this book.

One of the problems in the importance sampling filter (ISF) is as follows. For approximation of the filtering density functions (the target density functions), a researcher has to choose another density function appropriately, called the importance density, which is quite *ad hoc*. In order to avoid such an *ad hoc* assumption, furthermore, Tanizaki and Mariano (1995a, 1995b) proposed the density-based Monte-Carlo filter (DMF), where the method of Monte-Carlo stochastic simulations is utilized as an alternative solution to the nonlinear and nonnormal state-space model. The features of the algorithm proposed by Tanizaki and Mariano (1995a, 1995b) are that the random numbers of the sate-vector are generated from the transition equation for all time, and that the algorithm requires the density function from the measurement equation. That is, the measurement equation is utilized for deriving the density function while the transition equation is used to generate the random numbers of the state-variable. We call the nonlinear estimator the density-based Monte-Carlo filter (DMF), which does not yield the recursive algorithm.

Also, Tanizaki (1995b), Tanizaki and Mariano (1995b), Mariano and Tanizaki (1996) proposed another nonlinear and nonnormal filter using a Monte-Carlo approach. For a solution to nonlinear and nonnormal state-space model, we use the random draws to obtain filtering estimates, where rejection sampling is utilized to generate random draws from the filtering densities. By rejection sampling, a recursive algorithm of the random draws are obtained, where the random draws are directly generated from the filtering density, which is called the rejection sampling filter (RSF) in this book. There, we do not need the functional forms of the probability density functions which are required for the numerical integration filter (NIF) and the importance sampling filter (ISF). Thus, the random draws of the state-vector are directly generated from the filtering density. Programming is also very easy compared with the other nonlinear filters.

The extended Kalman filter (EKF), the second-order nonlinear filter (SNF), the Monte-Carlo simulation filter (MSF), the single-stage iteration filter (SIF), the Gaussian sum filter (GSF), the numerical integration filter (NIF), the importance sampling filter (ISF), the density-based Monte-Carlo filter (DMF) and the rejection sampling filter (RSF) are introduced in this book, i.e., the first four filters (EKF, SNF, MSF and SIF, known as the traditional nonlinear filters) are in Chapter 3 and the last five filters (GSF, NIF, ISF, DMF and RSF, known as the density-based nonlinear filters) are in Chapter 4. Of the nine nonlinear filters, MSF, ISF, DMF and RSF are the simulation-based nonlinear filters.

As an application, we take an example of estimating permanent consumption, taking into account nonlinearity of the Euler equation, variable interest rate and transitory consumption. In the last decade, a large amount of macroeconomic research has been devoted to various aspects of the life cycle permanent income hypothesis under rational expectations. See, for example, Hall (1978), Flavin (1981), Hall and Mishkin (1982), Mankiw and Shapiro (1985), Campbell (1987), Deaton (1987), Campbell and Mankiw (1987, 1990), West (1988), Campbell and Deaton (1989) and Diebold and Rudebusch (1991). Moreover, Flavin (1981), Hall and Mishkin (1982) and Campbell and Mankiw (1987) examined the permanent income hypothesis taking into account transitory consumption. Hayashi (1985a, 1985b) considered testing it based on liquidity constraints and durable goods. Mankiw (1981) and Muellbauer (1983) introduced the variable interest rate into the model. For another research on asset pricing models, we have Hamori (1992a, 1992b, 1993), Hamori and Kitasaka (1996), Hamori, Kitasaka and Tanizaki (1996) and so on. Thus, the permanent income hypothesis has been tested from various aspects (for a concise and useful survey, see Diebold and Nerlove (1989)).

Hall (1978) proposed a new approach to the permanent income hypothesis. The essence is as follows. If consumption is based on the permanent income hypothesis, by which kind of model can a sequence of consumption be written? According to the approach proposed by Hall (1978), the permanent income hypothesis can be tested without estimating a consumption function as conventionally done.

In Chapter 6 of this book, the nonlinear filtering techniques are applied to testing the permanent income hypothesis in the following basic setup. Total consumption consists of a sum of permanent consumption and transitory consumption, which is represented by a linear relationship (i.e., identity equation). This identity equation is taken as the measurement equation, where transitory consumption is assumed to be independent of permanent consumption, permanent income and transitory income, and accordingly transitory consumption is taken as an independently distributed error term. Permanent consumption is related to the permanent income hypothesis. Therefore, the Euler equation is obtained by solving a representative utility maximization problem, which is regarded as the transition equation. Here, total consumption is observable while neither permanent nor transition consumption are observed. Thus, permanent consumption is taken as the state-variable and transitory consumption as the random shock. For the setup of the model, see Mariano and Tanizaki (1996).

However, the permanent income hypothesis proposed by Hall (1978) sometimes has the criticism that it holds only in the short run. In this empirical application, we take the consumption data from 1965 to 1994, which are quite a long term. In addition, it is well known that variables other than lagged consumption appear to play a significant role in the determination of current consumption (see Diebold and Nerlove (1989)). Therefore, it might be more appropriate to consider that consumption consists of three parts, i.e., permanent consumption, transitory consumption and the other factor which is independent of the permanent income hypothesis and depends on income or time trend. That is, the basic setup discussed above is extended to the case where consumption includes the other factor as well as permanent and transitory consumption. If the coefficients of income (i.e., the other factor independent to the permanent income hypothesis is assumed to be a function of income) are statistically zero, we can conclude that the life cycle permanent income hypothesis actually holds even in the long run. However, according to the empirical studies in Chapter 6, we see that consumption significantly depends on income by the likelihood ratio test. Therefore, it is concluded that the permanent income hypothesis does not hold in reality. Thus, given this setup, permanent consumption is estimated for the following countries: Canada, France, Germany, Italy, Japan, Spain, United Kingdom (U.K.), and United States (U.S.).

1.3 Outline of the Book

The outline of this book is as follows.

In Chapter 2, first, a time-varying parameter model, a seasonal component model, an ARMA model and so on are concisely illustrated as the applications of the Kalman filter model in the simple linear case. Next, there are many ways to derive the standard Kalman filter algorithm. Here, three derivations are discussed:

- the derivation procedure under normality assumption for the error terms,
- the mixed estimation method, or equivalently, the generalized least squares estimation, and
- the minimum mean square linear estimator.

All the three derivation procedures result in the same linear recursive Kalman filter algorithm but an interpretation of the Kalman filter is quite different depending on the derivation procedures.

For one derivation, the conventional Kalman filter algorithm is introduced under the assumptions of normally distributed error terms and linear measurement and transition equations. In this case, it is known that the linear recursive algorithm can be derived from the first- and the second-moments of the normal densities by the property of the normal distribution.

For another derivation, by combining the present sample and the prior information and applying the mixed estimation method (Goldberger-Theil estimation), the standard linear recursive Kalman filter algorithm can be derived, where normality assumption is not imposed. In this derivation process, the state-variable at time t is represented by the prior information. That is, the state-variable at time t consists of the expectation of the state-variable at time t given information available at time $t-1$ and the error term. The present sample at time t represents the relationship between the observed variable at time t and the unobserved state-variable at time t. From the two equations (one represents the prior information and another shows the present sample), the mixed estimation method is applied. The interpretation based on the mixed estimation is similar to the recursive generalized least squares estimation method.

As the third derivation, the Kalman filter is interpreted as the minimum mean square estimator. There, normality assumption is not required. Note that the Kalman filter under the normality assumption is the minimum mean square estimator while the Kalman filter without the normality assumption is known as the minimum mean square linear estimator (see Harvey (1989)). Based on the derivation from the minimum mean square estimator, we can easily handle the case where the error term in the measurement equation is correlated with that in the transition equation.

Thus, although the same linear recursive algorithm is obtained, we see that the interpretation of the Kalman filter is different, depending on the derivation procedures.

The nonlinear filters are developed in Chapters 3 and 4. In the case where the nonlinear measurement and transition equations are linearized, the approximated error terms included in the two equations (i.e., measurement and transition equations) are referred to as the residuals. In Chapter 3, first, it is discussed that the approximated error terms (i.e., residuals) have the following problems.

(i) Both the error terms (i.e., residuals) are not necessarily zero-means.
(ii) In the measurement equation, the state-vector at time t is correlated with the error term (i.e., residual) at time t, and similarly in the transition equation the state-variable at time $t-1$ is also correlated with the error (i.e., residual) at time t.
(iii) The error terms (i.e., residuals) are correlated with each other. That is, the error term (i.e., residual) in the measurement equation is correlated with that in the transition equation at the same time period.
(iv) Both the error terms (i.e., residuals) are not normal.

Next, based on the Taylor series expansions, four nonlinear filters are introduced, which are as follows.

- The extended Kalman filter (EKF), where the first-order Taylor series approximation is applied to the standard linear recursive Kalman filter algorithm.
- The second-order nonlinear filter (SNF), where the second-order Taylor series expansion is used for approximation of the nonlinear functions and the expanded nonlinear functions are directly applied to the standard Kalman filter algorithm.
- The Monte-Carlo simulation filter (MSF), in which each expectation of nonlinear functions is evaluated by random draws.
- The single-stage iteration filter (SIF), which uses the iterative procedure at each time period and are equivalent to the Newton-Raphson nonlinear optimization procedure.

Three out of the four estimation methods are derived from the same theoretical framework, which are the extended Kalman filter (EKF, or first-order nonlinear filter), the second-order nonlinear filter (SNF) and the Monte-Carlo simulation filter (MSF). The extended Kalman filter (EKF) and the second-order nonlinear filter (SNF) have all the problems (i) – (iv) on the error terms (i.e., residuals). It is shown that the Monte-Carlo simulation filter (MSF) is the asymptotically unbiased estimator because it has the problems (iii) and (iv) only (that is, the problems (i) and (ii) on the error terms (i.e., residuals) are improved). As another nonlinear filter using the Taylor series expansions, we can consider the single-stage iteration filter (SIF), which is also one of the less biased estimator than the extended Kalman filter (EKF) and the higher-order nonlinear filters. This approach (i.e., SIF) is distinguished from the other three nonlinear filters (i.e., EKF, SNF and MSF) in its interpretation, even though the single-stage iteration filter (SIF) is derived from the Taylor series approximation of the nonlinear functions. The main difference in the derivation procedure between the single-stage iteration filter (SIF) and the other three filters (EKF, SNF and MSF) is that all the above three filters have to be based on normality approximation for the error terms (i.e., residuals), but the single-stage iteration filter (SIF) can be derived from the conventional nonlinear least squares approach such as the Newton-Raphson nonlinear optimization procedure. There, we do not have to impose normality approximation for the error terms (i.e., residuals) in order to derive the algorithm of the single-stage iteration filter (SIF).

Moreover, two theorems proposed by Tanizaki (1991) and Tanizaki and Mariano (1996) are introduced, which are related to one of the approximations on the error terms (i.e., residuals). It is shown that the correlation between the error terms (i.e., residuals) in the expanded state-space model disappears under a certain functional form of the measurement equation or the transition equation. This implies that in the case where we choose the functional form we do not need to take into account one of the approximations. The theorems might be useful for the Monte-Carlo simulation filter (MSF), because it has already zero-mean errors uncorrelated with the state-vectors, i.e., it has less problems than the extended Kalman filter (EKF), the second-order nonlinear filter (SNF) and the other higher-order nonlinear filters. The extended Kalman filter (EKF) and the other higher-order nonlinear filters have the problem (iii) as well as the problems (i), (ii) and (iv). However, for the Monte-Carlo simulation filter (MSF), the error terms (i.e., residuals) are transformed to be asymptotically uncorrelated with zero means. Accordingly, when applying the theorems proposed in Tanizaki (1991) and Tanizaki and Mariano (1996), the Monte-Carlo simulation filter (MSF) has only one problem, i.e., (iv), while the extended Kalman filter (EKF) and the other higher-order nonlinear filters still have all the problems, i.e., (i) – (iv). Therefore, the theorems are more useful for the Monte-Carlo simulation filter (MSF), rather than the extended Kalman filter (EKF) and the other higher-order nonlinear filter.

Attempts are made in Chapter 4 to approximate the underlying density functions. For approximation of the densities, the following five nonlinear filters are introduced and developed.

- The Gaussian sum filter (GSF), where each density is approximated as the weighted average of normal distributions.
- The numerical integration filter (NIF), in which the nodes and the segments are chosen and each density is approximated by numerical integration.
- The importance sampling filter (ISF), where the Monte-Carlo integration technique with importance sampling is applied and a recursion of weight functions instead of density functions is obtained.
- The density-based Monte-Carlo filter (DMF), where the method of Monte-Carlo stochastic simulations is utilized as an alternative solution to the nonlinear and nonnormal state-space model, i.e., the measurement equation is utilized to derive the density function while the transition equation is used to generate the random numbers of the state-variable.

- The rejection sampling filter (RSF), in which random draws are directly generated from filtering densities and a recursive algorithm based on the random draws is obtained.

A natural extension of density approximation is a Gaussian sum approach. The Gaussian sum filter (GSF) is a nonlinear algorithm which involves collections of extended Kalman filters, and thereby become both more powerful and more complex then the extended Kalman filter. In this algorithm, the conditional densities are approximated by a sum of Gaussian density functions. The filtering algorithm by the Gaussian sum approach basically follows from the weighted average of the extended Kalman filter equations. See Sorenson and Alspach (1971), Alspach and Sorenson (1972), and Anderson and Moore (1979)) for the Gaussian sum filter (GSF).

For EKF, SNF, MSF, SIF and GSF, approximation of the nonlinear measurement and transition equations by Taylor series expansions is used to obtain the estimates of the state-variables to some extent. Kitagawa (1987) and Kramer and Sorenson (1988)) proposed the nonlinear filter, where each density function is approximated by a piecewise linear (first-order spline) function and numerical integration is used for the density evaluation. Each density is specified by number of segments, location of nodes, and the value at each node. Thus, in the numerical integration filter (NIF), the underlying probability density functions, not the nonlinear state-space model, are approximated.

As an alternative to the numerical integration procedure, a recursive algorithm on the weight functions, rather than the density functions, are derived, where the weight function is represented as a ratio of two density functions. Tanizaki (1991), Tanizaki and Mariano (1994) and Mariano and Tanizaki (1995) developed a nonlinear and nonnormal filter using Monte-Carlo integration with importance sampling, where a recursive algorithm of the density functions is converted to that of the weight functions. By Monte-Carlo integration with importance sampling, a recursive filtering algorithm of the weight functions is derived, which is called the importance sampling filter (ISF) in this book.

For the numerical integration filter (NIF) and the importance sampling filter (ISF), choice of the nodes and the importance density function is one of the critical problems, because precision of the filtering estimates depends on location of the nodes for the numerical integration filter (NIF) and choice of the importance density for the importance sampling filter (ISF). The density-based Monte-Carlo filter (DMF) does not requires *ad hoc* assumptions such as choice of the nodes and that of the importance density because the random numbers are generated from the transition equation (see Tanizaki and Mariano (1995a, 1995b) and Mariano and Tanizaki (1996)). In the case where the transition equation follows a random walk process, the state-variable α_t has no mean an no variance as time t goes to infinity. Under this situation, the density-based Monte-Carlo filter (DMF) does not work. If the range of the state variable is restricted, the density-based Monte-Carlo filter (DMF) performs quite better. However, in the case where the transition equation is a random walk process, it is known that the density-based Monte-Carlo filter (DMF) is extremely poor.

Also, the rejection sampling filter (RSF) proposed by Tanizaki (1995b), Tanizaki and Mariano (1995b) and Mariano and Tanizaki (1996) does not require *ad hoc* assumptions such as choice of the nodes and that of the importance density. In this sense, the rejection sampling filter (RSF) is similar to the density-based Monte-Carlo filter (DMF). However, the rejection sampling filter (RSF) shows a good performance even in the case where the transition equation follows a random walk process (remember that DMF does not work when the transition equation follows a random walk process). For a solution to nonlinear and nonnormal state-space model, the rejection sampling filter (RSF) uses the random draws to obtain the filtering estimates. Given the random draws from the filtering density at the past time, we consider generating the random numbers from the filtering density at the present time. That is, a recursive algorithm of the filtering random draws is derived.

Monte-Carlo experiments are performed in Chapter 5 to examine the nonlinear filters. Taking various types of nonlinear functions, all of the nonlinear filters introduced in this book are compared in Section 5.2. For comparison, the bias (BIAS) and the root mean square error (RMSE) are used as the criteria of goodness of the estimators. It is known that the extended Kalman filter (EKF) and the second-order nonlinear filters (SNF) give us the biased filtering estimates and the biased parameter estimates. We obtain this result from the experiments. It is shown that the Monte-Carlo simulation filter (MSF) gives us the asymptotically unbiased filtering estimates but large RMSE. Also, it is discussed that the numerical integration filter (NIF), the importance sampling filter (ISF) and the rejection sampling filter (RSF) have a superior performance over the other nonlinear filters.

In Chapter 6, as an application of the nonlinear filters, we take an example of estimating permanent consumption, taking into account nonlinearity of the Euler equation, the variable interest rate and transitory consumption. Total consumption consists of permanent consumption, transitory consumption and the other factor of consumption. The Euler equation derived from a utility maximization problem of the representative agent is based on the permanent income hypothesis, and accordingly the Euler equation is a function of permanent consumption, which is not observed. Transitory consumption is regarded as a random shock which is also unobservable. The other factor which is independent of the permanent income hypothesis is assumed to be a function of income. The measurement equation in this model is taken as the identity equation in which total consumption is represented by a sum of permanent consumption, the random shock related to transitory consumption and the other factor of consumption which is independent of the permanent income hypothesis. The Euler equation is taken as the transition equation in the state-space model. The filtering techniques can be applied to this problem, where permanent consumption is regarded as the state-variable. Given the above setup, we consider testing the permanent income hypothesis for some countries. We have the result that for almost all the countries the life cycle permanent income hypothesis is rejected. And also, a ratio of permanent consumption relative to total consumption is estimated for each country.

Until Chapter 6, we deal with the filtering problem only. In Chapter 7, we consider both prediction (i.e., L-step ahead prediction) and smoothing (i.e., fixed-interval smoothing) problems. The relationship among prediction, filtering and smoothing is as follows. Filtering is used to estimate the present state at the present time, prediction is helpful to estimate the future state at the present time, and smoothing is an estimation of the past state at the present time. Moreover, the one-step ahead prediction estimate is utilized to obtain the filtering estimate, and we need the filtering estimate to derive the smoothing estimate. The nonlinear filters discussed in Chapter 4 are extended to the prediction and smoothing algorithms. There, we discuss the prediction and smoothing algorithms by the numerical integration procedure, the importance sampling procedure, the density-based Monte-Carlo method and the rejection sampling approach. We have the three kinds of smoothing, i.e., the fixed-point smoothing, the fixed-lag smoothing and the fixed-interval smoothing. Of the three, we consider the fixed-interval smoothing, which might be the most useful tool in a field of economics. Thus, the density-based nonlinear filters are extended to prediction and smoothing algorithms.

Chapter 8 contains the summary of this book and the directions for further research.

2. State-Space Model in Linear Case

2.1 Introduction

There is a great deal of literature about applications of the Kalman filter model. Kalman (1960) and Kalman and Bucy (1961) developed this new approach to linear filtering and prediction, which has also been developed by engineers. Economic applications of the Kalman filter include a time-varying parameter model, an autoregressive-moving average process, estimation of seasonal components, prediction of final data given preliminary data and so on, which are introduced later in this chapter. Thus, the Kalman filter can be applied to various models, especially, the models that include unobservable variables.

As it is well known, there are many approaches to derive the Kalman filter algorithm. Here, three derivations are discussed; the first is based on normality assumption for error terms, the second is related to the mixed estimation approach, so-called Goldberger-Theil estimator, where we do not have to impose normality assumption for error terms, and the third is interpreted as minimum mean square linear estimator. All the derivations result in the same linear recursive algorithm, although the derivation procedure is different.

Finally, in this chapter, an estimation problem of unknown parameters is taken. In the case where measurement and transition equations include unknown parameters, the maximum likelihood method is usually used for estimation. The maximum likelihood estimation procedure is to maximize an innovation form of the likelihood function, which is very simple and requires no extra computation because all of the values used in the maximization procedure are obtained in the filtering algorithm.

Thus, we aim to introduce past research on the Kalman filter by way of a preface to this chapter.

In general, the model represented by the following two equations is called the state-space model.

$$\text{(Measurement equation)} \qquad y_t = Z_t \alpha_t + d_t + S_t \epsilon_t, \tag{2.1}$$

$$\text{(Transition equation)} \qquad \alpha_t = T_t \alpha_{t-1} + c_t + R_t \eta_t, \tag{2.2}$$

$$\begin{pmatrix} \epsilon_t \\ \eta_t \end{pmatrix} \sim N\left(\begin{pmatrix} 0 \\ 0 \end{pmatrix}, \begin{pmatrix} H_t & 0 \\ 0 & Q_t \end{pmatrix} \right),$$

$$\begin{array}{lllll} y_t : g \times 1, & Z_t : g \times k, & d_t : g \times 1, & S_t : g \times g, & \epsilon_t : g \times 1, \\ \alpha_t : k \times 1 & T_t : k \times k, & c_t : k \times 1, & R_t : k \times k, & \eta_t : k \times 1, \end{array}$$

where ϵ_t is a serially uncorrelated disturbance with mean zero and covariance matrix H_t, and α_t is a $k \times 1$ vector which is not observable and is assumed to be generated by the first-order Markov process, as shown in equation (2.2). T_t is a transition matrix and η_t is a random vector of serially uncorrelated disturbances with mean zero and covariance matrix Q_t. And also Z_t, d_t, S_t, T_t, c_t and R_t may depend on information available at time $t-1$ if we assume normality for the errors ϵ_t and η_t (see Harvey (1989)). Equation (2.1) is known as the measurement equation while equation (2.2) is called the transition equation. The state-space model requires the following two additional assumptions:

(i) The initial vector α_0 has a mean of a_0 and a covariance matrix of Σ_0, i.e., $\mathrm{E}(\alpha_0) = a_0$ and $\mathrm{Var}(\alpha_0) = \Sigma_0$.
(ii) The disturbances ϵ_t and η_s are uncorrelated with each other for all time periods, and uncorrelated with the initial state-variable, i.e., $\mathrm{E}(\epsilon_t \eta_s') = 0$ for all t and all s, and $\mathrm{E}(\epsilon_t \alpha_0') = \mathrm{E}(\eta_t \alpha_0') = 0$ for $t = 1, \cdots, T$.

Notable points are as follows:

(1) The assumption (ii) guarantees no correlation between ϵ_t and α_t, and no correlation between η_t and α_{t-1}, i.e., $\mathrm{E}(\epsilon_t \alpha_t') = 0$ and $\mathrm{E}(\eta_t \alpha_{t-1}') = 0$ for all t.
(2) Z_t, d_t, S_t, T_t, c_t and R_t may depend on an unknown parameter vector, say θ. In this case, θ has to be estimated. We consider this estimation problem later in this chapter.
(3) The error terms ϵ_t and η_t are usually assumed to be normal, but we do not necessarily need normality assumption, depending on the derivation procedure of the linear recursive algorithm. In the case of the derivations by mixed estimation and minimum mean square linear estimator, we do not need to assume any distribution for the errors ϵ_t and η_t. However, if we derive the standard linear recursive algorithm based on density functions, normality assumption is required. This is discussed in Section 2.3.

2.2 Applications of Kalman Filter

Some applications of the state-space model are shown below. In this section, we take the following examples to Kalman filter; Time-Varying Parameter Model (Section 2.2.1), Autoregressive-Moving Average Process (Section 2.2.2), Seasonal Component Model (Section 2.2.3), Prediction of Final

Data Based on Preliminary Data (Section 2.2.4) and Estimation of Permanent Consumption (Section 2.2.5).

2.2.1 Time-Varying Parameter Model

In the case where we deal with time series data, we can usually write the regression model as follows:

$$y_t = x_t\beta + u_t,$$

for $t = 1, \cdots, T$, where y_t is a dependent variable, x_t denotes a $1 \times k$ vector of the explanatory variables, a $k \times 1$ vector of unknown parameters to be estimated is given by β, and u_t is the error term distributed with mean zero and variance σ^2. There are some methods to estimate the equation above, for example, ordinary least squares, generalized least squares, instrumental variable estimation, etc. In any case, the estimated parameters are constant over time. This model is known as the fixed-parameter model. However, structural changes (for example, the first- and the second-oil shocks), specification errors, nonlinearities, proxy variables and aggregation are all the sources of parameter variation; see Sarris (1973), Belsley (1973), Belsley and Kuh (1973) and Cooley and Prescott (1976). Therefore, we need to consider the model such that the parameter is a function of time, which is called the time-varying parameter model. Using the state-space form, the model is represented as the following two equations:

$$\text{(Measurement equation)} \qquad y_t = x_t\beta_t + u_t, \tag{2.3}$$

$$\text{(Transition equation)} \qquad \beta_t = \Psi\beta_{t-1} + v_t, \tag{2.4}$$

where the movement of the parameter is generally assumed to be the first-order autoregressive process. The error term v_t, independent of u_t, is a white noise with mean zero and variance R. Here, equations (2.3) and (2.4) are referred to as the measurement equation and the transition equation, respectively. Note that it is possible to consider that β_t follows an AR(p) model for the transition equation. The time-varying parameter β_t is unobservable, which is estimated using the observed data y_t and x_t. Clearly, we have $Z_t = x_t$, $d_t = c_t = 0$, $S_t = R_t = I_k$, $\alpha_t = \beta_t$, $T_t = \Psi$ and $g = 1$ in equations (2.1) and (2.2).

As an example of the time-varying parameter model, Tanizaki (1987, 1993c) estimated a small macro-econometric model of Japan, using the Kalman filtering technique, where a consumption function, an investment function, an export function, an import function, a money demand function and so on are estimated equation-by-equation, and the movements of each coefficient are analyzed. There are numerous other papers which deal with the time-varying parameter model, for example, Cooley (1977), Cooper (1973), Cooley, Rosenberg and Wall (1977), Dziechciar (1989), Johnston (1984), Nicholls and Pagan (1985), Laumas and Mehra (1976), Garbade (1976), Pagan (1980), Rosenberg (1977), Tanizaki (1989, 1993b) and Watanabe (1985).

2.2.2 Autoregressive-Moving Average Process

It is well-known that any autoregressive-moving average process (ARMA) can be written as the state-space form. See, for example, Aoki (1987), Gardner, Harvey and Phillips (1980), Hannan and Deistler (1988), Harvey (1981, 1989), Kirchen (1988) and Burridge and Wallis (1988). First of all, consider the following ARMA(p, q) process.

$$y_t = a_1 y_{t-1} + \cdots + a_p y_{t-p} + \epsilon_t + b_1 \epsilon_{t-1} + \cdots + b_q \epsilon_{t-q},$$

where ϵ_t is a white noise. The model above is rewritten as:

$$y_t = a_1 y_{t-1} + \cdots + a_m y_{t-m} + \epsilon_t + b_1 \epsilon_{t-1} + \cdots + b_{m-1} \epsilon_{t-m+1},$$

where $m = \max(p, q+1)$ and some of the coefficients $a_1, \cdots, a_m, b_1, \cdots, b_{m-1}$ can be zero. As it is well known, the ARMA process above is represented as:

$$\begin{aligned} &\text{(Measurement equation)} \qquad && y_t = z\alpha_t, \\ &\text{(Transition equation)} \qquad && \alpha_t = A\alpha_{t-1} + B\epsilon_t, \end{aligned} \tag{2.5}$$

where z, A and B are defined as:

$$\underset{1\times m}{z = (1, 0, \cdots, 0)}, \qquad A = \underset{m\times m}{\left(\begin{array}{c|c} a_1 & \\ \vdots & I_{m-1} \\ a_{m-1} & \\ \hline a_m & 0 \end{array} \right)}, \qquad B = \underset{m\times 1}{\begin{pmatrix} 1 \\ b_1 \\ \vdots \\ b_{m-1} \end{pmatrix}},$$

Thus, the state-space form is obtained by setting $Z_t = z$, $d_t = S_t = c_t = 0$, $T_t = A$, $R_t = B$, $g = 1$ and $k = m$ in equations (2.1) and (2.2).

As an extension of this model (2.5), we can consider the ARMA(p, q) with a constant term, where the above transition equation is replaced as the following two equations:

$$\begin{aligned} &\text{(Measurement equation)} \qquad && y_t = z\alpha_t, \\ &\text{(Transition equation)} \qquad && \alpha_t - \overline{\alpha} = C\beta_t, \\ & && \beta_t = D\beta_{t-1} + E\eta_t. \end{aligned}$$

C represents a selection matrix consisting of zeros and ones with the characteristic: $CC' = I_m$, where I_m is the unity matrix. Or equivalently, the two equations related to the transition equation are compactly rewritten as:

$$\alpha_t - \overline{\alpha} = CDC'(\alpha_{t-1} - \overline{\alpha}) + CE\eta_t,$$

where $A = CDC'$ and $B = CE$ in the transition equation of the system (2.5). Kirchen (1988) concluded that this extension yields better results than the transition equation in the system (2.5).

2.2.3 Seasonal Component Model

As another application, we can take an estimation of seasonal components. A time-series consists of seasonal, cyclical, and irregular components. Each component is unobservable. The Kalman filter is applied to estimate each component separately.

The following suggestion by Pagan (1975) is essentially a combination of an econometric model for the cyclical components with the filtering and estimation of the seasonal components formulated in a state-space form. See Chow (1983).

Assume, first, that an endogenous variable y_t is the sum of cyclical, seasonal, and irregular components, as given by:

$$y_t = y_t^c + y_t^s + v_t,$$

and, second, that the cyclical component y_t^c is represented as the following model:

$$y_t^c = Ay_{t-1}^c + Cx_t + u_t,$$

where x_t is a $k \times 1$ vector of exogenous variables and u_t denotes a random disturbance. Finally, an autoregressive seasonal model is assumed for the seasonal component, i.e.,

$$y_t^s = By_{t-m}^s + w_t,$$

where w_t represents a random residual and m can be 4 for a quarterly model and 12 for a monthly model. Combining these three equations above, we can construct the following state-space form:

$$\begin{aligned} &\text{(Measurement equation)} \qquad && y_t = z\alpha_t + v_t, \\ &\text{(Transition equation)} && \alpha_t = M\alpha_{t-1} + Nx_t + \eta_t, \end{aligned} \tag{2.6}$$

where z, α_t, M, N and η_t are given by:

$$\underset{1\times(m+1)}{z = (1,1,0,\cdots,0)}, \qquad \underset{(m+1)\times 1}{\alpha_t = \begin{pmatrix} y_t^c \\ y_t^s \\ \alpha_{2t} \\ \vdots \\ \alpha_{mt} \end{pmatrix}}, \qquad \underset{(m+1)\times(m+1)}{M = \left(\begin{array}{c|cc} A & & 0 \\ \hline & 0 & I_{m-1} \\ 0 & B & 0 \end{array}\right)},$$

$$\underset{(m+1)\times k}{N = \begin{pmatrix} C \\ 0 \end{pmatrix}}, \qquad \underset{(m+1)\times 1}{\eta_t = \begin{pmatrix} u_t \\ w_1 \\ 0 \end{pmatrix}}.$$

In this example, α_t is the unobservable variable to be estimated by the Kalman filter. In the system (2.6), we can take each variable as $Z_t = z$, $d_t = 0$, $S_t = 1$, $\epsilon_t = v_t$, $T_t = M$, $c_t = Nx_t$, and $R_t = I$.

2.2.4 Prediction of Final Data Based on Preliminary Data

It is well known that economic indicators are usually reported according to the following two steps: (i) first of all, the preliminary data are reported, and (ii) then we can obtain the final or revised data after a while (see Table 2.1). The problem is how to estimate the final data (or the revised data) while only the preliminary data are available.

In the case of annual data on the U.S. national accounts, the preliminary data at the present time are reported at the beginning of the next year. The revision process is performed over a few years and every decade, which is shown in Table 2.1, taking an example of the nominal gross domestic product data (GDP, billion dollars). In Table 2.1, the preliminary data of 1988 and 1992 are taken from *Survey of Current Business* (January, 1989 and January, 1993). The rest of the data in Table 2.1 are from *Economic Report of the President* (ERP), published from 1984 to 1995. Each column indicates the year when ERP is published, while each row represents the data of the corresponding year. The superscript p denotes the preliminary data, the superscript r implies the data revised in the year corresponding to each column, and NA indicates that the data are not available (i.e., the data have not yet been published). For instance, take the GDP data of 1984 (see the corresponding row in Table 2.1). The preliminary GDP data of 1984 was reported in 1985 (i.e., 3616.3), and it was revised in 1986 for the first time (i,e., 3726.7). In 1987 and 1988, the second and third revised data were published, respectively (i.e., 3717.5 and 3724.8). Since it was not revised in 1989, the GDP data of 1984 published in 1989 is given by 3724.8. Moreover, the GDP data of 1984 was revised as 3777.2 in 1992. Also, as another example, the GDP data of 1975 was revised from 1531.9 to 1580.9 in 1986 and to 1585.9 in 1992.

Thus, each data series is revised every year for the first few years and thereafter less frequently. This implies that we cannot really know the true final data, because the data are revised forever while the preliminary data are reported only once. Therefore, it might be possible to consider that the final data are unobservable, which leads to estimation of the final data given the preliminary data.

There is a wide literature dealing with the data revision process. Conrad and Corrado (1979) applied the Kalman filter to improve upon published preliminary estimates of monthly retail sales, using an ARIMA model. Howrey (1978, 1984) used the preliminary data in econometric forecasting and obtained the substantial improvements in forecast accuracy if the preliminary and revised data are used optimally.

In the context of the revision process, the Kalman filter is used as follows. There is some relationship between the final and preliminary data, because they are originally same data (see, for example, Conrad and Corrado (1979)). This relationship is referred to as the measurement equation, where the final

Table 2.1. Revision Process of U.S. National Accounts (Nominal GDP)

	1984	1985	1986	1987	1988	1989
1975	1531.9	1531.9	1580.9^r	1580.9	1580.9	1580.9
1976	1697.5	1697.5	1761.7^r	1761.7	1761.7	1761.7
1977	1894.9	1894.9	1965.1^r	1965.1	1965.1	1965.1
1978	2134.3	2134.3	2219.1^r	2219.1	2219.1	2219.1
1979	2375.2	2375.2	2464.4^r	2464.4	2464.4	2464.4
1980	2586.4^r	2586.4	2684.4^r	2684.4	2684.4	2684.4
1981	2904.5^r	2907.5^r	3000.5^r	3000.5	3000.5	3000.5
1982	3025.7^r	3021.3^r	3114.8^r	3114.8	3114.8	3114.8
1983	3263.4^p	3256.5^r	3350.9^r	3355.9^r	3355.9	3355.9
1984	NA	3616.3^p	3726.7^r	3717.5^r	3724.8^r	3724.8
1985	NA	NA	3951.8^p	3957.0^r	3970.5^r	3974.1^r
1986	NA	NA	NA	4171.2^p	4201.3^r	4205.4^r
1987	NA	NA	NA	NA	4460.2^p	4497.2^r
1988	NA	NA	NA	NA	NA	4837.8^p

	1990	1991	1992	1993	1994	1995
1975	1580.9	1580.9	1585.9^r	1585.9	1585.9	1585.9
1976	1761.7	1761.7	1768.4^r	1768.4	1768.4	1768.4
1977	1965.1	1965.1	1974.1^r	1974.1	1974.1	1974.1
1978	2219.1	2219.1	2232.7^r	2232.7	2232.7	2232.7
1979	2464.4	2464.4	2488.6^r	2488.6	2488.6	2488.6
1980	2684.4	2684.4	2708.0^r	2708.0	2708.0	2708.0
1981	3000.5	3000.5	3030.6^r	3030.6	3030.6	3030.6
1982	3114.8	3114.8	3149.6^r	3149.6	3149.6	3149.6
1983	3355.9	3355.9	3405.0^r	3405.0	3405.0	3405.0
1984	3724.8	3724.8	3777.2^r	3777.2	3777.2	3777.2
1985	3974.1	3974.1	4038.7^r	4038.7	4038.7	4038.7
1986	4197.2^r	4197.2	4268.6^r	4268.6	4268.6	4268.6
1987	4493.8^r	4486.7^r	4539.9^r	4539.9	4539.9	4539.9
1988	4847.3^r	4840.2^r	4900.4^r	4900.4	4900.4	4900.4
1989	5199.6^p	5163.2^r	5244.0^r	5250.8^r	5250.8	5250.8
1990	NA	5424.4^p	5513.8^r	5522.2^r	5546.1^r	5546.1
1991	NA	NA	5671.8^p	5677.5^r	5722.8^r	5724.8^r
1992	NA	NA	NA	5945.7^p	6038.5^r	6020.2^r
1993	NA	NA	NA	NA	6374.0^p	6343.3^r
1994	NA	NA	NA	NA	NA	6736.9^p

data is unobservable but the preliminary data is observed. The equation obtained by some economic theory is related to the final data, rather than the preliminary data. This equation is taken as the transition equation. We can represent this problem with the state-space form:

$$\begin{aligned} &\text{(Measurement equation)} \quad && y_t^p = \gamma y_t^f + u_t, \\ &\text{(Transition equation)} \quad && y_t^f = \theta_1 y_{t-1}^f + \theta_2 x_t + v_t, \end{aligned} \tag{2.7}$$

where y_t^p and y_t^f denote the preliminary data and the final data, respectively. y_t^f represents the state-variable. For $t = 1, \cdots, T$, the preliminary data y_t^p are observed, while we do not have the final data y_t^f. x_t is assumed to be exogenous, nonstochastic and available for $t = 1, \cdots, T$. θ_1, θ_2 and γ are the parameters to be estimated. u_t and v_t are error terms, which are mutually and independently distributed. Suppose that the relationship between y_t and x_t:

$$y_t = \theta_1 y_{t-1} + \theta_2 x_t + v_t,$$

which is derived from an economic theory. The equation obtained by some economic theory is related to the final data, rather than the preliminary data. Therefore, y_t should be replaced by the final data, not the preliminary data, because the preliminary data include the measurement errors. If we estimate with such data, the appropriate results cannot be obtained. Therefore, as in the system (2.7), the transition equation in this example is given by

$$y_t^f = \theta_1 y_{t-1}^f + \theta_2 x_t + v_t.$$

Mariano and Tanizaki (1995) also examined the above problem, where several predictors are introduced and compared using the U.S. national accounts.

2.2.5 Estimation of Permanent Consumption

The next application is concerned with an estimation of permanent consumption. Total consumption consists of permanent and transitory consumption. This relationship is represented by an identity equation, which is taken as the measurement equation where total consumption is observable but both permanent and transitory consumption are unobservable. Permanent consumption depends on life-time income expected in the future, which is defined as permanent income. Solving a dynamic optimization problem, maximizing a quadratic utility function of the representative agent with respect to permanent consumption for each time and assuming that the discount rate is equal to the reciprocal of the gross rate of return on savings, it is shown that permanent consumption follows a random walk, i.e., this implies solving the following utility maximization problem:

$$\max_{\{c_t^p\}} \quad \mathrm{E}_0\left(\sum_t \beta^t u(c_t^p)\right), \qquad \text{subject to} \qquad A_{t+1} = R_t(A_t + y_t - c_t),$$

where $0 < \beta < 1$, $u(c_t^p) = -\frac{1}{2}(\bar{c} - c_t^p)^2$, $c_t = c_t^p + c_t^T$ and $\beta R_t = 1$. c_t, c_t^p, c_t^f, R_t, A_t, y_t, β, $u(\cdot)$ and $\mathrm{E}_t(\cdot)$ denote per capita total consumption, per capita permanent consumption, per capita transitory consumption, the real gross rate of return on savings between periods t and $t+1$, the stock of assets at the beginning of period t, per capita labor income, the discount rate, the representative utility function and the mathematical expectation given information up to t, respectively. See Hall (1978).

Next, the assumption on transitory consumption is as follows. In a cross sectional framework,

$$\sum_i C_{it}^T = 0,$$

which is assumed by Friedman (1957), where C_{it}^T denotes transitory consumption of the i-th household at time t. Moreover, Friedman (1957) assumed that C_{it}^T is independent of permanent income, transitory income and permanent consumption. Also, see Branson (1979). Note that the relationship between c_t^T and C_{it}^T is:

$$c_t^T = \frac{1}{L_t}\sum_i C_{it}^T,$$

where L_t is the number of population at time t. We can approximate as

$$\frac{1}{L_t}\sum_i C_{it}^T \approx \mathrm{E}(C_{it}^T),$$

which is equal to zero. Here, assuming that $C_{it}^T \equiv \epsilon_{it}$ is identically distributed with mean zero and variance σ_ϵ^2, transitory consumption of the representative agent (i.e., c_t^T) is given by a random shock with mean zero and variance σ_ϵ^2/L_t. Thus, the model to this problem is given by:

$$\begin{aligned} &\text{(Measurement equation)} \qquad c_t = c_t^p + c_t^T, \\ &\text{(Transition equation)} \qquad c_t^p = c_{t-1}^p + \eta_t, \end{aligned} \tag{2.8}$$

where the error terms $c_t^T = \epsilon_t$ and η_t are mutually and independently distributed. ϵ_t has a distribution with mean zero and variance σ_ϵ^2/L_t, and η_t is a random variable with mean zero and variance σ_η^2, i.e.,

$$\begin{pmatrix} \epsilon_t \\ \eta_t \end{pmatrix} \sim N\left(\begin{pmatrix} 0 \\ 0 \end{pmatrix}, \begin{pmatrix} \sigma_\epsilon^2/L_t & 0 \\ 0 & \sigma_\eta^2 \end{pmatrix}\right)$$

In the state-space model (2.8), c_t^p is regarded as the state-variable which is estimated by the Kalman filtering technique. Thus, we can estimate permanent and transitory consumption separately. In Chapter 6, we consider this example with the nonlinear transition equation, where the utility function of the representative agent is assumed to be a constant relative risk aversion type of utility function.

For the other applications, we can find estimation of the rational expectation models (for example, see Burmeister and Wall (1982), Engle and Watson (1987) and McNelis and Neftci (1983)). See Harvey (1987) for a survey of applications of the Kalman filter model.

2.3 Derivations of Kalman Filter Algorithm

In this chapter, we derive the filtering algorithm of the state-space form for the linear case given by equations (2.1) and (2.2). One may consider the problem of estimating α_t using information up to time s, i.e., $Y_s = \{y_1, y_2, \cdots, y_s\}$. Denote by $\mathrm{E}(\alpha_t|Y_s) \equiv a_{t|s}$ the conditional expectation of α_t given Y_s. The evaluation of $a_{t|s}$ is known as filtering if $t = s$, smoothing if $t < s$, and prediction if $t > s$, respectively. In this book we focus only on the filtering problem. Finally, $\mathrm{Cov}(\alpha_t|Y_s) \equiv \Sigma_{t|s}$ is defined as the conditional covariance matrix of α_t given Y_s.

The filtering algorithm in this case is known as the Kalman filter, which is given by the following equations:

$$a_{t|t-1} = T_t a_{t-1|t-1} + c_t, \tag{2.9}$$

$$\Sigma_{t|t-1} = T_t \Sigma_{t-1|t-1} T_t' + R_t Q_t R_t', \tag{2.10}$$

$$y_{t|t-1} = Z_t a_{t|t-1} + d_t, \tag{2.11}$$

$$F_{t|t-1} = Z_t \Sigma_{t|t-1} Z_t' + S_t H_t S_t', \tag{2.12}$$

$$K_t = \Sigma_{t|t-1} Z_t' F_{t|t-1}^{-1}, \tag{2.13}$$

$$\Sigma_{t|t} = \Sigma_{t|t-1} - K_t F_{t|t-1} K_t', \tag{2.14}$$

$$a_{t|t} = a_{t|t-1} + K_t(y_t - y_{t|t-1}), \tag{2.15}$$

for $t = 1, \cdots, T$, where Z_t, d_t, S_t, T_t, c_t, R_t, Q_t, H_t, $a_{0|0}$ and $\Sigma_{0|0}$ are known for now. See Anderson and Moore (1979), Gelb (1974) and Jazwinski (1970) for the Kalman filter algorithm.

Since equations (2.9) and (2.10) essentially predict the parameter α_t using the information up to time $t-1$, they are called the prediction equations. On the other hand, equations (2.14) and (2.15) play the role of combining the new observation obtained at time t (i.e., y_t) and the past information up to time $t-1$ (i.e., Y_{t-1}). They are known as the updating equations. Moreover, equation (2.13) is called the Kalman gain. An interpretation on the Kalman gain K_t depends on the derivation process. When the filtering algorithm is derived from the minimum mean square linear estimator, K_t is chosen such that the filtering estimate $a_{t|t}$ has minimum variance (see Section 2.3.3). In this case, K_t is referred to as the Kalman gain. Equations (2.11) and (2.12) represent the prediction estimate and its variance of the observed data y_t.

According to the Kalman filter algorithm shown above, $a_{1|0}$ and $\Sigma_{1|0}$ are readily calculated from prediction equations (2.9) and (2.10) since we assume that $a_{0|0}$ and $\Sigma_{0|0}$ are known. $a_{1|1}$ and $\Sigma_{1|1}$ are obtained by substituting $a_{1|0}$ and $\Sigma_{1|0}$ with K_1 into equations (2.14) and (2.15). Thus, $a_{t|t}$ and $\Sigma_{t|t}$, $t = 1, \cdots, T$, are calculated recursively from (2.9) – (2.15), once Q_t, H_t, $a_{0|0}$ and $\Sigma_{0|0}$ are assumed to be known.

There are several ways to interpret the Kalman filter algorithm. Their interpretations are related to the derivations of the algorithm, i.e., the derivation under the assumption of normality (Harvey (1989)), by orthogonal projection (Brockwell and Davis (1987) and Chow (1983)), from the mixed estimation (Cooley (1977), Harvey (1981) and Diderrich (1985)), from the generalized least squares method (Sant (1977)), and by the minimum mean square linear estimator (Burridge and Wallis (1988) and Harvey (1989)). Here, we discuss three of these derivations, i.e., derivation under normality assumption for the error terms (Section 2.3.1), use of mixed estimation procedure (Section 2.3.2) and interpretation of minimum mean square linear estimator (Section 2.3.3).

2.3.1 Derivation under Normality Assumption

A derivation of the Kalman filter algorithm under normality assumption comes from a recursive Bayesian estimation. If ϵ_t and η_t are mutually, independently and normally distributed and both measurement and transition equations are linear in the state-vector and the error term as in equations (2.1) and (2.2), then the conventional linear recursive algorithm represented by equations (2.9) – (2.15) can be obtained. The derivation is shown as follows.

Let $P(\cdot|\cdot)$ be a conditional density function. The prediction and updating equations, based on the probability densities, are given as follows (see, for example, Kitagawa (1987), Kramer and Sorenson (1988), and Harvey (1989)):

(Prediction equation)

$$\begin{aligned} P(\alpha_t|Y_{t-1}) &= \int P(\alpha_t, \alpha_{t-1}|Y_{t-1}) \mathrm{d}\alpha_{t-1} \\ &= \int P(\alpha_t|\alpha_{t-1}, Y_{t-1}) P(\alpha_{t-1}|Y_{t-1}) \mathrm{d}\alpha_{t-1} \\ &= \int P(\alpha_t|\alpha_{t-1}) P(\alpha_{t-1}|Y_{t-1}) \mathrm{d}\alpha_{t-1}, \end{aligned} \tag{2.16}$$

(Updating equation)

$$\begin{aligned} P(\alpha_t|Y_t) &= P(\alpha_t|y_t, Y_{t-1}) \\ &= \frac{P(\alpha_t, y_t|Y_{t-1})}{P(y_t|Y_{t-1})} \end{aligned}$$

$$= \frac{P(\alpha_t, y_t|Y_{t-1})}{\displaystyle\int P(\alpha_t, y_t|Y_{t-1}) d\alpha_t}$$

$$= \frac{P(y_t|\alpha_t, Y_{t-1}) P(\alpha_t|Y_{t-1})}{\displaystyle\int P(y_t|\alpha_t, Y_{t-1}) P(\alpha_t|Y_{t-1}) d\alpha_t}$$

$$= \frac{P(y_t|\alpha_t) P(\alpha_t|Y_{t-1})}{\displaystyle\int P(y_t|\alpha_t) P(\alpha_t|Y_{t-1}) d\alpha_t}, \tag{2.17}$$

for $t = 1, \cdots, T$. See Appendix A2.1 for the derivation in detail. Equations (2.16) and (2.17) are already a recursive algorithm of the density functions, which is called the density-based filtering algorithm in this book. $P(\alpha_t|\alpha_{t-1})$ is obtained from the transition equation (2.2) if the distribution of the error term η_t is specified, and also $P(y_t|\alpha_t)$ is derived from the measurement equation (2.1) given the specific distribution of the error term ϵ_t. Thus, the filtering density $P(\alpha_t|Y_t)$ is obtained recursively, given the distributions $P(\alpha_t|\alpha_{t-1})$ and $P(y_t|\alpha_t)$.

Now, we derive the filtering algorithm represented by equations (2.9) – (2.15), i.e., density-based filtering algorithm.

If α_0, ϵ_t and η_t are normally distributed, it is known that $P(\alpha_t|Y_s)$, $s = t, t-1$, is expressed by the Gaussian distribution:

$$P(\alpha_t|Y_s) = \Phi(\alpha_t - a_{t|s}, \Sigma_{t|s}),$$

where $\Phi(\alpha_t - a_{t|s}, \Sigma_{t|s})$ denotes the normal density with mean $a_{t|s}$ and variance $\Sigma_{t|s}$, i.e.,

$$\begin{aligned}&\Phi(\alpha_t - a_{t|s}, \Sigma_{t|s}) \\ &= (2\pi)^{-k/2} |\Sigma_{t|s}|^{-1/2} \exp\left(-\frac{1}{2}(\alpha_t - a_{t|s})' \Sigma_{t|s}^{-1} (\alpha_t - a_{t|s})\right).\end{aligned}$$

Note as follows:

$$\Phi(\alpha_t - a_{t|s}, \Sigma_{t|s}) \equiv N(a_{t|s}, \Sigma_{t|s}).$$

Then, from equation (2.16), one-step ahead prediction density $P(\alpha_t|Y_{t-1})$ is given by:

$$\begin{aligned}&P(\alpha_t|Y_{t-1}) \\ &= \int P(\alpha_t|\alpha_{t-1}) P(\alpha_{t-1}|Y_{t-1}) d\alpha_{t-1} \\ &= \int \Phi(\alpha_t - T_t\alpha_{t-1} - c_t, R_tQ_tR_t')\, \Phi(\alpha_{t-1} - a_{t-1|t-1}, \Sigma_{t-1|t-1}) d\alpha_{t-1}\end{aligned}$$

$$= \Phi(\alpha_t - T_t a_{t-1|t-1} - c_t, T_t \Sigma_{t-1|t-1} T_t' + R_t Q_t R_t')$$

$$\equiv \Phi(\alpha_t - a_{t|t-1}, \Sigma_{t|t-1}), \tag{2.18}$$

where $P(\alpha_t|\alpha_{t-1})$ and $P(\alpha_{t-1}|Y_{t-1})$ are rewritten as follows:

$$P(\alpha_t|\alpha_{t-1}) = \Phi(\alpha_t - T_t\alpha_{t-1} - c_t, R_t Q_t R_t'),$$
$$P(\alpha_{t-1}|Y_{t-1}) = \Phi(\alpha_{t-1} - a_{t-1|t-1}, \Sigma_{t-1|t-1}).$$

The third equality in equation (2.18) is proved in Proof I of Appendix A2.2.

Thus, comparing each argument (moment) in the two normal densities given by the fourth line and the fifth line in equation (2.18), we can derive the prediction equations (2.9) and (2.10).

Next, from equation (2.17), to derive the updating equations, equation (2.17) is calculated as follows:

$$P(\alpha_t|Y_t) = \frac{P(y_t|\alpha_t)P(\alpha_t|Y_{t-1})}{\displaystyle\int P(y_t|\alpha_t)P(\alpha_t|Y_{t-1})\mathrm{d}\alpha_t}$$

$$= \frac{\Phi(y_t - Z_t\alpha_t - d_t, S_t H_t S_t')\ \Phi(\alpha_t - a_{t|t-1}, \Sigma_{t|t-1})}{\displaystyle\int \Phi(y_t - Z_t\alpha_t - d_t, S_t H_t S_t')\ \Phi(\alpha_t - a_{t|t-1}, \Sigma_{t|t-1})\mathrm{d}\alpha_t}$$

$$= \Phi\big(\alpha_t - a_{t|t-1} - K_t(y_t - y_{t|t-1}), \Sigma_{t|t-1} - K_t F_{t|t-1} K_t'\big)$$

$$\equiv \Phi(\alpha_t - a_{t|t}, \Sigma_{t|t}), \tag{2.19}$$

where

$$y_{t|t-1} \equiv \mathrm{E}(y_t|Y_{t-1}) = Z_t a_{t|t-1} + d_t,$$
$$F_{t|t-1} \equiv \mathrm{Var}(y_t|Y_{t-1}) = Z_t \Sigma_{t|t-1} Z_t' + S_t H_t S_t',$$
$$K_t = \Sigma_{t|t-1} Z_t' F_{t|t-1}^{-1}.$$

Note that $P(y_t|\alpha_t)$ and $P(\alpha_t|Y_{t-1})$ are rewritten as:

$$P(y_t|\alpha_t) = \Phi(y_t - Z_t\alpha_t - d_t, S_t H_t S_t'),$$
$$P(\alpha_t|Y_{t-1}) = \Phi(\alpha_t - a_{t|t-1}, \Sigma_{t|t-1}).$$

The third equality in equation (2.19) is proved in Proof II of Appendix A2.2.

Therefore, the updating equations (2.13), (2.14) and (2.15) are obtained, comparing each argument in equation (2.19). Thus, the filtering algorithm is represented by (2.9) – (2.15).

There are two notable points:

(1) Even if the lagged dependent variables are contained in the measurement and transition equations, i.e., even if Z_t, d_t, S_t, T_t, c_t and R_t in equations (2.1) and (2.2) depend on the lagged dependent variables $y_{t-1}, \cdots, y_1$, we

can derive the same algorithm as equation (2.9) – (2.15) under normality assumption for the error terms ϵ_t and η_t (see Harvey, 1989, p.156). In this case, the probability densities $P(y_t|\alpha_t)$ and $P(\alpha_t|\alpha_{t-1})$ in (2.16) and (2.17) are replaced by:

$$P(y_t|\alpha_t, y_{t-1}, \cdots, y_1) \qquad \text{(i.e., } P(y_t|\alpha_t, Y_{t-1}) \text{ in general)},$$

$$P(\alpha_t|\alpha_{t-1}, y_{t-1}, \cdots, y_1) \qquad \text{(i.e., } P(\alpha_t|\alpha_{t-1}, Y_{t-1}) \text{ in general)}.$$

Also, see Appendix A2.1.

(2) Even if the system is nonlinear, i.e., even if the measurement equation (2.1) and the transition equation (2.2) are given by:

$$\text{(Measurement equation)} \qquad q_t(y_t, \alpha_t) = \epsilon_t, \tag{2.20}$$

$$\text{(Transition equation)} \qquad f_t(\alpha_t, \alpha_{t-1}) = \eta_t, \tag{2.21}$$

respectively, the prediction and updating equations represented by the densities (2.16) and (2.17) are unchanged (see Appendix A2.1). In the proceeding chapters, we deal with a nonlinear system as the general case of the state-space model (2.20) and (2.21).

Thus, under the normality and linearity assumptions, from equations (2.16) and (2.17), we can derive the linear recursive Kalman filter algorithm given by equations (2.9) – (2.15).

The following is an alternative derivation based on normality assumption, which is very simple and easy. The prediction equations (2.9) and (2.10) are easily obtained by taking the conditional expectation and variance given information up to time $t-1$ on both sides of equation (2.2).

Since the transition equation (2.2) is linear and α_{t-1} and η_t are normal, α_t (which is a sum of α_{t-1} and η_t) is also normal. Therefore, when the initial value α_0 is normal, α_t for $t = 1, \cdots, T$ are normal.

In order to obtain the updating equations (2.13) – (2.15), consider the joint probability density function of α_t and y_t, i.e., $P(\alpha_t, y_t|Y_{t-1})$, which are normal. For the measurement equation, when α_t and ϵ_t are normal, y_t (which is a sum of α_t and ϵ_t) is also normal. Accordingly, given information Y_{t-1}, the joint density function $P(\alpha_t, y_t|Y_{t-1})$ is as follows:

$$\begin{pmatrix} \alpha_t \\ y_t \end{pmatrix} \sim N\left(\begin{pmatrix} a_{t|t-1} \\ y_{t|t-1} \end{pmatrix}, \begin{pmatrix} \Sigma_{t|t-1} & \Sigma_{t|t-1}Z_t' \\ Z_t\Sigma_{t|t-1} & F_{t|t-1} \end{pmatrix} \right). \tag{2.22}$$

Note that each component of conditional mean and variance is computed as:

$$\begin{aligned} \mathrm{E}(\alpha_t|Y_{t-1}) &\equiv a_{t|t-1} \\ &= T_t a_{t|t-1} + c_t, \end{aligned}$$

$$\begin{aligned} \mathrm{E}(y_t|Y_{t-1}) &\equiv y_{t|t-1} \\ &= Z_t a_{t|t-1} + d_t, \end{aligned}$$

$$\mathrm{E}\big((\alpha_t - a_{t|t-1})(\alpha_t - a_{t|t-1})'|Y_{t-1}\big) \equiv \Sigma_{t|t-1},$$

$$\begin{aligned} &\mathrm{E}\big((y_t - y_{t|t-1})(\alpha_t - a_{t|t-1})'|Y_{t-1}\big) \\ &= Z_t \mathrm{E}\big((\alpha_t - a_{t|t-1})(\alpha_t - a_{t|t-1})'|Y_{t-1}\big) \\ &= Z_t \Sigma_{t|t-1}, \end{aligned}$$

$$\mathrm{E}\big((y_t - y_{t|t-1})(y_t - y_{t|t-1})'|Y_{t-1}\big) \equiv F_{t|t-1}.$$

Here, applying Lemma 3 to equation (2.22), the conditional distribution $P(\alpha_t|y_t, Y_{t-1}) \equiv P(\alpha_t|Y_t)$ is obtained as:

$$\alpha_t \sim N(a_{t|t-1} - K_t(y_t - y_{t|t-1}),\ \Sigma_{t|t-1} - K_t F_{t|t-1} K_t'),$$

where K_t is given by equation (2.13). Since $P(\alpha_t|Y_t)$ implies

$$\alpha_t \sim N(a_{t|t},\ \Sigma_{t|t}).$$

Comparing each argument, the following updating equations can be obtained.

$$\begin{aligned} a_{t|t} &= a_{t|t-1} - K_t(y_t - y_{t|t-1}), \\ \Sigma_{t|t} &= \Sigma_{t|t-1} - K_t F_{t|t-1} K_t', \end{aligned}$$

which are given by equations (2.14) and (2.15).

Thus, based on normality assumption, two derivations were introduced in this section. When the measurement and transition equations are linear and the error terms in the two equations are normal, the Kalman filter algorithm can be obtained from equations (2.16) and (2.17) to equations (2.9) – (2.15).

In the following two sections, the Kalman filter algorithm is derived without normality assumption.

2.3.2 Derivation by Mixed Estimation

Next, we derive the Kalman filter algorithm by utilizing the mixed estimation method (so-called Goldberger-Theil estimation), which is introduced by Harvey (1981), Diderrich (1985) and Tanizaki (1993b). According to this estimation method, we do not have to give a specific distribution to the error terms ϵ_t and η_t because it is based on the generalized least squares estimation.

The prediction equations are simply derived by taking the conditional expectation and variance given information up to time $t-1$ on both sides of equation (2.2). They are given by (2.9) and (2.10). α_t is rewritten based on prior information as follows. α_t consists of one-step ahead prediction $a_{t|t-1}$ and error term ξ_t, i.e.,

$$\alpha_t = a_{t|t-1} + \xi_t, \tag{2.23}$$

where mean and variance of ξ_t are given by:

$$\mathrm{E}(\xi_t|Y_{t-1}) = 0$$
$$\mathrm{Var}(\xi_t|Y_{t-1}) = \Sigma_{t|t-1}.$$

Again, we write the measurement equation (2.1) together with equation (2.23) as follows:

$$y_t = Z_t\alpha_t + d_t + S_t\epsilon_t, \tag{2.1}$$

$$a_{t|t-1} = \alpha_t - \xi_t. \tag{2.23}$$

Equation (2.23) represents the prior information and the present sample is described by equation (2.1). Accordingly, equations (2.1) and (2.23) hold all the information available at time t. By the assumption of no correlation between ϵ_t and η_t, the error terms ϵ_t and ξ_t are independently distributed as follows:

$$\mathrm{E}\begin{pmatrix}\epsilon_t\\ \xi_t\end{pmatrix} = \begin{pmatrix}0\\ 0\end{pmatrix}, \qquad \mathrm{Var}\begin{pmatrix}\epsilon_t\\ \xi_t\end{pmatrix} = \begin{pmatrix}H_t & 0\\ 0 & \Sigma_{t|t-1}\end{pmatrix},$$

where we do not have to assume a functional form of the distribution of the error terms ϵ_t and ξ_t.

Equations (2.1) and (2.23) are rewritten as:

$$\begin{pmatrix}y_t - d_t\\ a_{t|t-1}\end{pmatrix} = \begin{pmatrix}Z_t\\ I_k\end{pmatrix}\alpha_t + \begin{pmatrix}S_t & 0\\ 0 & -I_k\end{pmatrix}\begin{pmatrix}\epsilon_t\\ \xi_t\end{pmatrix}. \tag{2.24}$$

Let $a_{t|t}$ be the generalized least squares estimate of the state-variable α_t in equation (2.24). Apply the generalized least squares method (GLS) to (2.24), and we can obtain $a_{t|t}$ which is the filtering estimate of α_t.

$$\begin{aligned}
a_{t|t} &= \left((Z_t' \;\; I_k)\begin{pmatrix}S_tH_tS_t' & 0\\ 0 & \Sigma_{t|t-1}\end{pmatrix}^{-1}\begin{pmatrix}Z_t\\ I_k\end{pmatrix}\right)^{-1}\\
&\qquad\times(Z_t' \;\; I_k)\begin{pmatrix}S_tH_tS_t' & 0\\ 0 & \Sigma_{t|t-1}\end{pmatrix}^{-1}\begin{pmatrix}y_t - d_t\\ a_{t|t-1}\end{pmatrix}\\
&= \left(Z_t'(S_tH_tS_t')^{-1}Z_t + \Sigma_{t|t-1}^{-1}\right)^{-1}\\
&\qquad\times\left(Z_t'(S_tH_tS_t')^{-1}(y_t - d_t) + \Sigma_{t|t-1}^{-1}a_{t|t-1}\right).
\end{aligned} \tag{2.25}$$

Note that we have:

$$\mathrm{Var}\left(\begin{pmatrix}S_t & 0\\ 0 & -I_k\end{pmatrix}\begin{pmatrix}\epsilon_t\\ \xi_t\end{pmatrix}\right) = \begin{pmatrix}S_tH_tS_t' & 0\\ 0 & \Sigma_{t|t-1}\end{pmatrix}.$$

The above GLS estimate of α_t obtained from equation (2.24), i.e., $a_{t|t}$ in equation (2.25), implies $\mathrm{E}(\alpha_t|Y_t)$, because the present sample y_t and the past information Y_{t-1} are included in equation (2.24).

And it is easily shown that:

$$\begin{aligned}\Sigma_{t|t} &= (Z_t'(S_tH_tS_t')^{-1}Z_t + \Sigma_{t|t-1}^{-1})^{-1} \\ &= \Sigma_{t|t-1} - \Sigma_{t|t-1}Z_t'F_{t|t-1}^{-1}Z_t\Sigma_{t|t-1} \\ &= \Sigma_{t|t-1} - K_tF_{t|t-1}K_t', \end{aligned} \tag{2.26}$$

where

$$K_t = \Sigma_{t|t-1}Z_t'F_{t|t-1}^{-1}.$$

Note that Lemma 1 in Appendix A2.2 is used for the second equality in equation (2.26). Substituting equation (2.26) into equation (2.25), we can rewrite equation (2.25) as:

$$\begin{aligned}a_{t|t} &= \left(Z_t'(S_tH_tS_t')^{-1}Z_t + \Sigma_{t|t-1}^{-1}\right)^{-1} \\ &\qquad \times\left(Z_t'(S_tH_tS_t')^{-1}(y_t - d_t) + \Sigma_{t|t-1}^{-1}a_{t|t-1}\right) \\ &= \left(\Sigma_{t|t-1} - \Sigma_{t|t-1}Z_t'F_{t|t-1}^{-1}Z_t\Sigma_{t|t-1}\right) \\ &\qquad \times\left(Z_t'(S_tH_tS_t')^{-1}(y_t - d_t) + \Sigma_{t|t-1}^{-1}a_{t|t-1}\right) \\ &= a_{t|t-1} + \Sigma_{t|t-1}Z_t'F_{t|t-1}^{-1}(y_t - Z_ta_{t|t-1} - d_t) \\ &= a_{t|t-1} + K_t(y_t - y_{t|t-1}). \end{aligned} \tag{2.27}$$

Thus, the algorithm of Kalman filter is represented by equations (2.9) – (2.15).

(2.26) and (2.27) represent the updating equations. The updating problem is to utilize the added sample with the prior information to improve the parameter estimates.

The advantage of this approach is that we do not have to assume any distribution function for the error terms ϵ_t and η_t in equations (2.1) and (2.2).

2.3.3 Minimum Mean Square Linear Estimator

The Kalman filter as minimum mean square linear estimator do not require normality assumption for the error terms. Prediction equations (2.9) and (2.10) are similarly obtained by taking the conditional mean and variance given information up to time $t-1$ on both sides of the transition equation (2.2).

In order to derive the updating equations (2.13) – (2.15), first, note that $a_{t|t}$ is written as:

$$a_{t|t} = A_ta_{t-1|t-1} + B_tc_t + D_td_t + K_ty_t, \tag{2.28}$$

because $a_{t|t}$ denotes the conditional mean of α_t given information Y_t and the measurement and transition equations are linear (also, here we consider a linear estimator only). Therefore, As shown above, $a_{t|t}$ is represented by a linear function of the past information (i.e., $a_{t-1|t-1}$ which includes information up to time $t-1$, Y_{t-1}) and the present sample (i.e., c_t, d_t and y_t). We obtain A_t, B_t, D_t and K_t such that $a_{t|t}$ is the minimum mean square linear estimator.

Define $e_t = \alpha_t - a_{t|t}$. Then, e_t is transformed into:

$$\begin{aligned}
e_t &= \alpha_t - A_t a_{t-1|t-1} - B_t c_t - D_t d_t - K_t y_t \\
&= (T_t \alpha_{t-1} + c_t + R_t \eta_t) - A_t(\alpha_{t-1} - e_{t-1}) - B_t c_t - D_t d_t \\
&\qquad - K_t \Big(Z_t (T_t \alpha_{t-1} + c_t + R_t \eta_t) + d_t + S_t \epsilon_t \Big) \\
&= A_t e_{t-1} + (T_t - A_t - K_t Z_t T_t)\alpha_{t-1} + (I_k - B_t - K_t Z_t) c_t \\
&\qquad - (D_t + K_t) d_t + (I_k - K_t Z_t) R_t \eta_t - K_t S_t \epsilon_t.
\end{aligned}$$

In calculating, the followings are used.

$$\alpha_t = T_t \alpha_{t-1} + c_t + R_t \eta_t,$$

$$a_{t-1|t-1} = \alpha_{t-1} - e_{t-1},$$

$$\begin{aligned}
y_t &= Z_t \alpha_t + d_t + S_t \epsilon_t \\
&= Z_t (T_t \alpha_{t-1} + c_t + R_t \eta_t) + d_t + S_t \epsilon_t.
\end{aligned}$$

In order for $a_{t|t}$ to be unbiased, the expectation of e_t yields zero. Accordingly, we have:

$$\begin{aligned}
&T_t - A_t - K_t Z_t T_t = 0, \\
&I_k - B_t - K_t Z_t = 0, \\
&D_t + K_t = 0.
\end{aligned}$$

Therefore, eliminating A_t, B_t and D_t, e_t is rewritten as follows:

$$e_t = (I_k - K_t Z_t)(T_t e_{t-1} + R_t \eta_t) - K_t S_t \epsilon_t.$$

In the above equation, we still have K_t to be computed. Next, we have to obtain K_t such that e_t has minimum variance. Since e_{t-1}, η_t and ϵ_t are mutually independent, variance of e_t, i.e., $\Sigma_{t|t}$, is defined as:

$$\begin{aligned}
\Sigma_{t|t} &= (I_k - K_t Z_t)(T_t \Sigma_{t-1|t-1} T_t' + R_t Q_t R_t')(I_k - K_t Z_t)' + K_t S_t H_t S_t' K_t' \\
&= (I_k - K_t Z_t)\Sigma_{t|t-1}(I_k - K_t Z_t)' + K_t S_t H_t S_t' K_t' \qquad (2.29)
\end{aligned}$$

Note that variance of e_t can be defined as $\Sigma_{t|t}$ because e_t includes information up to time t, i.e., Y_t.

It is easily shown that e_t has minimum variance when K_t takes the following:

$$K_t = \Sigma_{t|t-1} Z_t' F_{t|t-1}^{-1},$$

where

$$F_{t|t-1} = Z_t \Sigma_{t|t-1} Z_t' + S_t H_t S_t',$$

which denotes the conditional variance of y_t given information at time $t-1$, Y_{t-1}. Taking into account equations (2.9) and (2.10), equations (2.28) and (2.29) are represented as:

$$a_{t|t} = a_{t|t-1} + K_t(y_t - Z_t a_{t|t-1} - d_t),$$

$$\Sigma_{t|t} = \Sigma_{t|t-1} - K_t F_{t|t-1} K_t'.$$

Thus, the Kalman filter algorithm is derived. K_t is called the Kalman gain, because it is obtained such that $a_{t|t}$ has minimum variance.

In this section, we focused on derivation of the Kalman filter algorithm. The same algorithm was derived from the three different ways; derivation based on normal density in Section 2.3.1, and distribution-free derivations in Sections 2.3.2 and 2.3.3. In Section 2.3.3, the Kalman filter is derived as the minimum mean square estimator. Using this derivation, we can easily obtain the filtering algorithm even when the error terms ϵ_t and η_t in the state-space model represented by equations (2.1) and (2.10) are correlated with each other. See Burridge and Wallis (1988) and Harvey (1989).

2.4 Estimation of Unknown Parameters

Finally, we discuss estimation of unknown parameters, using the maximum likelihood function procedure. In this section, the likelihood function to be maximized is constructed by assuming normality for the error terms ϵ_t and η_t.

In the case where Z_t, d_t, H_t, T_t, c_t, R_t and Q_t depend on the unknown parameter vector to be estimated, say θ, the following innovation form of the likelihood function is maximized with respect to θ (see Chow (1983) and Watson and Engle (1983)).

$$\begin{aligned} &P(y_T, y_{T-1}, \cdots, y_1) \\ &= P(y_T|Y_{T-1}) P(y_{T-1}|Y_{T-2}) \cdots P(y_2|y_1) P(y_1) \\ &= \prod_{t=1}^{T} P(y_t|Y_{t-1}) \\ &= \prod_{t=1}^{T} (2\pi)^{-g/2} |F_{t|t-1}|^{-1/2} \exp\left(-\frac{1}{2}(y_t - y_{t|t-1})' F_{t|t-1}^{-1} (y_t - y_{t|t-1}) \right). \end{aligned} \tag{2.30}$$

Note that, in equation (2.30), the initial density function $P(y_1)$ is approximated as $P(y_1|Y_0)$. According to this maximization procedure, we cannot obtain a solution in closed form. The simplicity of using the innovation form is that $y_{t|t-1}$ and $F_{t|t-1}$ are computed in the filtering algorithm (2.9) – (2.15), where no extra algorithm for the maximization procedure is required. Therefore, the innovation form (2.30) is broadly used for the maximum likelihood procedure.

For another estimation method, the expectation of the log-likelihood function is maximized. This estimation is called the EM algorithm, where unobservable variables are replaced by the conditionally expected values given all the observed data $\{y_1, y_2, \cdots, y_T\}$, i.e., Y_T. Since the state-space model is originally developed for estimation of unobservable variables, the Kalman filter model is consistent with the EM algorithm in the concept and therefore the EM algorithm is often used for estimation of unknown parameters in the state-space form. See Shumway and Stoffer (1982), Tanizaki (1989, 1993b) and Watson and Engle (1983) for an application of the EM algorithm to the Kalman filter model and Dempster, Laird and Rubin (1977) and Rund (1991) for the EM algorithm. It is known that according to the EM algorithm the convergence is very slow, but it quickly searches the neighborhood of the true parameter. In the case where Z_t, d_t, H_t, T_t and c_t are linear in the unknown parameter vector θ, a solution of unknown parameters can be obtained explicitly by a function of the smoothed estimates. See Watson and Engle (1983). Thus, the expected log-likelihood function is maximized with respect to the unknown parameter vector (i.e., θ), when θ is included in the measurement and transition equations. The problem of using the EM algorithm is that we need the smoothed estimates and therefore the extra algorithm is necessary, which implies a greater computational burden.

In the proceeding chapters, the innovation form (2.30), rather than the expected likelihood function, is maximized in order to obtain parameter estimates, because of the reduced computational burden.

2.5 Summary

In this chapter, by the way of introduction to the Kalman filter, we have discussed some applications of the state-space model, the conventional linear recursive algorithm of Kalman filter, its derivations and the estimation problem of unknown parameters.

For the derivations, there are several approaches. Here we took three derivations; the first is based on normality assumption, the second is based on the mixed estimation and the third is taken as the minimum mean square linear estimator. The advantage of the mixed estimation approach and the minimum mean square linear estimator is that we do not have to assume normality for the error terms. The mixed estimation approach is useful for deriving the single-stage iteration filtering algorithm, which is one of the nonlinear filters introduced in the next chapter. The approach under the assumption of normality is used for deriving the nonlinear and/or nonnormal filtering algorithms based on the density functions, which are discussed in Chapter 4.

Finally, in the case where the unknown parameters are included in the measurement and transition equations, the maximum likelihood method is used. Here we consider maximizing the innovation form of the likelihood function.

A2 Appendix

A2.1 Density-Based Filtering Algorithm

We prove the third equality in (2.16) and the fifth equality in (2.17), where $P(\alpha_t|\alpha_{t-1}, Y_{t-1})$ is replaced by $P(\alpha_t|\alpha_{t-1})$ in equation (2.16) and $P(y_t|\alpha_t, Y_{t-1})$ is by $P(y_t|\alpha_t)$ in equation (2.17).

We give the proof more generally by using the nonlinear forms given by equations (2.20) and (2.21), which are the following state-space model:

$$\text{(Measurement equation)} \qquad q_t(y_t, \alpha_t) = \epsilon_t, \tag{2.20}$$

$$\text{(Transition equation)} \qquad f_t(\alpha_t, \alpha_{t-1}) = \eta_t. \tag{2.21}$$

Let $P_\epsilon(\epsilon_t)$ and $P_\eta(\eta_t)$ be the probability density functions of ϵ_t and η_t. Then, the joint density function of $\eta_t, \cdots, \eta_1$, $\epsilon_{t-1}, \cdots, \epsilon_1$ is given by:

$$P(\eta_t, \cdots, \eta_1, \epsilon_{t-1}, \cdots, \epsilon_1) = P_\eta(\eta_t) \cdots P_\eta(\eta_1) P_\epsilon(\epsilon_{t-1}) \cdots P_\epsilon(\epsilon_1),$$

because $\eta_t, \cdots, \eta_1$, $\epsilon_{t-1}, \cdots, \epsilon_1$ are mutually independent. By transforming the variables to obtain the joint density of $\alpha_t, \cdots, \alpha_1$, $y_{t-1}, \cdots, y_1$, we have

$$\begin{aligned}
&P(\alpha_t, \cdots, \alpha_1, y_{t-1}, \cdots, y_1) \\
&= P(\alpha_t, \cdots, \alpha_1, Y_{t-1}) \\
&= \prod_{s=1}^{t-1} \left| \frac{\partial q_s(y_s, \alpha_s)}{\partial y_s'} \right| P_\epsilon\big(q_s(y_s, \alpha_s)\big) \prod_{s=1}^{t} \left| \frac{\partial f_s(\alpha_s, \alpha_{s-1})}{\partial \alpha_s'} \right| P_\eta\big(f_s(\alpha_s, \alpha_{s-1})\big).
\end{aligned}$$

Here, we can take as:

$$P(y_s|\alpha_s) = \left| \frac{\partial q_s(y_s, \alpha_s)}{\partial y_s'} \right| P_\epsilon\big(q_s(y_s, \alpha_s)\big),$$

which is a conditional density of y_s, given α_s. Similarly, $P(\alpha_s|\alpha_{s-1})$ is given as follows:

$$P(\alpha_s|\alpha_{s-1}) = \left| \frac{\partial f_s(\alpha_s, \alpha_{s-1})}{\partial \alpha_s'} \right| P_\eta\big(f_s(\alpha_s, \alpha_{s-1})\big),$$

which is derived from the density $P_\eta(\eta_s)$.

Therefore, the joint density of $\alpha_t, \cdots, \alpha_1, y_{t-1}, \cdots, y_1$ is:

$$
\begin{aligned}
&P(\alpha_t, \cdots, \alpha_1, Y_{t-1}) \\
&= P(\alpha_t|\alpha_{t-1}) \cdots P(\alpha_1|\alpha_0) P(y_{t-1}|\alpha_{t-1}) \cdots P(y_1|\alpha_1),
\end{aligned}
$$

where α_0 is given. Next, to obtain $P(\alpha_t, \alpha_{t-1}, Y_{t-1})$, the above joint-density function is integrated with respect to $\alpha_{t-2}, \cdots, \alpha_1$.

$$
\begin{aligned}
&P(\alpha_t, \alpha_{t-1}, Y_{t-1}) \\
&= \int \cdots \int P(\alpha_t, \cdots, \alpha_1, Y_{t-1}) \mathrm{d}\alpha_{t-2} \cdots \mathrm{d}\alpha_1 \\
&= \int \cdots \int P(\alpha_t|\alpha_{t-1}) \cdots P(\alpha_1|\alpha_0) \\
&\qquad \times P(y_{t-1}|\alpha_{t-1}) \cdots P(y_1|\alpha_1) \mathrm{d}\alpha_{t-2} \cdots \mathrm{d}\alpha_1 \\
&= P(\alpha_t|\alpha_{t-1}) P(y_{t-1}|\alpha_{t-1}) \\
&\qquad \times \int \cdots \int P(\alpha_{t-1}|\alpha_{t-2}) \cdots P(\alpha_1|\alpha_0) \\
&\qquad\qquad \times P(y_{t-2}|\alpha_{t-2}) \cdots P(y_1|\alpha_1) \mathrm{d}\alpha_{t-2} \cdots \mathrm{d}\alpha_1 \\
&= P(\alpha_t|\alpha_{t-1}) P(y_{t-1}|\alpha_{t-1}) P(\alpha_{t-1}, Y_{t-2}),
\end{aligned}
$$

where the joint density $P(\alpha_{t-1}, Y_{t-2})$ is given by:

$$
\begin{aligned}
&P(\alpha_{t-1}, Y_{t-2}) \\
&= \int \cdots \int P(\alpha_t|\alpha_{t-1}) \cdots P(\alpha_1|\alpha_0) \\
&\qquad \times P(y_{t-1}|\alpha_{t-1}) \cdots P(y_1|\alpha_1) \mathrm{d}\alpha_{t-2} \cdots \mathrm{d}\alpha_1.
\end{aligned}
$$

To obtain the joint density of Y_{t-1}, $P(\alpha_t, \alpha_{t-1}, Y_{t-1})$ is integrated with respect to α_t and α_{t-1}, i.e.,

$$
\begin{aligned}
P(Y_{t-1}) &= \iint P(\alpha_t|\alpha_{t-1}) P(y_{t-1}|\alpha_{t-1}) P(\alpha_{t-1}, Y_{t-2}) \mathrm{d}\alpha_t \mathrm{d}\alpha_{t-1} \\
&= \int P(y_{t-1}|\alpha_{t-1}) P(\alpha_{t-1}, Y_{t-2}) \mathrm{d}\alpha_{t-1}.
\end{aligned}
$$

Therefore, the conditional density function of α_t and α_{t-1} is as follows:

$$
\begin{aligned}
P(\alpha_t, \alpha_{t-1}|Y_{t-1}) &= \frac{P(\alpha_t, \alpha_{t-1}, Y_{t-1})}{P(Y_{t-1})} \\
&= \frac{P(\alpha_t|\alpha_{t-1}) P(y_{t-1}|\alpha_{t-1}) P(\alpha_{t-1}, Y_{t-2})}{\int P(y_{t-1}|\alpha_{t-1}) P(\alpha_{t-1}, Y_{t-2}) \mathrm{d}\alpha_{t-1}} \\
&= P(\alpha_t|\alpha_{t-1}) \frac{P(y_{t-1}|\alpha_{t-1}) P(\alpha_{t-1}, Y_{t-2})}{\int P(y_{t-1}|\alpha_{t-1}) P(\alpha_{t-1}, Y_{t-2}) \mathrm{d}\alpha_{t-1}} \\
&= P(\alpha_t|\alpha_{t-1}) \frac{P(y_{t-1}|\alpha_{t-1}) P(\alpha_{t-1}|Y_{t-2})}{\int P(y_{t-1}|\alpha_{t-1}) P(\alpha_{t-1}|Y_{t-2}) \mathrm{d}\alpha_{t-1}}
\end{aligned}
$$

$$= P(\alpha_t|\alpha_{t-1})P(\alpha_{t-1}|Y_{t-1}),$$

where we can consider $P(\alpha_{t-1}|Y_{t-1})$ as follows:

$$P(\alpha_{t-1}|Y_{t-1}) = \frac{P(y_{t-1}|\alpha_{t-1})P(\alpha_{t-1}|Y_{t-2})}{\displaystyle\int P(y_{t-1}|\alpha_{t-1})P(\alpha_{t-1}|Y_{t-2})\mathrm{d}\alpha_{t-1}},$$

which is equivalent to equation (2.17). Accordingly, in equation (2.16), we have:

$$P(\alpha_t|Y_{t-1}) = P(\alpha_t|\alpha_{t-1})P(\alpha_{t-1}|Y_{t-1}).$$

Thus, the third equality in (2.16) and the fifth equality in (2.17) are proved, i.e., $P(\alpha_t|\alpha_{t-1}, Y_{t-1})$ is replaced by $P(\alpha_t|\alpha_{t-1})$ in equation (2.16) and $P(y_t|\alpha_t, Y_{t-1})$ is by $P(y_t|\alpha_t)$ in equation (2.17).

A2.2 Conditional Normal Distribution

Preliminaries: Useful Lemmas.

Lemma 1: Let A and C be $n \times n$ and $m \times m$ nonsingular matrices, and B be a $n \times m$ matrix. Then, we have the following formula.

$$(A + BCB')^{-1} = A^{-1} - A^{-1}B(C^{-1} + B'A^{-1}B)^{-1}B'A^{-1}.$$

Lemma 2: Let A, B and D be $k \times k$, $k \times g$ and $g \times g$ matrices, where A and D are nonsingular. Then, we have the following formula.

$$\begin{pmatrix} A & B \\ B' & D \end{pmatrix}^{-1} = \begin{pmatrix} X & Y \\ Y' & Z \end{pmatrix},$$

where X, Y and Z are as follows:

$$X = (A - BD^{-1}B')^{-1}$$
$$= A^{-1} + A^{-1}B(D - B'A^{-1}B)^{-1}B'A^{-1},$$

$$Y = -(A - BD^{-1}B')^{-1}BD^{-1}$$
$$= -A^{-1}B(D - B'A^{-1}B)^{-1},$$

$$Z = (D - B'A^{-1}B)^{-1}$$
$$= D^{-1} + D^{-1}B'(A - BD^{-1}B')^{-1}BD^{-1}.$$

Note that Lemma 1 is used to obtain the second equality in X and Z.

Lemma 3: Let y and x be $k \times 1$ and $g \times 1$ vectors of random variables, which are distributed as the following bivariate normal density:

$$\begin{pmatrix} y \\ x \end{pmatrix} \sim N\left(\begin{pmatrix} \mu_y \\ \mu_x \end{pmatrix}, \begin{pmatrix} \Sigma_{yy} & \Sigma_{yx} \\ \Sigma_{xy} & \Sigma_{xx} \end{pmatrix} \right),$$

which implies a joint distribution of x and y, i.e., $P(x,y)$.

Then, the conditional density of y given x (i.e., $P(y|x)$) follows a normal distribution with mean

$$\mu_y + \Sigma_{yx}\Sigma_{xx}^{-1}(x-\mu_x)$$

and variance

$$\Sigma_{yy} - \Sigma_{yx}\Sigma_{xx}^{-1}\Sigma_{xy}.$$

Moreover, the marginal density of x (i.e., $P(x)$) is also a normal distribution with mean μ_x and variance Σ_{xx}. Accordingly, we have the following:

$$P(y|x) = \Phi\big(y-\mu_y-\Sigma_{yx}\Sigma_{xx}^{-1}(x-\mu_x), \Sigma_{yy}-\Sigma_{yx}\Sigma_{xx}^{-1}\Sigma_{xy}\big),$$

$$P(x) = \Phi(x-\mu_x, \Sigma_{xx}).$$

Note that $P(x,y) = P(y|x)P(x)$.

Lemma 4: Let A, B, C and D be $k \times k$, $k \times g$, $g \times k$ and $g \times g$ matrices, where A and D are nonsingular. Then, the following formula on the determinant is obtained.

$$\begin{vmatrix} A & B \\ C & D \end{vmatrix} = |D||A - BD^{-1}C|.$$

Applying the above formula to variance in Lemma 3, the following relationship holds:

$$\begin{vmatrix} \Sigma_{yy} & \Sigma_{yx} \\ \Sigma_{xy} & \Sigma_{xx} \end{vmatrix} = |\Sigma_{xx}||\Sigma_{yy} - \Sigma_{yx}\Sigma_{xx}^{-1}\Sigma_{xy}|,$$

where A, B, C and D correspond to Σ_{yy}, Σ_{yx}, Σ_{xy} and Σ_{xx}, respectively.

Proof I. In this appendix, we prove the following equality:

$$\int \Phi(\alpha_t - T_t\alpha_{t-1} - c_t, R_tQ_tR_t')\Phi(\alpha_{t-1} - a_{t-1|t-1}, \Sigma_{t-1|t-1})d\alpha_{t-1}$$
$$= \Phi(\alpha_t - T_ta_{t-1|t-1} - c_t, T_t\Sigma_{t-1|1-1}T_t' + R_tQ_tR_t'),$$

which is in equation (2.18).

For simplicity of discussion, each variable is re-defined as follows:

$$
\begin{aligned}
&x = \alpha_{t-1} - a_{t-1|t-1}, \\
&\Sigma_{xx} = \Sigma_{t-1|t-1}, \\
&y = \alpha_t - T_t a_{t-1|t-1} - c_t, \\
&\Sigma_{yy} = R_t Q_t R_t', \\
&A = T_t.
\end{aligned}
$$

Substituting each variable, the equality to be proved is given by:

$$\int \Phi(y - Ax, \Sigma_{yy})\Phi(x, \Sigma_{xx})\mathrm{d}x = \Phi(y, A\Sigma_{xx}A' + \Sigma_{yy}).$$

The two normal distributions $\Phi(y - Ax, \Sigma_{yy})$ and $\Phi(x, \Sigma_{xx})$ are written as:

$$\Phi(y - Ax, \Sigma_{yy}) = (2\pi)^{-g/2}|\Sigma_{yy}|^{-1/2} \exp\left(-\frac{1}{2}(y - Ax)'\Sigma_{yy}^{-1}(y - Ax)\right),$$

$$\Phi(x, \Sigma_{xx}) = (2\pi)^{-k/2}|\Sigma_{xx}|^{-1/2} \exp\left(-\frac{1}{2}x'\Sigma_{xx}^{-1}x\right).$$

The dimensions of y and x are given by $g \times 1$ and $k \times 1$, respectively. A denotes a $g \times k$ matrix. Note that $g = k$ in Proof I but $g \neq k$ in Proof II.

A product of these two normal densities is transformed into:

$$
\begin{aligned}
&\Phi(y - Ax, \Sigma_{yy})\Phi(x, \Sigma_{xx}) \\
&= (2\pi)^{-g/2}|\Sigma_{yy}|^{-1/2} \exp\left(-\frac{1}{2}(y - Ax)'\Sigma_{yy}^{-1}(y - Ax)\right) \\
&\times (2\pi)^{-k/2}|\Sigma_{xx}|^{-1/2} \exp\left(-\frac{1}{2}x'\Sigma_{xx}^{-1}x\right) \\
&= (2\pi)^{-(g+k)/2}|\Sigma_{yy}|^{-1/2}\ |\Sigma_{xx}|^{-1/2} \\
&\times \exp\left(-\frac{1}{2}\begin{pmatrix} y - Ax \\ x \end{pmatrix}'\begin{pmatrix} \Sigma_{yy} & 0 \\ 0 & \Sigma_{xx} \end{pmatrix}^{-1}\begin{pmatrix} y - Ax \\ x \end{pmatrix}\right) \\
&= (2\pi)^{-(g+k)/2}|\Sigma_{yy}|^{-1/2}\ |\Sigma_{xx}|^{-1/2} \\
&\times \exp\left(-\frac{1}{2}\begin{pmatrix} y \\ x \end{pmatrix}'\begin{pmatrix} I_g & -A \\ 0 & I_k \end{pmatrix}'\begin{pmatrix} \Sigma_{yy} & 0 \\ 0 & \Sigma_{xx} \end{pmatrix}^{-1}\begin{pmatrix} I_g & -A \\ 0 & I_k \end{pmatrix}\begin{pmatrix} y \\ x \end{pmatrix}\right) \\
&= (2\pi)^{-(g+k)/2}|\Sigma_{yy}|^{-1/2}\ |\Sigma_{xx}|^{-1/2} \\
&\times \exp\left(-\frac{1}{2}\begin{pmatrix} y \\ x \end{pmatrix}'\begin{pmatrix} I_g & A \\ 0 & I_k \end{pmatrix}'^{-1}\begin{pmatrix} \Sigma_{yy} & 0 \\ 0 & \Sigma_{xx} \end{pmatrix}^{-1}\begin{pmatrix} I_g & A \\ 0 & I_k \end{pmatrix}^{-1}\begin{pmatrix} y \\ x \end{pmatrix}\right) \\
&= (2\pi)^{-(g+k)/2}|\Sigma_{yy}|^{-1/2}\ |\Sigma_{xx}|^{-1/2}
\end{aligned}
$$

$$\times \exp\left(-\frac{1}{2}\begin{pmatrix} y \\ x \end{pmatrix}' \left(\begin{pmatrix} I_g & A \\ 0 & I_k \end{pmatrix}\begin{pmatrix} \Sigma_{yy} & 0 \\ 0 & \Sigma_{xx} \end{pmatrix}\begin{pmatrix} I_g & A \\ 0 & I_k \end{pmatrix}'\right)^{-1}\begin{pmatrix} y \\ x \end{pmatrix}\right)$$

$$= (2\pi)^{-(g+k)/2}|\Sigma_{yy}|^{-1/2}\,|\Sigma_{xx}|^{-1/2}$$

$$\times \exp\left(-\frac{1}{2}\begin{pmatrix} y \\ x \end{pmatrix}' \begin{pmatrix} A\Sigma_{xx}A' + \Sigma_{yy} & A\Sigma_{xx} \\ \Sigma_{xx}A' & \Sigma_{xx} \end{pmatrix}^{-1}\begin{pmatrix} y \\ x \end{pmatrix}\right).$$

Note that, in deriving the above equation, the inverse of the following matrix is used.

$$\begin{pmatrix} I_g & -A \\ 0 & I_k \end{pmatrix}^{-1} = \begin{pmatrix} I_g & A \\ 0 & I_k \end{pmatrix}.$$

Furthermore, using Lemma 4, we have:

$$\begin{vmatrix} A\Sigma_{xx}A' + \Sigma_{yy} & A\Sigma_{xx} \\ \Sigma_{xx}A' & \Sigma_{xx} \end{vmatrix} = |\Sigma_{xx}||\Sigma_{yy}|.$$

Accordingly, the joint distribution of x and y is represented by the following bivariate normal distribution.

$$\begin{pmatrix} y \\ x \end{pmatrix} \sim N\left(\begin{pmatrix} 0 \\ 0 \end{pmatrix}, \begin{pmatrix} A\Sigma_{xx}A' + \Sigma_{yy} & A\Sigma_{xx} \\ \Sigma_{xx}A' & \Sigma_{xx} \end{pmatrix}\right), \tag{2.31}$$

From Lemma 3, the marginal density of y is given by:

$$P(y) = \Phi(y, A\Sigma_{xx}A' + \Sigma_{yy}),$$

which implies that

$$P(\alpha_t|Y_{t-1}) = \Phi(\alpha_t - T_t a_{t-1|t-1} - c_t, T_t\Sigma_{t-1|1-1}T_t' + R_tQ_tR_t').$$

Thus, the third equality in equation (2.18) is derived.

Proof II. We prove that the following equation holds.

$$\begin{aligned} &\Phi(y_t - Z_t\alpha_t - d_t, S_tH_tS_t')\Phi(\alpha_t - a_{t|t-1}, \Sigma_{t|t-1}) \\ &= \Phi\big(\alpha_t - a_{t|t-1} - K_t(y_t - y_{t|t-1}), \Sigma_{t|t-1} - K_tF_{t|t-1}K_t'\big) \\ &\qquad \times\Phi(y_t - y_{t|t-1}, F_{t|t-1}), \end{aligned}$$

which is utilized in equation (2.19).

Similarly, we re-define each variable as follows:

$$\begin{aligned} &x = \alpha_t - a_{t|t-1}, \\ &\Sigma_{xx} = \Sigma_{t|t-1}, \\ &y = y_t - Z_t a_{t|t-1} - d_t, \\ &\Sigma_{yy} = S_tH_tS_t', \\ &A = Z_t, \end{aligned}$$

where x is a $k \times 1$ vector, y is a $g \times 1$ vector, and A is a $g \times k$ matrix.

By taking into account the following three equations,

$$
\begin{aligned}
y_{t|t-1} &= Z_t a_{t|t-1} + d_t, \\
F_{t|t-1} &= Z_t \Sigma_{t|t-1} Z_t' + S_t H_t S_t', \\
K_t &= \Sigma_{t|t-1} Z_t' F_{t|t-1}^{-1},
\end{aligned}
$$

the equation to be proved is represented by:

$$
\begin{aligned}
&\Phi(y - Ax, \Sigma_{yy}) \Phi(x, \Sigma_{xx}) \\
&= \Phi\big(x - \Sigma_{xx} A' (A \Sigma_{xx} A' + \Sigma_{yy})^{-1} y, \\
&\qquad \Sigma_{xx} - \Sigma_{xx} A' (A \Sigma_{xx} A' + \Sigma_{yy})^{-1} A \Sigma_{xx}\big) \Phi(y, A \Sigma_{xx} A' + \Sigma_{yy}).
\end{aligned}
$$

We can prove the above equation in the exactly same fashion as Proof I, and accordingly the joint density of x and y (i.e., $P(x, y)$) follows the same distribution as equation (2.31).

Using Lemma 3, the conditional density of x given y (i.e., $P(x|y)$) and the marginal density of y (i.e., $P(y)$) are derived as:

$$
\begin{aligned}
&P(x, y) \\
&= P(x|y) P(y) \\
&= \Phi\big(x - \Sigma_{xx} A' (A \Sigma_{xx} A' + \Sigma_{yy})^{-1} y, \\
&\qquad \Sigma_{xx} - \Sigma_{xx} A' (A \Sigma_{xx} A' + \Sigma_{yy})^{-1} A \Sigma_{xx}\big) \Phi(y, A \Sigma_{xx} A' + \Sigma_{yy}),
\end{aligned}
$$

where

$$
\begin{aligned}
P(x|y) &= \Phi\big(x - \Sigma_{xx} A' (A \Sigma_{xx} A' + \Sigma_{yy})^{-1} y, \\
&\qquad \Sigma_{xx} - \Sigma_{xx} A' (A \Sigma_{xx} A' + \Sigma_{yy})^{-1} A \Sigma_{xx}\big), \\
P(y) &= \Phi(y, A \Sigma_{xx} A' + \Sigma_{yy}).
\end{aligned}
$$

Therefore, the following equation is obtained.

$$
\begin{aligned}
&\Phi(y_t - Z_t \alpha_t - d_t, S_t H_t S_t') \Phi(\alpha_t - a_{t|t-1}, \Sigma_{t|t-1}) \\
&= \Phi\big(\alpha_t - a_{t|t-1} - K_t (y_t - y_{t|t-1}), \Sigma_{t|t-1} - K_t F_{t|t-1} K_t'\big) \\
&\qquad \times \Phi(y_t - y_{t|t-1}, F_{t|t-1}).
\end{aligned}
$$

Note that we have the following:

$$
\Phi(y_t - y_{t|t-1}, F_{t|t-1}) = \int P(y_t|\alpha_t) P(\alpha_t|Y_{t-1}) \mathrm{d}\alpha_t.
$$

Thus, we can obtain the third equality of equation (2.19).

3. Traditional Nonlinear Filters

3.1 Introduction

We have been dealing with the linear transition and measurement equations. The equations derived from economic theories, however, are nonlinear in general. Unless the distributions are normal and the measurement and transition equations are linear, we cannot derive the explicit expression for the filtering algorithm. Therefore, some approximation is necessary for estimation. In Chapter 3, the nonlinear filtering algorithms are derived by approximating the nonlinear measurement and transition equations based on the Taylor series expansions, which are known as the traditional nonlinear filters.

In this chapter, first we aim to analyze what is approximated in order to obtain the algorithms of the extended Kalman filter (EKF) and the second-order nonlinear filter (SNF). It is shown that some approximations have to be made for derivation of the nonlinear filtering algorithms; one is approximating the nonnormal error terms (i.e., residuals) to be normal and another approximation is assuming that the error terms are mutually independent even if they are clearly correlated. Also, an additional approximation has to be imposed on the nonlinear transition and measurement equations to evaluate the expectations included in the algorithms.

Next, the nonlinear filter proposed in Tanizaki and Mariano (1996) is discussed, which is a combination of the extended Kalman filter (EKF) and the Monte-Carlo simulation method. We call this nonlinear filter the Monte-Carlo simulations applied to the first-order (extended) Kalman filter, or simply, the Monte-Carlo simulation filter (MSF). As pointed out in Mariano and Brown (1983, 1989) and Brown and Mariano (1984, 1989), approximating the expectation of a nonlinear function by the Taylor series expansion gives us the biased estimate, and accordingly, we might have a better approximation of the expectation if random numbers are artificially generated for the error terms. This Monte-Carlo approach is applied to the extended Kalman filter (EKF). Moreover, it is shown that, under certain conditions, the error terms treated as the residuals are uncorrelated with each other. Two theorems are introduced.

Finally, it is shown that the single-stage iteration filter (SIF) is derived by the mixed estimation approach, which gives us a different interpretation from the other three filters introduced in this chapter. The extended Kalman filter (EKF), the second-order nonlinear filter (SNF) and the Monte-Carlo simulation filter (MSF) are based on normality assumption for the approximated error terms (i.e., residuals), where the error terms (i.e., residuals) are approximated as the normal error terms even if they are not normal. The single-stage iteration filter (SIF) can be derived without assuming any distribution for the error terms, which is based on the nonlinear generalized least squares estimation method. Also, in Wishner, Tabaczynski and Athans (1969) and Gelb (1974), the single-stage iteration filter (SIF) was introduced. However, their explanation was not clear about the derivation (see Appendix A3.3). Therefore, in this chapter, the single-stage iteration filter (SIF) is reinterpreted by the mixed estimation method, so-called Goldberger-Theil estimator.

We treat the nonlinear filtering problem, where the state-space model is specified as follows:

$$\text{(Measurement equation)} \qquad y_t = h_t(\alpha_t, \epsilon_t), \tag{3.1}$$

$$\text{(Transition equation)} \qquad \alpha_t = g_t(\alpha_{t-1}, \eta_t), \tag{3.2}$$

$$y_t\colon g \times 1, \quad \epsilon_t\colon g \times 1, \quad \alpha_t\colon k \times 1, \quad \eta_t\colon k \times 1,$$

where functions $h_t(\cdot,\cdot)$ and $g_t(\cdot,\cdot)$ may depend on the other exogenous variables (here, we omit them for simplicity). Assume in this chapter that ϵ_t and η_t are independently distributed as random vectors, i.e.,

$$\begin{pmatrix} \epsilon_t \\ \eta_t \end{pmatrix} \sim \left(\begin{pmatrix} 0 \\ 0 \end{pmatrix}, \begin{pmatrix} H_t & 0 \\ 0 & Q_t \end{pmatrix} \right),$$

which are not necessarily normal.

Moreover, we assume that the functional forms of $h_t(\cdot,\cdot)$ and $g_t(\cdot,\cdot)$ are known and that they are invertible, i.e.,

$$y_t = h_t(\alpha_t, \epsilon_t) \quad \text{implies} \quad q_t(y_t, \alpha_t) = \epsilon_t,$$

$$\alpha_t = g_t(\alpha_{t-1}, \eta) \quad \text{implies} \quad f_t(\alpha_t, \alpha_{t-1}) = \eta_t.$$

Originally, $q_t(y_t, \alpha_t) = \epsilon_t$ and $f_t(\alpha_t, \alpha_{t-1}) = \eta_t$ are the equations which are derived from economic theory and are referred to as the structural equations. The reduced form given by equations (3.1) and (3.2), rather than the structural from given by equations (2.20) and (2.21), however, is used for the derivations of nonlinear filters, because it is easier to handle the problem with the reduced form.

Given the above setup, several nonlinear filtering algorithms are derived in this and the next chapters. The most heuristic method to derive the nonlinear filtering algorithm utilizes the Taylor series expansions. See Anderson and Moore (1979), Gelb (1974) and Wishner, Tabaczynski and Athans (1969). However, it is well-known that the estimate of α_t, given Y_t, is biased because of nonlinearity of the functions $h_t(\cdot,\cdot)$ and $g_t(\cdot,\cdot)$. To avoid this problem, we can exclude the bias by generating random numbers for the error terms (see Tanizaki and Mariano (1996)). This method is known as the Monte-Carlo stochastic simulation method and it is applied to the nonlinear filter which is called the Monte-Carlo simulation filter (MSF). It is known that the bias disappears asymptotically according to the stochastic simulation (see, for example, Brown and Mariano (1984, 1989) and Mariano and Brown (1983, 1989)).

In this chapter, we derive the filtering algorithms using the first- and second-order Taylor series approximations. Another approach, discussed in Chapter 4, comes from approximating the probability density functions, i.e., $P(\alpha_t|Y_{t-1})$ and $P(\alpha_t|Y_t)$. The algorithms derived from probability densities are more essential than those from the Taylor series expansions, and their estimates are more precise in the sense that they give us asymptotically unbiased filtering estimates.

3.2 Re-consideration of Taylor Series Expansions

Approaches using the Taylor series expansions are common in the literature. A lot of references deal with the measurement and transition equations which are nonlinear in the state-vector but linear in the error term (see Appendix A3.1). Here, we consider the measurement and transition equations which are nonlinear in both the state-variable and the error term, i.e., equations (3.1) and (3.2).

In order to apply the standard linear recursive Kalman filter algorithm represented by equations (2.9) – (2.15), the nonlinear functions $h_t(\alpha_t, \epsilon_t)$ and $g_t(\alpha_{t-1}, \eta_t)$ have to be linearized with respect to (α_t, ϵ_t) and (α_{t-1}, η_t), respectively. Therefore, we first consider the following state-space model which represent the exact relationships, not approximations.

(Measurement equation)

$$\begin{aligned} y_t &= h_t(\alpha_t, \epsilon_t) \\ &= c_{t|t-1} + Z_{t|t-1}(\alpha_t - a_{t|t-1}) + u_t, \end{aligned} \tag{3.3}$$

(Transition equation)

$$\begin{aligned} \alpha_t &= g_t(\alpha_{t-1}, \eta_t) \\ &= d_{t|t-1} + T_{t|t-1}(\alpha_{t-1} - a_{t-1|t-1}) + v_t, \end{aligned} \tag{3.4}$$

where $c_{t|t-1}$, $Z_{t|t-1}$, $d_{t|t-1}$ and $T_{t|t-1}$ depend on the information set Y_{t-1}. The measurement equation (3.3) is transformed to the linear function of α_t, depending on Y_{t-1}, and the transition equation (3.4) is also linear in α_{t-1}. Here, we can take any functions of Y_{t-1} for $c_{t|t-1}$, $Z_{t|t-1}$, $d_{t|t-1}$ and $T_{t|t-1}$ in the above equations (3.3) and (3.4). Note that the expectations of u_t and v_t are not necessarily zero. Thus, the somewhat *ad hoc* but more general linearization is given in equations (3.3) and (3.4). Such a linearization is taken in order to analyze what has to be approximated in the case of applying the Taylor series expansions to the nonlinear state-space model.

Example of $c_{t|t-1}$, $Z_{t|t-1}$ and u_t. As an example of the above transformation, consider the function:

$$h_t(\alpha_t, \epsilon_t) = \exp(\alpha_t)\exp(\epsilon_t).$$

(1) Then, based on equation (3.3), we can transform the function as follows:

$$\begin{aligned} h_t(\alpha_t, \epsilon_t) &= \exp(\alpha_t)\exp(\epsilon_t) \\ &= \exp(a_{t|t-1}) + \exp(a_{t|t-1})(\alpha_t - a_{t|t-1}) + u_t, \end{aligned}$$

where we may choose:

$$\begin{aligned} c_{t|t-1} &= \exp(a_{t|t-1}), \\ Z_{t|t-1} &= \exp(a_{t|t-1}), \\ u_t &= \exp(\alpha_t + \epsilon_t) - \exp(a_{t|t-1}) - \exp(a_{t|t-1})(\alpha_t - a_{t|t-1}). \end{aligned}$$

(2) Also, we can transform as:

$$\begin{aligned} h_t(\alpha_t, \epsilon_t) &= \exp(\alpha_t)\exp(\epsilon_t) \\ &= \exp(a_{t|t-1} + \frac{1}{2}\Sigma_{t|t-1} + \frac{1}{2}\sigma_\epsilon^2) \\ &\quad + \exp(a_{t|t-1} + \frac{1}{2}\Sigma_{t|t-1} + \frac{1}{2}\sigma_\epsilon^2)(\alpha_t - a_{t|t-1}) + u_t, \end{aligned}$$

where wc can take:

$$\begin{aligned} c_{t|t-1} &= \exp(a_{t|t-1} + \frac{1}{2}\Sigma_{t|t-1} + \frac{1}{2}\sigma_\epsilon^2), \\ Z_{t|t-1} &= \exp(a_{t|t-1} + \frac{1}{2}\Sigma_{t|t-1} + \frac{1}{2}\sigma_\epsilon^2), \\ u_t &= \exp(\alpha_t + \epsilon_t) - \exp(a_{t|t-1} + \frac{1}{2}\Sigma_{t|t-1} + \frac{1}{2}\sigma_\epsilon^2) \\ &\quad - \exp(a_{t|t-1} + \frac{1}{2}\Sigma_{t|t-1} + \frac{1}{2}\sigma_\epsilon^2)(\alpha_t - a_{t|t-1}). \end{aligned}$$

(3) Moreover, the following transformation is also possible.

$$\begin{aligned} h_t(\alpha_t, \epsilon_t) &= \exp(\alpha_t)\exp(\epsilon_t) \\ &= 1 + \exp(a_{t|t-1})(\alpha_t - a_{t|t-1}) + u_t, \end{aligned}$$

where

$$\begin{aligned} c_{t|t-1} &= 1, \\ Z_{t|t-1} &= \exp(a_{t|t-1}), \\ u_t &= \exp(\alpha_t + \epsilon_t) - 1 - \exp(a_{t|t-1})(\alpha_t - a_{t|t-1}). \end{aligned}$$

The three examples of the transformation are taken. Thus, we can take any functions for $c_{t|t-1}$, $Z_{t|t-1}$ and u_t, which are not unique.

Note that the linear recursive algorithm as in equations (2.9) – (2.15) can be obtained under the condition that u_t and v_t are normal with zero means and uncorrelated with each other, even if $c_{t|t-1}$, $Z_{t|t-1}$, $d_{t|t-1}$ and $T_{t|t-1}$ are functions of Y_{t-1}, which is the lagged endogenous variable, not the exogenous one. See Harvey (1989). Clearly, however, both error terms u_t and v_t are not normally distributed, because the error terms are represented as the residuals, i.e.,

$$\begin{aligned} u_t &= y_t - c_{t|t-1} - Z_{t|t-1}(\alpha_t - a_{t|t-1}) \\ &= h_t(\alpha_t, \epsilon_t) - c_{t|t-1} - Z_{t|t-1}(\alpha_t - a_{t|t-1}), \end{aligned} \tag{3.5}$$

$$\begin{aligned} v_t &= \alpha_t - d_{t|t-1} - T_{t|t-1}(\alpha_{t-1} - a_{t-1|t-1}) \\ &= g_t(\alpha_{t-1}, \eta_t) - d_{t|t-1} - T_{t|t-1}(\alpha_{t-1} - a_{t-1|t-1}). \end{aligned} \tag{3.6}$$

As we can see, it is clear from the structure of equations (3.5) and (3.6) that u_t and v_t are not necessarily normal even if ϵ_t and η_t are normal.

In equations (3.3) and (3.4), there are some other problems. First of all, clearly, both u_t and v_t do not necessarily have zero means. Next, the basic assumptions on the error terms are: in the measurement equation the state-vector is uncorrelated with the error term, and similarly, in the transition equation the lagged state-vector is not correlated with the error term, i.e., $\mathrm{E}(\alpha_t\epsilon_t')$ $=\mathrm{E}(\alpha_{t-1}\eta_t') = 0$ in equations (2.1) and (2.2). This is not the case, however, because u_t is a function of $(\alpha_t - a_{t|t-1})$ and also v_t is a function of $(\alpha_{t-1} - a_{t-1|t-1})$. Thus, u_t is correlated with $(\alpha_t - a_{t|t-1})$, and in addition, v_t is also correlated with $(\alpha_{t-1} - a_{t-1|t-1})$. Furthermore, u_t is correlated with v_t. Summarizing the above facts, we have the following problems on u_t and v_t:

(i) $\mathrm{E}(u_t|Y_{t-1}) \neq 0$ and $\mathrm{E}(v_t|Y_{t-1}) \neq 0$,
(ii) $\mathrm{E}\big(u_t(\alpha_t - a_{t|t-1})'|Y_{t-1}\big) \neq 0$ and $\mathrm{E}\big(v_t(\alpha_{t-1} - a_{t|t-1})'|Y_{t-1}\big) \neq 0$,
(iii) $\mathrm{E}(u_t v_t'|Y_{t-1}) \neq 0$,
(iv) u_t and v_t are not normal.

Thus, all of the assumptions required for the recursive algorithm are violated in the equations (3.3) and (3.4).

Now, we consider eliminating the correlation between u_t and $(\alpha_t - a_{t|t-1})$ and the correlation between v_t and $(\alpha_{t-1} - a_{t|t-1})$, transforming the error terms u_t and v_t . The following theorem found in a usual textbook is helpful for the transformation.

Theorem: Let x and y be vectors of random variables, which are distributed as

$$\mathrm{E}\begin{pmatrix} x \\ y \end{pmatrix} = \begin{pmatrix} \mu_x \\ \mu_y \end{pmatrix}, \qquad \mathrm{Var}\begin{pmatrix} x \\ y \end{pmatrix} = \begin{pmatrix} \Sigma_{xx} & \Sigma_{xy} \\ \Sigma_{xy}' & \Sigma_{yy} \end{pmatrix}.$$

Define $\widetilde{x}$ as:

$$\widetilde{x} = x - \Sigma_{xy}\Sigma_{yy}^{-1}(y - \mu_y).$$

Then, $(\widetilde{x} - \mu_x)$ is uncorrelated with $(y - \mu_y)$, i.e.,

$$\mathrm{E}\big((\widetilde{x} - \mu_x)(y - \mu_y)\big) = 0.$$

For the measurement equation, to exclude the correlation between u_t and α_t, we transform u_t using the above theorem. Since the expectations corresponding to μ_x, μ_y, Σ_{xy} and Σ_{yy} are as follows:

$$\begin{aligned}
\mu_x &= \mathrm{E}(u_t|Y_{t-1}) = y_{t|t-1} - c_{t|t-1},\\
\mu_y &= \mathrm{E}(\alpha_t|Y_{t-1}) = \mathrm{E}\big(g_t(\alpha_{t-1},\eta_t)|Y_{t-1}\big) = a_{t|t-1},\\
\Sigma_{xy} &= \mathrm{Cov}(u_t,\alpha_t|Y_{t-1}) = M_{t|t-1} - Z_{t|t-1}\Sigma_{t|t-1},\\
\Sigma_{yy} &= \mathrm{Var}(\alpha_t|Y_{t-1}) = \Sigma_{t|t-1},
\end{aligned}$$

where

$$\begin{aligned}
M_{t|t-1} &= \mathrm{E}\big((y_t - y_{t|t-1})(\alpha_t - a_{t|t-1})'|Y_{t-1}\big),\\
y_{t|t-1} &= \mathrm{E}(y_t|Y_{t-1}) = \mathrm{E}\big(h_t(\alpha_t,\epsilon_t)|Y_{t-1}\big).
\end{aligned}$$

The transformed error term $\widetilde{u}_t$ can be represented as:

$$\begin{aligned}
\widetilde{u}_t &= u_t - y_{t|t-1} + c_{t|t-1} - (M_{t|t-1} - Z_{t|t-1}\Sigma_{t|t-1})\Sigma_{t|t-1}^{-1}(\alpha_t - a_{t|t-1})\\
&= y_t - y_{t|t-1} - M_{t|t-1}\Sigma_{t|t-1}^{-1}(\alpha_t - a_{t|t-1}).
\end{aligned}$$

Now, by performing the transformation, clearly we have:

$$\begin{aligned}
&\mathrm{E}(\widetilde{u}_t|Y_{t-1}) = 0,\\
&\mathrm{Cov}(\widetilde{u}_t,\alpha_t|Y_{t-1}) = 0.
\end{aligned}$$

By the transformation shown in Theorem, we have the error term with zero-mean and no correlation.

Similarly, for the transition equation, we eliminate the correlation between α_{t-1} and v_t. Each expectation is given by:

$$\begin{aligned}
\mu_x &= \mathrm{E}(v_t|Y_{t-1}) = a_{t|t-1} - d_{t|t-1},\\
\mu_y &= \mathrm{E}(\alpha_{t-1}|Y_{t-1}) = a_{t-1|t-1},\\
\Sigma_{xy} &= \mathrm{Cov}(v_t,\alpha_{t-1}|Y_{t-1}) = N_{t|t-1} - T_{t|t-1}\Sigma_{t-1|t-1},\\
\Sigma_{yy} &= \mathrm{Var}(\alpha_{t-1}|Y_{t-1}) = \Sigma_{t-1|t-1},
\end{aligned}$$

where

$$N_{t|t-1} = \mathrm{E}\big((\alpha_t - a_{t|t-1})(\alpha_{t-1} - a_{t-1|t-1})'|Y_{t-1}\big).$$

The transformed error term $\widetilde{v}_t$ is represented as:

$$\begin{aligned}\widetilde{v}_t &= v_t - a_{t|t-1} + d_{t|t-1} \\ &\quad - (N_{t|t-1} - T_{t|t-1}\Sigma_{t-1|t-1})\Sigma_{t-1|t-1}^{-1}(\alpha_{t-1} - a_{t-1|t-1}) \\ &= \alpha_t - a_{t|t-1} - N_{t|t-1}\Sigma_{t-1|t-1}^{-1}(\alpha_{t-1} - a_{t-1|t-1}).\end{aligned}$$

Now, by performing the transformation, clearly we have:

$$\mathrm{E}(\widetilde{v}_t|Y_{t-1}) = 0,$$
$$\mathrm{Cov}(\widetilde{v}_t, \alpha_{t-1}|Y_{t-1}) = 0.$$

Thus, we have the error term with zero-mean and no correlation.

From the transformation shown above, $\widetilde{u}_t$ is not correlated with $(\alpha_t - a_{t|t-1})$ and at the same time $(\alpha_{t-1} - a_{t-1|t-1})$ is uncorrelated with $\widetilde{v}_t$. Moreover, we have: $\mathrm{E}(\widetilde{u}_t|Y_{t-1}) = \mathrm{E}(\widetilde{v}_t|Y_{t-1}) = 0$.

The measurement equation and the transition equation in equations (3.3) and (3.4) can then be modified as the following state-space model:

(Measurement equation)

$$y_t = y_{t|t-1} + M_{t|t-1}\Sigma_{t|t-1}^{-1}(\alpha_t - a_{t|t-1}) + \widetilde{u}_t, \tag{3.7}$$

(Transition equation)

$$\alpha_t = a_{t|t-1} + N_{t|t-1}\Sigma_{t-1|t-1}^{-1}(\alpha_{t-1} - a_{t-1|t-1}) + \widetilde{v}_t, \tag{3.8}$$

where

$$\begin{aligned} y_{t|t-1} &= \mathrm{E}\big(h_t(\alpha_t, \epsilon_t)|Y_{t-1}\big), \\ M_{t|t-1} &= \mathrm{E}\big((y_t - y_{t|t-1})(\alpha_t - a_{t|t-1})'|Y_{t-1}\big), \\ N_{t|t-1} &= \mathrm{E}\big((\alpha_t - a_{t|t-1})(\alpha_{t-1} - a_{t-1|t-1})'|Y_{t-1}\big). \end{aligned}$$

The conditional variances of $\widetilde{u}_t$ and $\widetilde{v}_t$ are given by:

$$\mathrm{Var}(\widetilde{u}_t|Y_{t-1}) = F_{t|t-1} - M_{t|t-1}\Sigma_{t|t-1}^{-1}M'_{t|t-1},$$
$$\mathrm{Var}(\widetilde{v}_t|Y_{t-1}) = \Sigma_{t|t-1} - N_{t|t-1}\Sigma_{t-1|t-1}^{-1}N'_{t|t-1},$$

where

$$F_{t|t-1} = \mathrm{E}\big((y_t - y_{t|t-1})(y_t - y_{t|t-1})'|Y_{t-1}\big).$$

Hereafter, we call $\widetilde{u}_t$ and $\widetilde{v}_t$ the residuals to distinguish them from the error terms ϵ_t and η_t.

$\widetilde{u}_t$ and $\widetilde{v}_t$ have zero mean and no correlation with the state-vector In general, however, $\widetilde{u}_t$ is still correlated with $\widetilde{v}_t$ in equations (3.7) and (3.8). Furthermore, the residuals $\widetilde{u}_t$ and $\widetilde{v}_t$ are not normally distributed.

Summarizing, the new residuals $\widetilde{u}_t$ and $\widetilde{v}_t$ have the fewer problems than the original residuals u_t and v_t, which are:

(i) $\mathrm{E}(\widetilde{u}_t|Y_{t-1}) = 0$ and $\mathrm{E}(\widetilde{v}_t|Y_{t-1}) = 0$,
(ii) $\mathrm{E}\big(\widetilde{u}_t(\alpha_t - a_{t|t-1})'|Y_{t-1}\big) = 0$ and $\mathrm{E}\big(\widetilde{v}_t(\alpha_{t-1} - a_{t|t-1})'|Y_{t-1}\big) = 0$,

(iii) $\mathrm{E}(\widetilde{u}_t \widetilde{v}_t' | Y_{t-1}) \neq 0$,
(iv) $\widetilde{u}_t$ and $\widetilde{v}_t$ are not normal.

The problems (i) and (ii) are improved but we still have the problems (iii) and (iv). Ignoring these nonnormal error terms and their correlation, we apply equations (3.7) and (3.8) to the standard Kalman filter algorithm (2.9) – (2.15) introduced in Chapter 2, and the following filtering algorithm can be obtained.

$$a_{t|t-1} = \mathrm{E}\big(g_t(\alpha_{t-1}, \eta_t) | Y_{t-1}\big), \tag{3.9}$$
$$\Sigma_{t|t-1} = \mathrm{E}\big((\alpha_t - a_{t|t-1})(\alpha_t - a_{t|t-1})' | Y_{t-1}\big), \tag{3.10}$$
$$y_{t|t-1} = \mathrm{E}\big(h_t(\alpha_t, \epsilon_t) | Y_{t-1}\big), \tag{3.11}$$
$$F_{t|t-1} = \mathrm{E}\big((y_t - y_{t|t-1})(y_t - y_{t|t-1})' | Y_{t-1}\big), \tag{3.12}$$
$$M_{t|t-1} = \mathrm{E}\big((y_t - y_{t|t-1})(\alpha_t - a_{t|t-1})' | Y_{t-1}\big), \tag{3.13}$$
$$K_t = M_{t|t-1}' F_{t|t-1}, \tag{3.14}$$
$$\Sigma_{t|t} = \Sigma_{t|t-1} - K_t F_{t|t-1} K_t', \tag{3.15}$$
$$a_{t|t} = a_{t|t-1} + K_t (y_t - y_{t|t-1}), \tag{3.16}$$

for $t = 1, \cdots, T$.

Note that we have the standard Kalman filter algorithm if the nonlinear functions $h_t(\alpha_t, \epsilon_t)$ and $g_t(\alpha_{t-1}, \eta_t)$ are linear in the state-variables and the error terms. In the case of nonlinear measurement and transition equations, however, the above algorithm is not operational, because the expectations of the nonlinear functions are included in the algorithm (3.9) – (3.16). Therefore, next we need to approximate the nonlinear measurement and transition functions in order to evaluate the expectations (3.9) – (3.13).

Before evaluating the expectations, first, consider an estimation problem in the case where an unknown parameter is included in $h_t(\cdot,\cdot)$ and $g_t(\cdot,\cdot)$, i.e., $h_t = h_t(\alpha_t, \epsilon_t; \theta)$ and $g_t = g_t(\alpha_{t-1}, \eta_t; \theta)$. Since we assume in equations (3.7) and (3.8) that the residuals $\widetilde{u}_t$ and $\widetilde{v}_t$ are distributed as approximately normal random variables, y_t is also approximately normal with mean $y_{t|t-1}$ and variance $F_{t|t-1}$. Using Bayes' formula, we have the log-likelihood function shown in equation (2.30), which is the innovation form of likelihood function. Note that in equation (2.30) $y_{t|t-1}$ and $F_{t|t-1}$ are approximated by the first-order Taylor series expansion (Section 3.2.1), and the second-order Taylor series expansion (Section 3.2.2) or the Monte-Carlo stochastic simulations (Section 3.2.3). The unknown parameter θ is included in $y_{t|t-1}$ and $F_{t|t-1}$. Thus, the approximated likelihood function (2.30) is maximized with respect to the unknown parameter θ, and we can obtain their estimates.

Now, in the above algorithm (3.9) – (3.16) and the likelihood function (2.30), we need to evaluate the expectations given by equations (3.9) – (3.13).

Until now, we have made some approximations on $\widetilde{u}_t$ and $\widetilde{v}_t$. One is assuming that the residuals $\widetilde{u}_t$ and $\widetilde{v}_t$ are approximately normally distributed, and another approximation is no correlation between them. Moreover, we make one additional approximation on the above expectations included in equations (3.9) – (3.13). Three nonlinear filters are derived by approximating the expectations (see Appendix A3.2) as follows.

3.2.1 Extended Kalman Filter

In the case where we approximate the two nonlinear functions $h_t(\alpha_t, \epsilon_t)$ and $g_t(\alpha_{t-1}, \eta_t)$ around $(\alpha_t, \epsilon_t) = (a_{t|t-1}, 0)$ and $(\alpha_{t-1}, \eta_t) = (a_{t-1|t-1}, 0)$ by the first-order Taylor series expansion in order to evaluate the expectations (3.9) – (3.13) explicitly, the extended Kalman filter (EKF, or the first-order nonlinear filter) can be obtained by applying the linearized nonlinear functions to the algorithm (3.9) – (3.16). See Anderson and Moore (1979), Gelb (1974), Harvey (1989), Tanizaki and Mariano (1996) and Wishner, Tabaczynski and Athans (1969) for the extended Kalman filter (EKF, or the first-order nonlinear filter).

In order to approximate the expectations included in equations (3.9) – (3.16), the measurement equation and the transition equation are approximated as the following state-space model:

(Measurement equation)

$$\begin{aligned} y_t &= h_t(\alpha_t, \epsilon_t) \\ &\approx h_{t|t-1} + Z_{t|t-1}(\alpha_t - a_{t|t-1}) + S_{t|t-1}\epsilon_t, \end{aligned} \tag{3.17}$$

(Transition equation)

$$\begin{aligned} \alpha_t &= g_t(\alpha_{t-1}, \eta_t) \\ &\approx g_{t|t-1} + T_{t|t-1}(\alpha_{t-1} - a_{t-1|t-1}) + R_{t|t-1}\eta_t, \end{aligned} \tag{3.18}$$

where

$$h_{t|t-1} = h_t(a_{t|t-1}, 0),$$

$$Z_{t|t-1} = \left.\frac{\partial h_t(\alpha_t, \epsilon_t)}{\partial \alpha_t'}\right|_{(\alpha_t, \epsilon_t) = (a_{t|t-1}, 0)},$$

$$S_{t|t-1} = \left.\frac{\partial h_t(\alpha_t, \epsilon_t)}{\partial \epsilon_t'}\right|_{(\alpha_t, \epsilon_t) = (a_{t|t-1}, 0)},$$

$$g_{t|t-1} = g_t(a_{t-1|t-1}, 0),$$

$$T_{t|t-1} = \left.\frac{\partial g_t(\alpha_{t-1}, \eta_t)}{\partial \alpha_{t-1}'}\right|_{(\alpha_{t-1}, \eta_t) = (a_{t-1|t-1}, 0)},$$

$$R_{t|t-1} = \left.\frac{\partial g_t(\alpha_{t-1}, \eta_t)}{\partial \eta_t'}\right|_{(\alpha_{t-1}, \eta_t) = (a_{t-1|t-1}, 0)}.$$

All of the functions $h_{t|t-1}$, $Z_{t|t-1}$, $a_{t|t-1}$, $S_{t|t-1}$, $g_{t|t-1}$, $T_{t|t-1}$, $a_{t-1|t-1}$ and $R_{t|t-1}$ depend on the information available at time $t-1$, i.e., Y_{t-1}. When these linearized measurement and transition equations, i.e., equations (3.17) and (3.18), are used to evaluate the expectations in equations (3.9) – (3.13), we can obtain the following filtering algorithm:

$$a_{t|t-1} = g_{t|t-1}, \tag{3.19}$$
$$\Sigma_{t|t-1} = T_{t|t-1}\Sigma_{t-1|t-1}T'_{t|t-1} + R_{t|t-1}Q_tR'_{t|t-1}, \tag{3.20}$$
$$y_{t|t-1} = h_{t|t-1}, \tag{3.21}$$
$$F_{t|t-1} = Z_{t|t-1}\Sigma_{t|t-1}Z'_{t|t-1} + S_{t|t-1}H_tS'_{t|t-1}, \tag{3.22}$$
$$M_{t|t-1} = Z_{t|t-1}\Sigma_{t|t-1}, \tag{3.23}$$
$$K_t = M'_{t|t-1}F^{-1}_{t|t-1}, \tag{3.14}$$
$$\Sigma_{t|t} = \Sigma_{t|t-1} - K_tF_{t|t-1}K'_t, \tag{3.15}$$
$$a_{t|t} = a_{t|t-1} + K_t(y_t - y_{t|t-1}), \tag{3.16}$$

for $t = 1, \cdots, T$.

The filtering algorithm is given by equations (3.19) – (3.23) and (3.14) – (3.16). In the likelihood function (2.30), $y_{t|t-1}$ and $F_{t|t-1}$ are given by equations (3.21) and (3.22).

It is known that the filtering estimate of α_t (i.e., $a_{t|t}$) is biased because one-step ahead prediction of α_t (i.e., $a_{t|t-1}$) is biased, which are caused by nonlinearity of the functions $h_t(\cdot,\cdot)$ and $g_t(\cdot,\cdot)$. Recall that we have $\mathrm{E}\big(g(x)\big) \neq g\big(\mathrm{E}(x)\big)$ for a nonlinear function $g(\cdot)$ and a random variable x. To reduce this problem and obtain the more precise filtered estimates, the second-order approximation can be applied to the nonlinear functions $h_t(\cdot,\cdot)$ and $g_t(\cdot,\cdot)$.

3.2.2 Second-Order Nonlinear Filter

Next, consider deriving the nonlinear filter algorithm by the second-order approximation of the functions $h_t(\cdot,\cdot)$ and $g_t(\cdot,\cdot)$. The estimates $a_{t|t}$ and $a_{t|t-1}$ give us more precise estimates compared with the first-order approximation, since we utilize extra information, i.e., the second-order terms which are ignored in the extended Kalman filter (EKF). See Gelb (1974), and Wishner, Tabaczynski and Athans (1969) for the second-order nonlinear filter.

Expanding the two nonlinear functions $h_t(\alpha_t, \epsilon_t)$ and $g_t(\alpha_{t-1}, \eta_t)$ around $(\alpha_t, \epsilon_t) = (a_{t|t-1}, 0)$ and $(\alpha_{t-1}, \eta_t) = (a_{t-1|t-1}, 0)$ by the second-order Taylor series expansion, we can obtain the following state-space model:

(Measurement equation)

$$\begin{aligned} y_t &= h_t(\alpha_t, \epsilon_t) \\ &\approx h_{t|t-1} + Z_{t|t-1}(\alpha_t - a_{t|t-1}) + S_{t|t-1}\epsilon_t \\ &\quad + \frac{1}{2}\sum_{j=1}^{g} \Psi_j(\alpha_t - a_{t|t-1})'A^j_{t|t-1}(\alpha_t - a_{t|t-1}) \end{aligned}$$

$$
\begin{aligned}
&+\frac{1}{2}\sum_{j=1}^{g}\Psi_j\epsilon_t' B_{t|t-1}^j\epsilon_t \\
&+\sum_{j=1}^{g}\Psi_j(\alpha_t-a_{t|t-1})'C_{t|t-1}^j\epsilon_t,
\end{aligned}
\tag{3.24}
$$

(Transition equation)

$$
\begin{aligned}
\alpha_t &= g_t(\alpha_{t-1},\eta_t) \\
&\approx g_{t|t-1}+T_{t|t-1}(\alpha_{t-1}-a_{t-1|t-1})+R_{t|t-1}\eta_t \\
&+\frac{1}{2}\sum_{j=1}^{k}\Psi_j(\alpha_{t-1}-a_{t-1|t-1})'D_{t|t-1}^j(\alpha_{t-1}-a_{t-1|t-1}) \\
&+\frac{1}{2}\sum_{j=1}^{k}\Psi_j\eta_t' E_{t|t-1}^j\eta_t \\
&+\sum_{j=1}^{k}\Psi_j(\alpha_{t-1}-a_{t-1|t-1})'F_{t|t-1}^j\eta_t,
\end{aligned}
\tag{3.25}
$$

where

$$
\begin{aligned}
A_{t|t-1}^j &= \left.\frac{\partial^2 h_{jt}(\alpha_t,\epsilon_t)}{\partial\alpha_t\alpha_t'}\right|(\alpha_t,\epsilon_t)=(a_{t|t-1},0), \\
B_{t|t-1}^j &= \left.\frac{\partial^2 h_{jt}(\alpha_t,\epsilon_t)}{\partial\epsilon_t\epsilon_t'}\right|(\alpha_t,\epsilon_t)=(a_{t|t-1},0), \\
C_{t|t-1}^j &= \left.\frac{\partial^2 h_{jt}(\alpha_t,\epsilon_t)}{\partial\epsilon_t\alpha_t'}\right|(\alpha_t,\epsilon_t)=(a_{t|t-1},0), \\
D_{t|t-1}^j &= \left.\frac{\partial^2 g_{jt}(\alpha_{t-1},\eta_t)}{\partial\alpha_{t-1}\alpha_{t-1}'}\right|(\alpha_{t-1},\eta_t)=(a_{t-1|t-1},0), \\
E_{t|t-1}^j &= \left.\frac{\partial^2 g_{jt}(\alpha_{t-1},\eta_t)}{\partial\eta_t\eta_t'}\right|(\alpha_{t-1},\eta_t)=(a_{t-1|t-1},0), \\
F_{t|t-1}^j &= \left.\frac{\partial^2 g_{jt}(\alpha_{t-1},\eta_t)}{\partial\eta_t\alpha_{t-1}'}\right|(\alpha_{t-1},\eta_t)=(a_{t-1|t-1},0), \\
\Psi_j &= (0,\cdots,0,1,0,\cdots,0)'.
\end{aligned}
$$

↗ j-th element

where $h_{jt}(\cdot,\cdot)$ and $g_{jt}(\cdot,\cdot)$ are the j-th elements of the vector functions $h_t(\cdot,\cdot)$ and $g_t(\cdot,\cdot)$.

When we evaluate each expectation in equations (3.9) – (3.16) by using equations (3.24) and (3.25), the second-order nonlinear filtering algorithm can be obtained. In the derivation procedure, the third- and the fourth-moments of $(\alpha_t - a_{t|t-1})$ and $(\alpha_{t-1} - a_{t-1|t-1})$ are required. Note that the third-moment of a normal distribution is zero and the fourth-moment is three times the second-moment. The filtering algorithm in this case is given by:

$$\begin{aligned} a_{t|t-1} = g_{t|t-1} &+ \frac{1}{2}\sum_{j=1}^{k} \Psi_j \mathrm{tr}(D^j_{t|t-1}\Sigma_{t-1|t-1}) \\ &+ \frac{1}{2}\sum_{j=1}^{k} \Psi_j \mathrm{tr}(E^j_{t|t-1}Q_t), \end{aligned} \tag{3.26}$$

$$\Sigma_{t|t-1} = T_{t|t-1}\sigma_{t-1|t-1}T'_{t|t-1} + R_{t|t-1}Q_t R'_{t|t-1} + \widetilde{D}_t + \widetilde{E}_t + \widetilde{F}_t, \tag{3.27}$$

$$y_{t|t-1} = h_{t|t-1} + \frac{1}{2}\sum_{j=1}^{g} \Psi_j \mathrm{tr}(A^j_{t|t-1}\Sigma_{t|t-1}) + \frac{1}{2}\sum_{j=1}^{g} \Psi_j \mathrm{tr}(B^j_{t|t-1}H_t), \tag{3.28}$$

$$F_{t|t-1} = Z_{t|t-1}\Sigma_{t|t-1}Z'_{t|t-1} + S_{t|t-1}H_t S'_{t|t-1} + \widetilde{A}_t + \widetilde{B}_t + \widetilde{C}_t, \tag{3.29}$$

$$M_{t|t-1} = Z_{t|t-1}\Sigma_{t|t-1}, \tag{3.30}$$

$$K_t = M'_{t|t-1}F^{-1}_{t|t-1}, \tag{3.14}$$

$$\Sigma_{t|t} = \Sigma_{t|t-1} - K_t F_{t|t-1} K'_t, \tag{3.15}$$

$$a_{t|t} = a_{t|t-1} + K_t(y_t - y_{t|t-1}), \tag{3.16}$$

where the (i,j)-th elements of $\widetilde{A}_t$, $\widetilde{B}_t$, $\widetilde{C}_t$, $\widetilde{D}_t$, $\widetilde{E}_t$ and $\widetilde{F}_t$ are given by $\widetilde{A}^{ij}_t$, $\widetilde{B}^{ij}_t$, $\widetilde{C}^{ij}_t$, $\widetilde{D}^{ij}_t$, $\widetilde{E}^{ij}_t$ and $\widetilde{F}^{ij}_t$, i.e.,

$$\widetilde{A}^{ij}_t = \frac{1}{2}\mathrm{tr}(A^i_{t|t-1}\Sigma_{t|t-1}A^j_{t|t-1}\Sigma_{t|t-1}),$$

$$\widetilde{B}^{ij}_t = \frac{1}{2}\mathrm{tr}(B^i_{t|t-1}H_t B^j_{t|t-1}H_t),$$

$$\widetilde{C}^{ij}_t = \mathrm{tr}(C^i_{t|t-1}H_{t|t-1}C^{j\prime}_{t|t-1}\Sigma_{t|t-1}),$$

$$\widetilde{D}^{ij}_t = \frac{1}{2}\mathrm{tr}(D^i_{t|t-1}\Sigma_{t-1|t-1}D^j_{t|t-1}\Sigma_{t-1|t-1}),$$

$$\widetilde{E}^{ij}_t = \frac{1}{2}\mathrm{tr}(E^i_{t|t-1}Q_t E^j_{t|t-1}Q_t),$$

$$\widetilde{F}^{ij}_t = \mathrm{tr}(F^i_{t|t-1}Q_t F^{j\prime}_{t|t-1}\Sigma_{t-1|t-1}).$$

The filtering algorithm is given by equations (3.26) – (3.30) and (3.14) – (3.16).

The second-order nonlinear filter (SNF) might be a less biased estimator than the extended Kalman filter (EKF) because the bias correction terms (i.e., the second-order terms) are included in the approximated nonlinear measurement and transition equations.

In the likelihood function (2.30), $y_{t|t-1}$ and $F_{t|t-1}$ are given by equations (3.28) and (3.29).

As discussed above, we have derived the extended Kalman filter (EKF) and the second-order nonlinear filter (SNF), approximating the nonlinear measurement and transition equations by the Taylor series expansions with respect to the state-vectors and the error terms. It is possible to consider the higher-order nonlinear filters, i.e., the third-order nonlinear filter, the fourth-order nonlinear filter, and so on (see Gelb (1974)). In the exactly same fashion, the higher-order nonlinear filters can be derived.

Even if the higher-order nonlinear filters give us less biased filtering estimates than the extended Kalman filter (EKF), the filtering estimates obtained from the higher-order nonlinear filters are still biased because the nonlinear functions are approximated ignoring the other higher-order terms. The Monte-Carlo stochastic simulation approach which follows might be less biased than the above two filters and the other higher-order nonlinear filters. There, for the error terms and the state-variables, the random numbers are artificially generated, based on the algorithm (3.9) – (3.16).

Also, note the following important point: when we approximate the expectations included in the algorithm (3.9) – (3.16) by the Taylor series expansion (for example, the first-order approximation, the second-order expansion and so on), the residuals $\widetilde{u}_t$ and $\widetilde{v}_t$ in equations (3.7) and (3.8) turn out to have non-zero means and be correlated with each other. Therefore, the problems of the above two nonlinear filters turn out to be:

(i) $\mathrm{E}(\widetilde{u}_t|Y_{t-1}) \neq 0$ and $\mathrm{E}(\widetilde{v}_t|Y_{t-1}) \neq 0$,

(ii) $\mathrm{E}\big(\widetilde{u}_t(\alpha_t - a_{t|t-1})'|Y_{t-1}\big) \neq 0$ and $\mathrm{E}\big(\widetilde{v}_t(\alpha_{t-1} - a_{t|t-1})'|Y_{t-1}\big) \neq 0$,

(iii) $\mathrm{E}(\widetilde{u}_t\widetilde{v}_t'|Y_{t-1}) \neq 0$,

(iv) $\widetilde{u}_t$ and $\widetilde{v}_t$ are not normal,

which are the exactly same as u_t and v_t in equations (3.3) and (3.4).

Thus we still have the exactly same problems as the errors u_t and v_t in equations (3.3) and (3.4), which are not improved at all. Accordingly, when we use the extended Kalman filter (EKF) and the other higher-order nonlinear filters, $\widetilde{u}_t$ and $\widetilde{v}_t$ have to be assumed as well-behaved error terms in order to apply the standard linear recursive algorithm even if they are not well-behaved.

In the following nonlinear filter, the above problems (i) and (ii) are asymptotically improved because normal random numbers are used for evaluation of the expectations.

3.2.3 Monte-Carlo Simulation Filter

According to the extended Kalman filter (EKF), the second-order nonlinear filter (SNF) and the other higher-order nonlinear filters, we have to expand the nonlinear measurement and transition equations with respect to the state-vectors α_t and α_{t-1} and the error terms ϵ_t and η_t to evaluate the expectations in equations (3.9) – (3.13).

Tanizaki (1991) and Tanizaki and Mariano (1996) proposed the new filtering algorithm based on the Monte-Carlo stochastic simulations, where the normal random numbers are generated for the error terms (ϵ_t and η_t) and the state-variables (α_t and α_{t-1}) to evaluate the expectations more precisely. This approach is also based on equations (3.7) and (3.8), which are linear in the state-variables α_t and α_{t-1}. Thus, the new algorithm is a combination of the extended Kalman filter and the Monte-Carlo stochastic simulations, which is called the first-order nonlinear filter with Monte-Carlo stochastic simulations or the Monte-Carlo simulation filter (MSF).

First, to obtain the algorithm of the nonlinear filter based on equations (3.9) – (3.16), consider approximating the nonlinear functions $h_t(\alpha_t, \epsilon_t)$ and $g_t(\alpha_{t-1}, \eta_t)$ by random numbers. Equations (3.9) – (3.13) are approximated by random draws. The filtering algorithm in this case is as follows:

$$a_{t|t-1} = \frac{1}{n}\sum_{i=1}^{n} a_{i,t|t-1}, \tag{3.31}$$

$$\begin{aligned}\Sigma_{t|t-1} &= \frac{1}{n}\sum_{i=1}^{n}(a_{i,t|t-1} - a_{t|t-1})(a_{i,t|t-1} - a_{t|t-1})' \\ &= \frac{1}{n}\sum_{i=1}^{n} a_{i,t|t-1}a'_{i,t|t-1} - a_{t|t-1}a'_{t|t-1},\end{aligned} \tag{3.32}$$

$$y_{t|t-1} = \frac{1}{n}\sum_{i=1}^{n} y_{i,t|t-1}, \tag{3.33}$$

$$\begin{aligned}F_{t|t-1} &= \frac{1}{n}\sum_{i=1}^{n}(y_{i,t|t-1} - y_{t|t-1})(y_{i,t|t-1} - y_{t|t-1})' \\ &= \frac{1}{n}\sum_{i=1}^{n} y_{i,t|t-1}y'_{i,t|t-1} - y_{t|t-1}y'_{t|t-1},\end{aligned} \tag{3.34}$$

$$\begin{aligned}M_{t|t-1} &= \frac{1}{n}\sum_{i=1}^{n}(y_{i,t|t-1} - y_{t|t-1})(a_{i,t|t-1} - a_{t|t-1})' \\ &= \frac{1}{n}\sum_{i=1}^{n} y_{i,t|t-1}a'_{i,t|t-1} - y_{t|t-1}a'_{t|t-1},\end{aligned} \tag{3.35}$$

$$K_t = M'_{t|t-1} F_{t|t-1}^{-1}, \tag{3.14}$$

$$\Sigma_{t|t} = \Sigma_{t|t-1} - K_t F_{t|t-1} K'_t, \tag{3.15}$$

$$a_{t|t} = a_{t|t-1} + K_t(y_t - y_{t|t-1}), \tag{3.16}$$

where

$$a_{i,t|t-1} = g_t(\alpha_{i,t-1|t-1}, \eta_{i,t}),$$

$$y_{i,t|t-1} = h_t(\alpha_{i,t|t-1}, \epsilon_{i,t}).$$

The random numbers $\alpha_{i,t-1|t-1}$, $\eta_{i,t}$, $\alpha_{i,t|t-1}$ and $\epsilon_{i,t}$ for $i = 1, \cdots, n$ are approximately generated from the following normal distributions:

$$\begin{pmatrix} \alpha_{i,t-1|t-1} \\ \eta_{i,t} \end{pmatrix} \sim N\left(\begin{pmatrix} a_{t-1|t-1} \\ 0 \end{pmatrix}, \begin{pmatrix} \Sigma_{t-1|t-1} & 0 \\ 0 & Q_t \end{pmatrix} \right),$$

$$\begin{pmatrix} \alpha_{i,t|t-1} \\ \epsilon_{i,t} \end{pmatrix} \sim N\left(\begin{pmatrix} a_{t|t-1} \\ 0 \end{pmatrix}, \begin{pmatrix} \Sigma_{t|t-1} & 0 \\ 0 & H_t \end{pmatrix} \right).$$

Approximating $\widetilde{u}_t$ and $\widetilde{v}_t$ to be normal is equivalent to approximating α_t and α_{t-1} given Y_{t-1} to be normal. Therefore, in the algorithm above, we may generate the normal random numbers for α_t and α_{t-1} to evaluate the expectations. We call this nonlinear filter the first-order nonlinear filter with Monte-Carlo stochastic simulations, or simply, the Monte-Carlo simulation filter (MSF). Thus, the filtering algorithm is given by equations (3.31) – (3.35) and (3.14) – (3.16).

Finally, for the likelihood function (2.30), $y_{t|t-1}$ and $F_{t|t-1}$ are given by equations (3.33) and (3.34).

Note that in equations (3.31) – (3.35) the following approximation is also possible:

$$a_{t|t-1} = \frac{1}{nm} \sum_{i=1}^{n} \sum_{j=1}^{m} a_{ij,t|t-1},$$

$$\begin{aligned} \Sigma_{t|t-1} &= \frac{1}{nm} \sum_{i=1}^{n} \sum_{i=1}^{m} (a_{ij,t|t-1} - a_{t|t-1})(a_{ij,t|t-1} - a_{t|t-1})' \\ &= \frac{1}{nm} \sum_{i=1}^{n} \sum_{i=1}^{m} a_{ij,t|t-1} a'_{ij,t|t-1} - a_{t|t-1} a'_{t|t-1}, \end{aligned}$$

$$y_{t|t-1} = \frac{1}{nm} \sum_{i=1}^{n} \sum_{i=1}^{m} y_{ij,t|t-1},$$

$$\begin{aligned} F_{t|t-1} &= \frac{1}{nm} \sum_{i=1}^{n} \sum_{i=1}^{m} (y_{ij,t|t-1} - y_{t|t-1})(y_{ij,t|t-1} - y_{t|t-1})' \\ &= \frac{1}{nm} \sum_{i=1}^{n} \sum_{i=1}^{m} y_{ij,t|t-1} y'_{ij,t|t-1} - y_{t|t-1} y'_{t|t-1}, \end{aligned}$$

$$M_{t|t-1} = \frac{1}{nm}\sum_{i=1}^{n}\sum_{i=1}^{m}(y_{ij,t|t-1} - y_{t|t-1})(a_{ij,t|t-1} - a_{t|t-1})'$$
$$= \frac{1}{nm}\sum_{i=1}^{n}\sum_{i=1}^{m} y_{ij,t|t-1}a'_{ij,t|t-1} - y_{t|t-1}a'_{t|t-1},$$

where

$$a_{ij,t|t-1} = g_t(\alpha_{i,t-1|t-1}, \eta_{j,t}),$$
$$y_{ij,t|t-1} = h_t(\alpha_{i,t|t-1}, \epsilon_{j,t}),$$

and one may use the normal random numbers for $\alpha_{i,t-1|t-1}$, $\alpha_{i,t|t-1}$, $\eta_{j,t}$ and $\epsilon_{j,t}$ for $i = 1, \cdots, n$ and $j = 1, \cdots, m$. n random draws are generated for the state-variables while m random draws are for the error terms. The above evaluation of expectations might be more precise but takes a longer time computationally.

According to this estimator, asymptotically, the problems (i) and (ii) on the residuals $\widetilde{u}_t$ and $\widetilde{v}_t$ are completely improved. However, we still have the problems (iii) and (iv).

The approximations for all of the algorithms introduced in this section are that we have assumed the distributions of $\widetilde{u}_t$ and $\widetilde{v}_t$ in equations (3.7) and (3.8) to be approximately normally distributed, and that we have ignored the correlation between $\widetilde{u}_t$ and $\widetilde{v}_t$. In order to apply the model (3.7) and (3.8) directly to the conventional Kalman filter algorithm, the residuals $\widetilde{u}_t$ and $\widetilde{v}_t$ must be conditionally normally distributed given Y_{t-1}. They are not normally distributed, however, because of nonlinearity of the functions $h_t(\cdot,\cdot)$ and $g_t(\cdot,\cdot)$. Moreover, $\widetilde{u}_t$ is correlated with $\widetilde{v}_t$. Approximation of the expectations in equations (3.9) – (3.13) is more appropriate for the Monte-Carlo simulation filter (MSF) than the extended Kalman filter (EKF) and the second-order nonlinear filter (SNF), because we have the four problems (i) – (iv) for the extended Kalman filter (EKF) and the other higher-order nonlinear filters but the two problems (iii) and (iv) for the Monte-Carlo simulation filter (MSF), which implies that the Monte-Carlo simulation filter (MSF) has less problems, compared with the extended Kalman filter (EKF), the second-order nonlinear filter (SNF) and the other higher-order nonlinear filters.

Next, we make clear the difference between the extended Kalman filter (EKF), the second-order nonlinear filter (SNF) and the Monte-Carlo simulation filter (MSF), taking an example of the following two expectations:

$$\mathrm{E}\big(h_t(\alpha_t, \epsilon_t)|Y_{t-1}\big)$$

and

$$\mathrm{E}\big(g_t(\alpha_{t-1}, \eta_t)|Y_{t-1}\big).$$

(1) The expectations in the extended Kalman filter (EKF) are approximated as:

$$\mathrm{E}\big(h_t(\alpha_t, \epsilon_t)|Y_{t-1}\big) \approx h_t(a_{t|t-1}, 0),$$
$$\mathrm{E}\big(g_t(\alpha_{t-1}, \eta_t)|Y_{t-1}\big) \approx g_t(a_{t-1|t-1}, 0).$$

Since $\mathrm{E}\big(g(x)\big) \neq g\big(\mathrm{E}(x)\big)$ for a nonlinear function $g(\cdot)$ and a random variable x, the equations above do not hold clearly, i.e.,

$$\mathrm{E}\big(h_t(\alpha_t, \epsilon_t)|Y_{t-1}\big) \neq h_t(a_{t|t-1}, 0),$$
$$\mathrm{E}\big(g_t(\alpha_{t-1}, \eta_t)|Y_{t-1}\big) \neq g_t(a_{t-1|t-1}, 0).$$

(2) In the second-order nonlinear filter (SNF), we approximate the expectations as follows:

$$\mathrm{E}\big(h_t(\alpha_t, \epsilon_t)|Y_{t-1}\big)$$
$$\approx h_{t|t-1} + \frac{1}{2}\sum_{j=1}^{g} \Psi_j \mathrm{tr}(A^j_{t|t-1}\Sigma_{t|t-1}) + \frac{1}{2}\sum_{j=1}^{g} \Psi_j \mathrm{tr}(B^j_{t|t-1}H_t),$$

$$\mathrm{E}\big(g_t(\alpha_{t-1}, \eta_t)|Y_{t-1}\big)$$
$$\approx g_{t-1|t-1} + \frac{1}{2}\sum_{j=1}^{k} \Psi_j \mathrm{tr}(D^j_{t|t-1}\Sigma_{t-1|t-1}) + \frac{1}{2}\sum_{j=1}^{k} \Psi_j \mathrm{tr}(E^j_{t|t-1}Q_t).$$

Since we have the bias correction terms (i.e., the second and the third terms in the right hand side of the above two equations), the estimates of the expectations are less biased than those of the extended Kalman filter. Since the nonlinear functions are approximated, however, evaluation of these expectations is still biased.

(3) According to the Monte-Carlo simulation filter (MSF), the expectations are approximated by:

$$\mathrm{E}\big(h_t(\alpha_t, \epsilon_t)|Y_{t-1}\big) \approx \frac{1}{n}\sum_{i=1}^{n} h_t(a_{i,t|t-1}, \epsilon_{i,t}),$$

$$\mathrm{E}\big(g_t(\alpha_{t-1}, \eta_t)|Y_{t-1}\big) \approx \frac{1}{n}\sum_{i=1}^{n} g_t(a_{i,t-1|t-1}, \eta_{i,t}).$$

Since each expectation is approximated by generating normal random numbers and evaluated more correctly, $a_{t|t}$ and $a_{t|t-1}$ might be less biased estimates than the extended Kalman filter and the other higher-order nonlinear filters.

The approximations for all the algorithms introduced in this section are that the distributions of $\widetilde{u}_t$ and $\widetilde{v}_t$ in equations (3.7) and (3.8) are assumed to be approximately normally distributed, and that correlation between $\widetilde{u}_t$ and $\widetilde{v}_t$ is ignored. In order to apply the model (3.7) and (3.8) directly to the algorithm (2.9) – (2.15), the residuals $\widetilde{u}_t$ and $\widetilde{v}_t$ must be conditionally normally distributed given Y_{t-1}. They are not normally distributed, however, because of nonlinearity of the functions $h_t(\cdot,\cdot)$ and $g_t(\cdot,\cdot)$. Moreover, $\widetilde{u}_t$ is correlated with $\widetilde{v}_t$. The error terms (i.e., residuals) in the extended Kalman filter (EKF) and the other higher-order nonlinear filters have non-zero means and are correlated with the state-variables, while those in the Monte-Carlo simulation approach do not have such problems. Therefore, strictly speaking, we cannot apply the model (3.7) and (3.8) to the linear recursive algorithm represented by equations (2.9) – (2.15). Here, however, $\widetilde{u}_t$ and $\widetilde{v}_t$ are approximated as mutually, independently and normally distributed error terms.

In the following section, we examine the condition under which we can relax one of the approximations. Under a certain functional form of either $h_t(\cdot,\cdot)$ or $g_t(\cdot,\cdot)$, it will be shown that $\widetilde{u}_t$ is uncorrelated with $\widetilde{v}_t$.

3.3 Theorems

In this section, two theorems proposed by Tanizaki (1991) and Tanizaki and Mariano (1996) are introduced, which are related to one of the approximations on the residuals $\widetilde{u}_t$ and $\widetilde{v}_t$.

It is shown that the correlation between the residuals $\widetilde{u}_t$ and $\widetilde{v}_t$ in equations (3.7) and (3.8) disappears under a certain functional form of either $h_t(\cdot,\cdot)$ or $g_t(\cdot,\cdot)$. This implies that in the case where we choose the functional form we do not need to take into account one of the approximations made clear in the previous section. This is useful for the Monte-Carlo simulation filter (MSF) introduced in the last section, because it has zero-mean errors uncorrelated with the state-vectors, i.e., it has less problems than the extended Kalman filter (EKF), the second-order nonlinear filter (SNF) and the other higher-order nonlinear filters. The extended Kalman filter (EKF) and the other higher-order nonlinear filters have the problem (iii) as well as the problems (i), (ii) and (iv) even if $\widetilde{u}_t$ and $\widetilde{v}_t$ are uncorrelated with each other, because the residuals $\widetilde{u}_t$ and $\widetilde{v}_t$ have non-zero means and therefore $\mathrm{E}(\widetilde{u}_t\widetilde{v}_t'|Y_{t-1}) \neq 0$. However, for the Monte-Carlo simulation filter (MSF), the residuals $\widetilde{u}_t$ and $\widetilde{v}_t$ are asymptotically zero means. Accordingly, when applying the theorems proposed in Tanizaki (1991) and Tanizaki and Mariano (1996), the Monte-Carlo simulation filter (MSF) has only one problem, i.e., (iv), while the extended Kalman filter (EKF) and the other higher-order nonlinear filters still have all the problems, i.e., (i) – (iv). Therefore, the theorems are more useful for the Monte-Carlo simulation filter (MSF), rather than the extended Kalman filter (EKF) and the other higher-order nonlinear filter.

To show no correlation between $\widetilde{u}_t$ and $\widetilde{v}_t$ under a certain functional form, consider the state-space model which consists of the following measurement and transition equations:

$$\text{(Measurement equation)} \qquad y_t = h_{1t}(\alpha_t)h_{2t}(\epsilon_t), \tag{3.36}$$

$$\text{(Transition equation)} \qquad \alpha_t = g_{1t}(\alpha_{t-1})g_{2t}(\eta_t). \tag{3.37}$$

In this case, $M_{t|t-1}$ defined in equation (3.14) is computed as:

$$\begin{aligned} M_{t|t-1} &= \mathrm{E}\big((y_t - y_{t|t-1})(\alpha_t - a_{t|t-1})'|Y_{t-1}\big) \\ &= \overline{h}_{2t}\mathrm{E}\big((h_{1t} - \overline{h}_{1t})(\alpha_t - a_{t|t-1})'|Y_{t-1}\big), \end{aligned}$$

because $(y_t - y_{t|t-1})$ is represented as:

$$y_t - y_{t|t-1} = h_{1t}(h_{2t} - \overline{h}_{2t}) + \overline{h}_{2t}(h_{1t} - \overline{h}_{1t}),$$

and $(h_{2t} - \overline{h}_{2t})$ is not correlated with $(\alpha_t - a_{t|t-1})$. h_{1t}, h_{2t}, $\overline{h}_{1t}$ and $\overline{h}_{2t}$ are defined as follows:

$$\begin{aligned} h_{1t} &\equiv h_{1t}(\alpha_t), \\ h_{2t} &\equiv h_{2t}(\epsilon_t), \\ \overline{h}_{1t} &\equiv \mathrm{E}_\alpha\big(h_{1t}(\alpha_t)\big), \\ \overline{h}_{2t} &\equiv \mathrm{E}_\epsilon\big(h_{2t}(\epsilon_t)\big), \end{aligned}$$

where $\overline{h}_{2t}$ denotes the expectation with respect to the error term ϵ_t, and $\overline{h}_{1t}$ represents the conditional expectation with respect to α_t, given information up to time $t-1$, i.e., Y_{t-1}. Note that $\mathrm{E}_\alpha(\cdot)$ denotes the expectation with respect to the state-variable while $\mathrm{E}_\epsilon(\cdot)$ and $\mathrm{E}_\eta(\cdot)$ represent the expectations with respect to the error terms.

Also, $N_{t|t-1}$ is obtained as:

$$\begin{aligned} N_{t|t-1} &= \mathrm{E}\big((\alpha_t - \alpha_{t|t-1})(\alpha_{t-1} - a_{t-1|t-1})'|Y_{t-1}\big) \\ &= \overline{g}_{2t}\mathrm{E}\big((g_{1t} - \overline{g}_{1t})(\alpha_{t-1} - a_{t-1|t-1})'|Y_{t-1}\big), \end{aligned}$$

because $(\alpha_t - a_{t|t-1})$ is represented as:

$$\alpha_t - a_{t|t-1} = g_{1t}(g_{2t} - \overline{g}_{2t}) + \overline{g}_{2t}(g_{1t} - \overline{g}_{1t}),$$

and $(g_{2t} - \overline{g}_{2t})$ is not correlated with $(\alpha_{t-1} - a_{t-1|t-1})$. g_{1t}, g_{2t}, $\overline{g}_{1t}$ and $\overline{g}_{2t}$ are defined as follows:

$$\begin{aligned} g_{1t} &\equiv g_{1t}(\alpha_{t-1}), \\ g_{2t} &\equiv g_{2t}(\eta_t), \\ \overline{g}_{1t} &\equiv \mathrm{E}_\alpha\big(g_{1t}(\alpha_{t-1})\big), \\ \overline{g}_{2t} &\equiv \mathrm{E}_\eta\big(g_{2t}(\eta_t)\big), \end{aligned}$$

where $\overline{g}_{2t}$ denotes the expectation with respect to the error term η_t and $\overline{g}_{1t}$ represents the conditional expectation with respect to α_{t-1}, given information up to time $t-1$, i.e., Y_{t-1}.

Therefore, transforming the error terms, $\widetilde{u}_t$ and $\widetilde{v}_t$ are represented by:

$$\widetilde{u}_t = h_{1t}(h_{2t} - \overline{h}_{2t}) + \overline{h}_{2t}(h_{1t} - \overline{h}_{1t})$$
$$- \overline{h}_{2t}\mathrm{E}\big((h_{1t} - \overline{h}_{1t})(\alpha_t - a_{t|t-1})'|Y_{t-1}\big)\Sigma_{t|t-1}^{-1}(\alpha_t - a_{t|t-1}), \tag{3.38}$$

$$\widetilde{v}_t = g_{1t}(g_{2t} - \overline{g}_{2t}) + \overline{g}_{2t}(g_{1t} - \overline{g}_{1t})$$
$$- \overline{g}_{2t}\mathrm{E}\big((g_{1t} - \overline{g}_{1t})(\alpha_{t-1} - a_{t-1|t-1})'|Y_{t-1}\big)\Sigma_{t-1|t-1}^{-1}(\alpha_{t-1} - a_{t-1|t-1}). \tag{3.39}$$

With this transformation, we have no correlation between $\widetilde{u}_t$ and $(\alpha_t - a_{t|t-1})$ and no correlation between $\widetilde{v}_t$ and $(\alpha_{t-1} - a_{t-1|t-1})$. $\widetilde{u}_t$ is, however, correlated with $\widetilde{v}_t$, because the third term in equation (3.39) is correlated with both the second and the third terms in equation (3.38). From the structure of the residuals $\widetilde{u}_t$ and $\widetilde{v}_t$, the following theorem can be derived.

Theorem 1: In the case of the state-space model given by equations (3.36) and (3.37), when one of the functions $h_{1t}(\alpha_t)$, $h_{2t}(\epsilon_t)$, $g_{1t}(\alpha_{t-1})$ and $g_{2t}(\eta_t)$ is linear in its argument, i.e., when $h_{1t}(\alpha_t) = Z_t\alpha_t$, $h_{2t}(\epsilon_t) = S_t\epsilon_t$, $g_{1t}(\alpha_{t-1}) = T_t\alpha_{t-1}$ or $g_{2t}(\eta_t) = R_t\eta_t$, then $\widetilde{u}_t$ is not correlated with $\widetilde{v}_t$. Note that Z_t, S_t, T_t and R_t may depend on Y_{t-1}.

Proof: First, consider the case where either $h_{1t}(\alpha_t)$ or $h_{2t}(\epsilon_t)$ is linear. Equation (3.38) reduces to:

$$\widetilde{u}_t = h_{1t}(h_{2t} - \overline{h}_{2t}).$$

Since $(h_{2t} - \overline{h}_{2t})$ is a function of ϵ_t, it is uncorrelated with η_t, α_t and α_{t-1}. Therefore, we have $\mathrm{E}(\widetilde{u}_t\widetilde{v}_t') = 0$.

Next, consider the case where either $g_{1t}(\alpha_{t-1})$ or $g_{2t}(\eta_t)$ is linear. Similarly, equation (3.39) reduces to:

$$\widetilde{v}_t = g_{1t}(g_{2t} - \overline{g}_{2t}),$$

which is not correlated with ϵ_t, α_t and α_{t-1}. In this case, we also have $\mathrm{E}(\widetilde{u}_t\widetilde{v}_t') = 0$.

Thus, in the case where one of the functions $h_{1t}(\alpha_t)$, $h_{2t}(\epsilon_t)$, $g_{1t}(\alpha_{t-1})$ and $g_{2t}(\eta_t)$ is linear, we do not have to take into account the correlation between the error terms (i.e., residuals). In this case, we might expect a better approximation. This theorem is more useful for the Monte-Carlo simulation filter (MSF), because this implies that the problems (i) – (iii), shown in Section 3.2, are avoided.

Next, consider whether one could not get rid of the correlation between $\widetilde{u}_t$ and $\widetilde{v}_t$ by starting with the following state-space model:

(Measurement equation) $\qquad y_t = h_{1t}(\alpha_t) + h_{2t}(\epsilon_t), \qquad (3.40)$

(Transition equation) $\qquad \alpha_t = g_{1t}(\alpha_{t-1}) + g_{2t}(\eta_t). \qquad (3.41)$

In this case, $M_{t|t-1}$ is computed as:

$$\begin{aligned} M_{t|t-1} &= \mathrm{E}\big((y_t - y_{t|t-1})(\alpha_t - a_{t|t-1})'|Y_{t-1}\big) \\ &= \mathrm{E}\big((h_{1t} - \overline{h}_{1t})(\alpha_t - a_{t|t-1})'|Y_{t-1}\big), \end{aligned}$$

because $(y_t - y_{t|t-1})$ is represented as:

$$y_t - y_{t|t-1} = (h_{1t} - \overline{h}_{1t}) + (h_{2t} - \overline{h}_{2t}),$$

and $(h_{2t} - \overline{h}_{2t})$ is not correlated with $(\alpha_t - a_{t|t-1})$.

Thus, $\widetilde{u}_t$ is written as follows:

$$\begin{aligned} \widetilde{u}_t &= (h_{1t} - \overline{h}_{1t}) + (h_{2t} - \overline{h}_{2t}) \\ &\quad - \mathrm{E}\big((h_{1t} - \overline{h}_{1t})(\alpha_t - a_{t|t-1})'|Y_{t-1}\big)\Sigma_{t|t-1}^{-1}(\alpha_t - a_{t|t-1}). \end{aligned}$$

Similarly, $\widetilde{v}_t$ is given by:

$$\begin{aligned} \widetilde{v}_t &= (g_{1t} - \overline{g}_{1t}) + (g_{2t} - \overline{g}_{2t}) \\ &\quad - \mathrm{E}\big((g_{1t} - \overline{g}_{1t})(\alpha_{t-1} - a_{t-1|t-1})'|Y_{t-1}\big)\Sigma_{t-1|t-1}^{-1}(\alpha_{t-1} - a_{t-1|t-1}). \end{aligned}$$

Here, the following theorem can be obtained.

Theorem 2: In the case of the state-space model given by equations (3.40) and (3.41), when either $h_{1t}(\alpha_t)$ or $g_{1t}(\alpha_{t-1})$ is linear in its argument, i.e., when either $h_{1t}(\alpha_t) = Z_t\alpha_t$ or $g_{1t} = T_t\alpha_{t-1}$, then $\widetilde{u}_t$ is not correlated with $\widetilde{v}_t$. Note that Z_t and T_t may depend on Y_{t-1}.

Proof: When we take $h_{1t}(\alpha_t) = Z_t\alpha_t$, $\widetilde{u}_t$ is rewritten as follows:

$$\widetilde{u}_t = h_{2t} - \overline{h}_{2t},$$

because

$$h_{1t} - \overline{h}_{1t} = Z_t(\alpha_t - a_{t|t-1}),$$

$$\begin{aligned} &\mathrm{E}\big((h_{1t} - \overline{h}_{1t})(\alpha_t - a_{t|t-1})'|Y_{t-1}\big)\Sigma_{t|t-1}^{-1}(\alpha_t - a_{t|t-1}) \\ &= Z_t(\alpha_t - a_{t|t-1}). \end{aligned}$$

Clearly, $\widetilde{u}_t = h_{2t} - \overline{h}_{2t}$ is independent of $\widetilde{v}_t$, because $\widetilde{u}_t$ depends on the error term ϵ_t only.

Similarly, if $g_{1t}(\alpha_{t-1}) = T_t\alpha_{t-1}$, $\widetilde{v}_t$ is given by:

$$\widetilde{v}_t = g_{2t} - \overline{g}_{2t},$$

which is independent of $\widetilde{u}_t$, because $\widetilde{v}_t$ is a function of the error term η_t only.

Thus, as shown in this section, under a certain functional form of either $h_t(\alpha_t, \epsilon_t)$ or $g_t(\alpha_{t-1}, \eta_t)$, the correlation between the residuals $\widetilde{u}_t$ and $\widetilde{v}_t$ in equations (3.7) and (3.8) disappears.

3.4 Single-Stage Iteration Filter

In Section 3.2, we have discussed as follows. In order to derive the algorithms of the extended Kalman filter (EKF), the second-order nonlinear filter (SNF) and the Monte-Carlo simulation filter (MSF), we need the approximation that the residuals u_t and v_t are mutually, independently and normally distributed with zero means.

The algorithm of the single-stage iteration filter (SIF) can be derived from the mixed estimation approach. See Harvey (1981), Diderrich (1985) and Tanizaki (1993b) for the mixed estimation approach. According to this approach, we do not have to assume any distribution for the error terms ϵ_t and η_t but we utilize the Taylor series expansions.

For the prediction equations, taking the conditional expectation on both sides of the transition equation $\alpha_t = g_t(\alpha_{t-1}, \eta_t)$, we have the equations (3.9) and (3.10). By approximating the nonlinear transition equation by the first-order Taylor series expansion, the prediction equations are given by equations (3.19) and (3.20). Possibly, the nonlinear transition equation can be approximated by the second-order Taylor series expansion of equations (3.26) and (3.27) or by the Monte-Carlo stochastic simulations of equations (3.31) and (3.32). Thus, in order to obtain the prediction equations, the nonlinear transition equation is approximated by the Taylor series expansions as discussed in Section 3.2. Thus, this approximation for the prediction equations is the exactly same as the extended Kalman filter (EKF), the second-order nonlinear filter (SNF) or the Monte-Carlo simulation filter (MSF).

For derivation of the updating equations, the mixed estimation method shown in Chapter 2 is adopted. Again, rewrite equations (3.1) and (2.23) as follows:

$$q_t(y_t, \alpha_t) = \epsilon_t, \tag{3.1}$$

$$a_{t|t-1} = \alpha_t - \xi_t. \tag{2.23}$$

where ϵ_t and ξ_t are distributed as:

$$\begin{pmatrix} \epsilon_t \\ \xi_t \end{pmatrix} \sim \left(\begin{pmatrix} 0 \\ 0 \end{pmatrix}, \begin{pmatrix} H_t & 0 \\ 0 & \Sigma_{t|t-1} \end{pmatrix} \right).$$

Let $a_{t|t}$ be α_t such that the following weighted sum of squared residuals is minimized.

$$q_t(y_t, \alpha_t)' H_t^{-1} q_t(y_t, \alpha_t) + (a_{t|t-1} - \alpha_t)' \Sigma_{t|t-1}^{-1} (a_{t|t-1} - \alpha_t). \tag{3.42}$$

Since, at the maximum point, the first-derivative of equation (3.42) with respect to α_t is equal to zero, we have the following equation:

$$Z_t' H_t^{-1} q_t(y_t, \alpha_t) - \Sigma_{t|t-1}^{-1} (a_{t|t-1} - \alpha_t) = 0, \tag{3.43}$$

where

$$Z_t = \frac{\partial q_t(y_t, \alpha_t)}{\partial \alpha_t'}.$$

Moreover, computing the second-derivative with respect to α_t and taking the expectation with respect to ϵ_t and ξ_t, the following expression can be obtained:

$$Z_t' H_t^{-1} Z_t + \Sigma_{t|t}^{-1}. \tag{3.44}$$

Let $a_{t|t}^{(i)}$ be the i-th iteration of the filtering estimate at time t. Then, from equations (3.43) and (3.44), using the Newton-Raphson nonlinear optimization method, we have the following expression:

$$\begin{aligned} a_{t|t}^{(i+1)} = a_{t|t}^{(i)} &- \left(Z_{t|t}^{(i)\prime} H_t^{-1} Z_{t|t}^{(i)} + \Sigma_{t|t-1}^{-1} \right)^{-1} \\ &\times \left(Z_{t|t}^{(i)\prime} H_t^{-1} q_t(y_t, a_{t|t}^{(i)}) - \Sigma_{t|t-1}^{-1} (a_{t|t-1} - a_{t|t}^{(i)}) \right), \end{aligned} \tag{3.45}$$

where $Z_{t|t}^{(i)}$ denotes Z_t evaluated at $a_{t|t}^{(i)}$. The covariance of α_t is given by:

$$\begin{aligned} \Sigma_{t|t} &= \left(Z_{t|t}^{(i)\prime} H_t^{-1} Z_{t|t}^{(i)} + \Sigma_{t|t-1}^{-1} \right)^{-1} \\ &= \Sigma_{t|t-1} - \Sigma_{t|t-1} Z_{t|t}^{(i)\prime} (Z_{t|t}^{(i)} \Sigma_{t|t-1} Z_{t|t}^{(i)\prime} + H_t)^{-1} Z_{t|t}^{(i)} \Sigma_{t|t-1} \\ &= \Sigma_{t|t-1} - K_t^{(i)} (Z_{t|t}^{(i)} \Sigma_{t|t-1} Z_{t|t}^{(i)\prime} + H_t)^{-1} K_t^{(i)\prime}, \end{aligned} \tag{3.46}$$

where

$$K_t^{(i)} = \Sigma_{t|t-1} Z_{t|t}^{(i)\prime} (Z_{t|t}^{(i)} \Sigma_{t|t-1} Z_{t|t}^{(i)\prime} + H_t)^{-1}. \tag{3.47}$$

Finally, substituting equation (3.46) into equation (3.45), we have the updating equation for α_t:

$$a_{t|t}^{(i+1)} = a_{t|t-1}^{(i)} - K_t^{(i)} \Big(q_t(y_t, a_{t|t}^{(i)}) + Z_{t|t}^{(i)} (a_{t|t-1} - a_{t|t}^{(i)}) \Big). \tag{3.48}$$

Thus, the filtering algorithm in this case is represented by:

- equations (3.19), (3.20) and (3.46) – (3.48),
- equations (3.26), (3.27) and (3.46) – (3.48), or
- equations (3.31), (3.32) and (3.46) – (3.48).

The iteration scheme is given at each time t because the right-hand side of equations (3.46) – (3.48) depends on $a_{t|t}$. In this sense, this algorithm is called the single-stage iteration filter (SIF); see Wishner, Tabaczynski and Athans (1969). Finally, note that there are some problems of the single-stage iteration filter (SIF) discussed in Wishner, Tabaczynski and Athans (1969) (see Appendix A3.3 in detail).

One of the applications dealing with the single-stage iteration filter (SIF) based on the approach introduced here is Tanizaki (1993b). There, the conventional Kalman filter model is combined with a qualitative response model such as the probit model and the logit model in the scheme of the time-varying parameter model, which results in a nonlinear filtering problem.

Note that the single-stage iteration filter (SIF) reduces to the extended Kalman filter (EKF) or the second-order nonlinear filter (SNF) if the measurement equation is linear in both the state-vector and the error term.

The advantage of the single-stage iteration filter (SIF) is that we do not have to assume any distribution for the error terms to derive the filtering algorithm. In this sense, this approach is different from the other three algorithms discussed in Section 3.2. According to the single-stage iteration filter (SIF), the bias in the filtered estimates arises from approximation of the transition equation by the Taylor series expansions or the Monte-Carlo stochastic simulations. The bias from approximation of the measurement equation might be much smaller, because the Newton-Raphson optimization procedure is taken in the iteration scheme.

3.5 Summary

In this chapter, we have discussed the following: when we use the nonlinear filters based on the Taylor series expansion, first, we must impose some approximations on the error terms (i.e., residuals) $\widetilde{u}_t$ and $\widetilde{v}_t$ and also we must linearize the nonlinear functions in order to evaluate the expectations in equations (3.9) – (3.13). Some of these approximations imply that the error terms (i.e., residuals) $\widetilde{u}_t$ and $\widetilde{v}_t$ are mutually, independently and normally distributed with zero means. The most heuristic approximation of nonlinear functions utilizes the Taylor series expansions. The extended Kalman filter (EKF) and the other higher-order nonlinear filters are obtained based on the Taylor series expansions. The problems of the extended Kalman filter (EKF) and the higher-order nonlinear filters are as follows:

(i) $\mathrm{E}(\widetilde{u}_t|Y_{t-1}) \neq 0$ and $\mathrm{E}(\widetilde{v}_t|Y_{t-1}) \neq 0$,
(ii) $\mathrm{E}\big(\widetilde{u}_t(\alpha_t - a_{t|t-1})'|Y_{t-1}\big) \neq 0$ and $\mathrm{E}\big(\widetilde{v}_t(\alpha_{t-1} - a_{t|t-1})'|Y_{t-1}\big) \neq 0$,
(iii) $\mathrm{E}(\widetilde{u}_t\widetilde{v}_t'|Y_{t-1}) \neq 0$,
(iv) $\widetilde{u}_t$ and $\widetilde{v}_t$ are not normal.

In order to use the Taylor series expansion approach, we have to approximate the error terms $\widetilde{u}_t$ and $\widetilde{v}_t$ to be well-behaved.

Second, using the Monte-Carlo stochastic simulation method, the nonlinear filter called the Monte-Carlo simulation filter (MSF) was discussed, where the problems (i) and (ii) are improved because we approximated the expectations by generating random numbers for the state-variables (i.e., α_t and α_{t-1}) and the error terms (i.e., ϵ_t and η_t). We might expect a better nonlinear filter by applying the Monte-Carlo simulation, because approximation of the expectations is evaluated more correctly.

Third, in the case where we take a certain functional form for either the measurement equation or the transition equation, we do not have to take into account the correlation between the error terms. When we take such a functional form for either measurement equation or transition equation, a better approximation is expected. Two theorems were introduced in Section 3.3, which are more useful for the Monte-Carlo simulation filter (MSF), because, if the theorems are applied, the Monte-Carlo simulation filter (MSF) remains only one problem (iv), while the extended Kalman filter (EKF) and the other higher-order nonlinear filter still have all the problems (i) – (iv).

Finally, an interpretation of the single-stage iteration filter (SIF) was made. The single-stage iteration filter (SIF) is similar to the filtering algorithms using the Taylor series expansions which are discussed in Section 3.2. However, the interpretation is quite different in the sense that the single-stage iteration filter (SIF) is not based on normality approximation for the error terms $\widetilde{u}_t$ and $\widetilde{v}_t$ (i.e., residuals). The single-stage iteration filter is (SIF) derived from the generalized least squares method, where the Newton-Raphson nonlinear optimization procedure is taken.

A3 Appendix

A3.1 State-Space Model

Difference between the model in this book and that in Wishner, Tabaczynski and Athans (1969) is as follows.

- In this book, we deal with the following transition and measurement equations:

$$
\begin{aligned}
y_t &= h_t(\alpha_t, \epsilon_t), \\
\alpha_t &= g_t(\alpha_{t-1}, \eta_t).
\end{aligned}
$$

  On the other hand, Wishner, Tabaczynski and Athans (1969) used the transition and measurement equations which are linear with respect to the error terms, i.e.,

$$
\begin{aligned}
y_t &= h_t(\alpha_t) + \epsilon_t, \\
\alpha_t &= g_t(\alpha_{t-1}) + \eta_t.
\end{aligned}
$$

  In such a case, the conditional density of y_t given α_t is normal if ϵ_t is assumed to be normal. Also the conditional density of α_t given α_{t-1} is normal, provided that η_t is normal. Normality approximation of the state-vectors is more acceptable in this situation.
- The model discussed by Wishner, Tabaczynski and Athans (1969), and Gelb (1974) is a continuous type. Here, however, we deal with a discrete one.

A3.2 Correlation between u_t and v_t

As shown above, u_t and v_t , u_t and $(\alpha_t - a_{t|t-1})$, and v_t and $(\alpha_{t-1} - a_{t-1|t-1})$ are clearly correlated, respectively. In the studies by Wishner, Tabaczynski and Athans (1969), and Gelb (1974), however, no discussions on this point are given.

A3.3 Single-Stage Iteration Filter

There are some problems with the single-stage iteration filter (SIF) introduced in Wishner, Tabaczynski and Athans (1969). First, note that the single-stage iteration filter (SIF) given by them is represented by the following equations (3.49) – (3.54):

$$a_{t|t-1}^{(i)} = g_{t|t}^{(i)} + T_{t|t}^{(i)}(a_{t-1|t-1} - a_{t-1|t}^{(i)}), \tag{3.49}$$

$$\Sigma_{t|t-1}^{(i)} = T_{t|t}^{(i)}\Sigma_{t-1|t-1}T_{t|t}^{(i)\prime} + R_{t|t}^{(i)}Q_t R_{t|t}^{(i)\prime}, \tag{3.50}$$

$$K_t^{(i)} = \Sigma_{t|t-1}^{(i)} Z_{t|t}^{(i)\prime}(Z_{t|t}^{(i)}\Sigma_{t|t-1}^{(i)}Z_{t|t}^{(i)\prime} + S_{t|t}^{(i)}H_t S_{t|t}^{(i)\prime})^{-1}, \tag{3.51}$$

$$\Sigma_{t|t}^{(i)} = \Sigma_{t|t-1}^{(i)} - K_t^{(i)}(Z_{t|t}^{(i)}\Sigma_{t|t-1}^{(i)}Z_{t|t}^{(i)\prime} + S_{t|t}^{(i)}H_t S_{t|t}^{(i)\prime})K_t^{(i)\prime}, \tag{3.52}$$

$$a_{t|t}^{(i+1)} = a_{t|t-1}^{(i)} + K_t^{(i)}\Big(h_{t|t}^{(i)} - Z_{t|t}^{(i)}(a_{t|t-1}^{(i)} - a_{t|t}^{(i)})\Big), \tag{3.53}$$

$$a_{t-1|t}^{(i+1)} = a_{t-1|t-1} + \Sigma_{t-1|t-1}T_{t|t}^{(i)}\Sigma_{t|t-1}^{(i)}{}^{-1}(a_{t|t}^{(i+1)} - a_{t|t-1}^{(i)}), \tag{3.54}$$

for $i = 1, 2, \cdots$ and $t = 1, \cdots, T$.

The starting conditions are $a_{t-1|t}^{(1)} = a_{t-1|t-1}$ and $a_{t|t}^{(1)} = a_{t|t-1}$, and the converged values of $a_{t|t}^{(i)}$, $a_{t-1|t}^{(i)}$ and $\Sigma_{t|t}^{(i)}$ are taken as the estimates $a_{t|t}$, $a_{t-1|t}$ and the covariance $\Sigma_{t|t}$, respectively. $Z_{t|t}^{(i)}$ and $S_{t|t}^{(i)}$ are the first-derivatives with respect to α_t and ϵ_t which are evaluated at $(\alpha_t, \epsilon_t) = (a_{t|t}^{(i)}, 0)$. Similarly, $T_{t|t}^{(i)}$ and $R_{t|t}^{(i)}$ are the first-derivatives with respect to α_{t-1} and η_t which are evaluated at $(\alpha_{t-1}, \eta_t) = (a_{t-1|t}^{(i)}, 0)$. Equation (3.54) represents the smoothing algorithm.

According to their approach, the two nonlinear functions $y_t = h_t(\alpha_t, \epsilon_t)$ and $\alpha_t = g_t(\alpha_{t-1}, \eta_t)$ are linearized around $(\alpha_t, \epsilon_t) = (a_{t|t}, 0)$ and $(\alpha_{t-1}, \eta_t) = (a_{t-1|t}, 0)$, respectively, which are represented as follows:

$$y_t \approx h_t(a_{t|t}, 0) + Z_{t|t}(\alpha_t - a_{t|t}) + S_{t|t}\epsilon_t, \tag{3.55}$$

$$\alpha_t \approx g_t(a_{t-1|t}, 0) + T_{t|t}(\alpha_{t-1} - a_{t-1|t}) + R_{t|t}\eta_t, \tag{3.56}$$

where $Z_{t|t}$ and $S_{t|t}$ are the first-derivatives with respect to α_t and ϵ_t which are evaluated at $(\alpha_t, \epsilon_t) = (a_{t|t}, 0)$. Similarly, $T_{t|t}$ and $R_{t|t}$ are the first-derivatives with respect to α_{t-1} and η_t which are evaluated at $(\alpha_{t-1}, \eta_t) = (a_{t-1|t}, 0)$. And they applied equations (3.55) and (3.56) to the conventional linear recursive Kalman filter algorithm given by equations (2.9) – (2.15), as if $a_{t|t}$ and $a_{t-1|t}$ were functions of Y_{t-1} and the other exogenous variables.

Specifically, the filtering estimate $a_{t|t}$ depends on the right-hand side of the measurement equation (3.55), which implies that $Z_{t|t}$ and $S_{t|t}$ are the functions of Y_t, i.e., $Y_t = \{y_1, y_2, \cdots, y_t\}$. Also, the smoothing estimate $a_{t-1|t}$ depends on the right-hand side of the transition equation (3.56). Unless the linearized points such as $a_{t|t}$ and $a_{t-1|t}$ are a function of Y_{t-1} only, the linear recursive algorithm cannot be obtained. To show this fact, we take the transition equation given by equation (3.56). Even if we take the conditional expectation and variance, given the information available at time $t-1$, from equation (3.56), we cannot obtain the following expressions:

$$a_{t|t-1} = g_t(a_{t-1|t}, 0) + T_{t|t}(a_{t-1|t-1} - a_{t-1|t}), \tag{3.57}$$

$$\Sigma_{t|t-1} = T_{t|t}\Sigma_{t-1|t-1}T_{t|t}{}' + R_{t|t}Q_t R_{t|t}{}', \tag{3.58}$$

because $g_t(a_{t-1|t}, 0)$, $T_{t|t}$, $R_{t|t}$ and $a_{t-1|t}$ depend on the information at time t. The expressions (3.57) and (3.58) can be derived by regarding the approximated points $a_{t|t}$ and $a_{t-1|t}$ as functions of Y_{t-1} or fixed (nonstochastic) values, but this is clearly inappropriate. $g_t(a_{t-1|t}, 0)$, $T_{t|t}$, $R_{t|t}$ and $a_{t-1|t}$ in equations (3.57) and (3.58) must be the expected values. However, Wishner, Tabaczynski and Athans (1969) did not take into account this point and they approximated the prediction equations as equations (3.57) and (3.58). In their paper, the same approximation was used for the updating equations.

Here, the iteration scheme is derived from the interpretation of the mixed estimation method and is related to the updating equations shown in equations (3.46) – (3.48).

4. Density-Based Nonlinear Filters

4.1 Introduction

In Chapter 3, we have analyzed which kinds of approximation have to be used in the nonlinear filters that utilize the Taylor series expansions. In addition to linearization of the nonlinear measurement and transition equations, we need to approximate the nonnormal error terms (ie.e, residuals) as the normal ones and the correlated error terms (i.e., residuals) as the uncorrelated ones. These approximations are very strong because it is known that Kalman filter models based on normality assumption are nonrobust (see Meinhold and Singpurwalla (1989)). Therefore, density approximation, rather than function approximation, has to be investigated to obtain less biased filtering estimates of the state-variables.

In this chapter, alternatively, we derive the filtering algorithms by approximating the conditional probability density functions of α_t, given Y_t and Y_{t-1}. In this approach, there are some methods; the density approximation as a sum of Gaussian distributions, the density approximation through numerical integration, the simulation-based density approximation using Monte-Carlo integration with importance sampling, and a recursive algorithm of random draws with use of rejection sampling.

The representative references about the Gaussian sum filter (GSF) are presented by Sorenson and Alspach (1971), Alspach and Sorenson (1972), and Anderson and Moore (1979). Numerical integration for the densities is found in Kitagawa (1987) and Kramer and Sorenson (1988), in which each density function is approximated by a piecewise linear (first-order spline) function, i.e., numerical integration filter (NIF). Carlin, Polson and Stoffer (1992) suggested applying an adaptive Monte-Carlo integration technique known as the Gibbs sampler to density approximation (see Appendix A4.1 for the further discussion, but we do not take the procedure in this book). An importance sampling procedure using weight functions is developed by Tanizaki (1991), Tanizaki and Mariano (1994), and Mariano and Tanizaki (1995), where each integration in the densities is evaluated by Monte-Carlo integration with importance sampling and a recursive algorithm of weight functions is derived, i.e., importance sampling filter (ISF). Moreover, Tanizaki and Mariano (1995a, 1995b) and Mariano and Tanizaki (1996) proposed the density-based Monte-Carlo filter (DMF), where the density function derived from the measurement equation and the random draws of the state-vector generated from the transition equation are utilized. Also, Tanizaki (1995b), Tanizaki and Mariano (1995a, 1995b) and Mariano and Tanizaki (1996) derived the rejection sampling filter (RSF), where the filtering estimate is recursively obtained generating random draws of the state vector directly from the filtering density. There, we do not evaluate any integration such as the numerical integration procedure and the Monte-Carlo integration approach.

Thus, we introduce the density-based nonlinear filters; the Gaussian sum filter (Section 4.2), the numerical integration filter (Section 4.3), the importance sampling filter (Section 4.4), the density-based Monte-Carlo filter (Section 4.5) and the rejection sampling filter (Section 4.6).

Let $P(\cdot|\cdot)$ be a conditional density function. As shown in Chapter 2, the prediction and updating equations, based on the probability densities, are given by equations (2.16) and (2.17). Writing the density-based filtering algorithm consisting of the two densities again,

$$P(\alpha_t|Y_{t-1}) = \int P(\alpha_t|\alpha_{t-1})P(\alpha_{t-1}|Y_{t-1})\mathrm{d}\alpha_{t-1}, \tag{2.16}$$

$$P(\alpha_t|Y_t) = \frac{P(y_t|\alpha_t)P(\alpha_t|Y_{t-1})}{\displaystyle\int P(y_t|\alpha_t)P(\alpha_t|Y_{t-1})\mathrm{d}\alpha_t}, \tag{2.17}$$

for $t = 1,\cdots,T$. $P(\alpha_t|\alpha_{t-1})$ is obtained from the transition equation (3.1), i.e., $\alpha_t = g_t(\alpha_{t-1},\eta_t)$, if the distribution of η_t is specified, and also $P(y_t|\alpha_t)$ is derived from the measurement equation (3.2), i.e., $y_t = h_t(\alpha_t,\epsilon_t)$, given a specific distribution for ϵ_t. Thus, the distributions $P(\alpha_t|Y_t)$ and $P(\alpha_t|Y_{t-1})$ are obtained recursively, given the distributions $P(\alpha_t|\alpha_{t-1})$ and $P(y_t|\alpha_t)$.

In the density-based filtering algorithm (2.16) and (2.17), since the conditional distribution of y_t given Y_{t-1} is given by the denominator of equation (2.17):

$$P(y_t|Y_{t-1}) = \int P(y_t|\alpha_t)P(\alpha_t|Y_{t-1})\mathrm{d}\alpha_t, \tag{4.1}$$

the likelihood function is given by:

$$\begin{aligned} P(y_T, y_{T-1},\cdots,y_1) &= \prod_{t=1}^{T} P(y_t|Y_{t-1}) \\ &= \prod_{t=1}^{T} \left(\int P(y_t|\alpha_t)P(\alpha_t|Y_{t-1})\mathrm{d}\alpha_t \right). \end{aligned} \tag{4.2}$$

Strictly speaking, we must have:

$$P(y_T, y_{T-1},\cdots,y_1) = P(y_1)\prod_{t=2}^{T} P(y_t|Y_{t-1}),$$

but we approximate the distribution $P(y_T, y_{T-1}, \cdots, y_1)$ as equation (4.2), i.e, $P(y_1) \approx P(y_1|Y_0)$.

In this chapter, the nonlinear filters are investigated by the density functions (2.16) and (2.17), and the likelihood function (4.2).

In the case where the functions $h_t(\alpha_t, \epsilon_t)$ and $g_t(\alpha_{t-1}, \eta_t)$ are linear and the error terms ϵ_t and η_t are normally distributed, equations (2.16) and (2.17) reduce to the conventional linear recursive Kalman filter algorithm which is shown in Chapter 2. When the system is nonlinear and/or the *a priori* distributions are non-Gaussian, however, two problems are encountered. First the integration in equations (2.16) and (2.17) cannot be accomplished in closed form, and second the moments are not easily obtained from the equations above. These problems lead to investigation of density approximation. The nonlinear filters introduced and developed in this chapter do not require normality approximation for the disturbances ϵ_t and η_t, because the distributions of the state-vector are exactly derived from the functional forms of the density functions of ϵ_t and η_t and the nonlinear measurement and transition equations (3.1) and (3.2). Therefore, they can be applied to the case where both the system is nonlinear and the error terms are non-Gaussian. In this chapter, five kinds of nonlinear filters are introduced; the Gaussian sum filter (GSF) in Section 4.2, the numerical integration filter (NIF) in Section 4.3, the importance sampling filter (ISF) in Section 4.4, the density-based Monte-Carlo filter (DMF) in Section 4.5 and the rejection sampling filter (RSF) in Section 4.6.

4.2 Gaussian Sum Filter

A natural extension of density approximation is a Gaussian sum approach. The Gaussian sum filter (GSF) is a nonlinear algorithm which involves collections of extended Kalman filters (EKF), and thereby become both more powerful and more complex then the extended Kalman filter (EKF) (see Sorenson and Alspach (1971), Alspach and Sorenson (1972) and Anderson and Moore (1979)). In this algorithm, the conditional densities $P(\alpha_t|Y_t)$ and $P(\alpha_{t-1}|Y_{t-1})$ are approximated by a sum of Gaussian density functions.

The filtering algorithm by the Gaussian sum approach basically follows from the weighted average of the extended Kalman filter (EKF) equations. In the derivation process, the nonlinear functions $h_t(\alpha_t, \epsilon_t)$ and $g_t(\alpha_{t-1}, \eta_t)$ are linearized around $(\alpha_t, \epsilon_t) = (a_{i,t|t-1}, 0)$ and $(\alpha_{t-1}, \eta_t) = (a_{i,t-1|t-1}, 0)$, respectively. See Appendix A4.2 for the difference between the Gaussian sum approach in this section and that in Anderson and Moore (1979).

Let us assume that the filtering density at time t, i.e., $P(\alpha_t|Y_t)$, is approximately expressed as the Gaussian sum:

$$P(\alpha_t|Y_t) \approx \sum_{i=1}^{n} \omega_{i,t} \Phi(\alpha_t - a_{i,t|t}, \Sigma_{i,t|t}),$$

where , as defined in Section 2.3.1, $\Phi(\alpha_t - a_{i,t|t}, \Sigma_{i,t|t})$ denotes the normal density with mean $a_{i,t|t}$ and variance $\Sigma_{i,t|t}$, i.e.,

$$\begin{aligned}&\Phi(\alpha_t - a_{i,t|t}, \Sigma_{i,t|t})\\ &= (2\pi)^{-k/2}|\Sigma_{i,t|t}|^{-1/2}\exp\left(-\frac{1}{2}(\alpha_t - a_{i,t|t})\Sigma_{i,t|t}^{-1}(\alpha_t - a_{i,t|t})'\right),\end{aligned}$$

and a sum of the weight function $\omega_{i,t}$ over i is equal to one, i.e.,

$$\sum_{i=1}^{n}\omega_{i,t} = 1.$$

Then, $a_{t|t} = \mathrm{E}(\alpha_t|Y_t)$ and $\Sigma_{t|t} = \mathrm{Var}(\alpha_t|Y_t)$ are readily calculated, and this is a crucial advantage of using Gaussian sums. The calculation is easy, yielding the following two equations:

$$a_{t|t} = \sum_{i=1}^{n}\omega_{i,t}a_{i,t|t}, \tag{4.3}$$

$$\Sigma_{t|t} = \sum_{i=1}^{n}\omega_{i,t}\big(\Sigma_{i,t|t} + (a_{t|t} - a_{i,t|t})(a_{t|t} - a_{i,t|t})'\big). \tag{4.4}$$

See Appendix A4.3 for derivation of equations (4.3) and (4.4).

Similarly, we assume that $P(\alpha_t|Y_{t-1})$ is given by the Gaussian sum:

$$\begin{aligned}&P(\alpha_t|Y_{t-1})\\ &= \int P(\alpha_t|\alpha_{t-1})P(\alpha_{t-1}|Y_{t-1})\mathrm{d}\alpha_{t-1}\\ &\approx \int\sum_{i=1}^{n}\omega_{i,t-1}\Phi\big(\alpha_t - g_{i,t|t-1} - T_{i,t|t-1}(\alpha_t - a_{i,t-1|t-1}),\\ &\qquad\qquad R_{i,t|t-1}Q_tR'_{i,t|t-1}\big)\\ &\qquad \times\Phi(\alpha_{t-1} - a_{i,t-1|t-1}, \Sigma_{i,t-1|t-1})\mathrm{d}\alpha_{t-1}\\ &= \sum_{i=1}^{n}\omega_{i,t-1}\int\Phi\big(\alpha_t - g_{i,t|t-1} - T_{i,t|t-1}(\alpha_t - a_{i,t-1|t-1}),\\ &\qquad\qquad R_{i,t|t-1}Q_tR'_{i,t|t-1}\big)\\ &\qquad \times\Phi(\alpha_{t-1} - a_{i,t-1|t-1}, \Sigma_{i,t-1|t-1})\mathrm{d}\alpha_{t-1}\\ &= \sum_{i=1}^{n}\omega_{i,t-1}\Phi(\alpha_t - g_{i,t|t-1}, T_{i,t|t-1}\Sigma_{i,t-1|t-1}T'_{i,t|t-1} + R_{i,t|t-1}Q_tR'_{i,t|t-1})\\ &\equiv \sum_{i=1}^{n}\omega_{i,t-1}\Phi(\alpha_t - a_{i,t|t-1}, \Sigma_{i,t|t-1}),\end{aligned} \tag{4.5}$$

where

$$g_{i,t|t-1} = g_t(a_{i,t-1|t-1}, 0),$$

$$T_{i,t|t-1} = \left.\frac{\partial g_t(\alpha_{t-1}, \eta_t)}{\partial \alpha'_{t-1}}\right|(\alpha_{t-1}, \eta_t) = (a_{i,t-1|t-1}, 0),$$

$$R_{i,t|t-1} = \left.\frac{\partial g_t(\alpha_{t-1}, \eta_t)}{\partial \eta'_t}\right|(\alpha_{t-1}, \eta_t) = (a_{i,t-1|t-1}, 0).$$

See Proof I in Appendix A2.2 for the fifth equality in equation (4.5). Note that, in the third line of equation (4.5), the nonlinear transition equation is linearized for each segment i. Thus, for each i, we can obtain the prediction equations:

$$a_{i,t|t-1} = g_{i,t|t-1}, \tag{4.6}$$

$$\Sigma_{i,t|t-1} = T_{i,t|t-1}\Sigma_{i,t-1|t-1}T'_{i,t|t-1} + R_{i,t|t-1}Q_t R'_{i,t|t-1}. \tag{4.7}$$

Moreover, $a_{t|t-1}$ and $\Sigma_{t|t-1}$ are represented as:

$$a_{t|t-1} = \sum_{i=1}^{n} \omega_{i,t-1} a_{i,t|t-1}, \tag{4.8}$$

$$\Sigma_{t|t-1} = \sum_{i=1}^{n} \omega_{i,t-1}\big(\Sigma_{i,t|t-1} + (a_{t|t-1} - a_{i,t|t-1})(a_{t|t-1} - a_{i,t|t-1})'\big). \tag{4.9}$$

Next, to derive the updating equations, we perform some manipulations as follows:

$$P(\alpha_t|Y_t)$$

$$= \frac{P(y_t|\alpha_t)P(\alpha_t|Y_{t-1})}{\displaystyle\int P(y_t|\alpha_t)P(\alpha_t|Y_{t-1})\mathrm{d}\alpha_t}$$

$$\approx \frac{\left(\begin{array}{l}\displaystyle\sum_{i=1}^{n} \omega_{i,t-1}\Phi\big(y - h_{i,t|t-1} - Z_{i,t|t-1}(\alpha_t - a_{i,t|t-1}), \\ \hfill S_{i,t|t-1}H_t S'_{i,t|t-1}\big) \\ \qquad \times \Phi(\alpha_t - a_{i,t|t-1}, \Sigma_{i,t|t-1})\end{array}\right)}{\left(\begin{array}{l}\displaystyle\int\sum_{i=1}^{n} \omega_{i,t-1}\Phi\big(y - h_{i,t|t-1} - Z_{i,t|t-1}(\alpha_t - a_{i,t|t-1}), \\ \hfill S_{i,t|t-1}H_t S'_{i,t|t-1}\big) \\ \qquad \times \Phi(\alpha_t - a_{i,t|t-1}, \Sigma_{i,t|t-1})\mathrm{d}\alpha_t\end{array}\right)}$$

$$= \frac{\left(\begin{array}{l}\displaystyle\sum_{i=1}^{n} \omega_{i,t-1}\Phi\big(\alpha_t - a_{i,t|t-1} - K_{i,t}(y_t - y_{i,t|t-1}), \\ \hfill \Sigma_{i,t|t-1} - K_{i,t}F_{i,t|t-1}K'_{i,t}\big) \\ \qquad \times \Phi(y - y_{i,t|t-1}, F_{i,t|t-1})\end{array}\right)}{\displaystyle\sum_{i=1}^{n} \omega_{i,t-1}\Phi(y - y_{i,t|t-1}, F_{i,t|t-1})}$$

$$= \sum_{i=1}^{n} \omega_{i,t} \Phi\big(\alpha_t - a_{i,t|t-1} - K_{i,t}(y_t - y_{i,t|t-1}), \Sigma_{i,t|t-1} - K_{i,t} F_{i,t|t-1} K'_{i,t}\big)$$

$$\equiv \sum_{i=1}^{n} \omega_{i,t} \Phi(\alpha_t - a_{i,t|t}, \Sigma_{i,t|t}), \tag{4.10}$$

where

$$h_{i,t|t-1} = h_t(a_{i,t|t-1}, 0),$$

$$Z_{i,t|t-1} = \left.\frac{\partial h_t(\alpha_t, \epsilon_t)}{\partial \alpha'_t}\right|_{(\alpha_t, \epsilon_t) = (a_{i,t|t-1}, 0)},$$

$$S_{i,t|t-1} = \left.\frac{\partial h_t(\alpha_t, \epsilon_t)}{\partial \epsilon'_t}\right|_{(\alpha_t, \epsilon_t) = (a_{i,t|t-1}, 0)},$$

$$y_{i,t|t-1} = h_{i,t|t-1},$$
$$F_{i,t|t-1} = Z_{i,t|t-1} \Sigma_{i,t|t-1} Z'_{i,t|t-1} + S_{i,t|t-1} H_t S'_{i,t|t-1},$$
$$K_{i,t} = \Sigma_{i,t|t-1} Z'_{i,t|t-1} F^{-1}_{i,t|t-1}.$$

See Proof I in Appendix A2.2 for the third equality in equation (4.10).

Also, note that, in the second line of (4.10), the nonlinear measurement equation is linearized for each segment i. From the fourth equality in equation (4.10), the relationship between $\omega_{i,t-1}$ and $\omega_{i,t}$ is given by:

$$\omega_{i,t} = \frac{\omega_{i,t-1} \Phi(y - y_{i,t|t-1}, F_{i,t|t-1})}{\sum_{i=1}^{n} \omega_{i,t-1} \Phi(y - y_{i,t|t-1}, F_{i,t|t-1})}. \tag{4.11}$$

Therefore, the updating equations are as follows:

$$y_{i,t|t-1} = h_{i,t|t-1}, \tag{4.12}$$
$$F_{i,t|t-1} = Z_{i,t|t-1} \Sigma_{i,t|t-1} Z'_{i,t|t-1} + S_{i,t|t-1} H_t S'_{i,t|t-1}, \tag{4.13}$$
$$K_{i,t} = \Sigma_{i,t|t-1} Z'_{i,t|t-1} F^{-1}_{i,t|t-1}, \tag{4.14}$$
$$\Sigma_{i,t|t} = \Sigma_{i,t|t-1} - K_{i,t} F_{i,t|t-1} K'_{i,t}, \tag{4.15}$$
$$a_{i,t|t} = a_{i,t|t-1} + K_{i,t}(y_t - y_{i,t|t-1}). \tag{4.16}$$

Thus, the filtering algorithm based on the Gaussian sum approximation is represented by (4.3), (4.4), (4.6) – (4.9) and (4.11) – (4.16). The estimation procedure of the Gaussian sum filter is as follows.

(i) Given the initial values $a_{i,0|0}$ and $\Sigma_{i,0|0}$, perform the recursive algorithm represented by equations (4.6), (4.7) and (4.12) – (4.16) and obtain $a_{i,t|s}$ and $\Sigma_{i,t|s}$ for $i = 1, \cdots, n$, $t = 1, \cdots, T$ and $s = t-1, t$

(ii) Given the initial weight $\omega_{i,0}$ and based on $y_{i,t|t-1}$ and $F_{i,t|t-1}$ (which are obtained in Step (i)), compute $\omega_{i,t}$ for $i = 1, \cdots, n$ and $t = 1, \cdots, T$.

(iii) Given $a_{i,t|s}$, $\Sigma_{i,t|s}$ and $\omega_{i,t}$, use equations (4.3), (4.4), (4.8) and (4.9) to obtain $a_{t|s}$ and $\Sigma_{t|s}$ for $s = t-1, t$.

Note that in equations (4.5) and (4.10) the nonlinear functions are linearly approximated as follows:

(Measurement equation)

$$\begin{aligned} y_t =& h_t(\alpha_t, \epsilon_t) \\ \approx& h_{i,t|t-1} + Z_{i,t|t-1}(\alpha_t - a_{i,t|t-1}) + S_{i,t|t-1}\epsilon_t, \end{aligned}$$

(Transition equation)

$$\begin{aligned} \alpha_t =& g_t(\alpha_{t-1}, \eta_t) \\ \approx& g_{i,t|t-1} + T_{i,t|t-1}(\alpha_{t-1} - a_{i,t-1|t-1}) + R_{i,t|t-1}\eta_t, \end{aligned}$$

where i is regarded as the segment or the node. Here, the nonlinear measurement and transition equations are divided by n segments. For each segment, the extended Kalman filter (EKF) algorithm is applied.

The problem of the Gaussian sum filter (GSF) is how to choose the initial values $a_{i,0|0}$ and $\Sigma_{i,0|0}$ for $i = 1, \cdots, n$. The nonlinear filter by Gaussian sum approach requires n nodes on the initial values $a_{i,0|0}$ and $\Sigma_{i,0|0}$, $i = 1, \cdots, n$. If $a_{i,0|0}$ and $\Sigma_{i,0|0}$ are same for all i, the Gaussian sum filter (GSF) is exactly equivalent to the extended Kalman filter (EKF).

Anderson and Moore (1979) made the following remarks. First, simulation results on some examples suggest that the above algorithm still works satisfactorily even when n is small. Second, in the case of $n = 1$, this algorithm reduces to the extended Kalman filter.

Finally, the denominator in the above equation (4.10) denotes the conditional distribution of y_t given Y_{t-1}, i.e., $P(y_t|Y_{t-1})$, which is represented as:

$$P(y_t|Y_{t-1}) = \sum_{i=1}^{n} \omega_{i,t-1}\Phi(y - y_{i,t|t-1}, F_{i,t|t-1}).$$

Therefore, the likelihood function (2.30) is approximated as follows:

$$P(y_T, y_{T-1}, \cdots, y_1) = \prod_{t=1}^{T}\left(\sum_{i=1}^{n} \omega_{i,t-1}\Phi(y - y_{i,t|t-1}, F_{i,t|t-1})\right). \qquad (4.17)$$

Thus, the likelihood function is also given by the weighted average of Gaussian distributions. When the unknown parameters are included in the measurement and transition equations, equation (4.17) is maximized with respect to the unknown parameters.

According to the Gaussian sum approach, the nonlinear measurement and transition equations are linearized by the first-order Taylor series expansion for each segment i and they are applied to the conventional linear recursive algorithm of the Kalman filter. Thus, the Gaussian sum approach is the weighted average of the extended Kalman filters (EKF), where the nonnormal error terms (i.e., residuals) are approximated as the normal error terms for each segment. This implies that the filtered estimates of the Gaussian sum filter (GSF) are still biased but might be less biased than the extended Kalman filter (EKF).

In the proceeding sections, the investigations on the recursive algorithm of the distributions based on equations (2.16) and (2.17) are performed. There, we utilize $P(y_t|\alpha_t)$ and $P(\alpha_t|\alpha_{t-1})$ directly without approximating the density functions as the normal densities, because $P(y_t|\alpha_t)$ and $P(\alpha_t|\alpha_{t-1})$ are derived from the measurement and transition equations by assuming the specific distributions for the error terms ϵ_t and η_t. Moreover, approximation of the conditional densities $P(\alpha_t|Y_{t-1})$ and $P(\alpha_t|Y_t)$ is made by numerical integration (Section 4.3), or Monte-Carlo integration with importance sampling (Section 4.4).

4.3 Numerical Integration Filter

Until now, some approximation of the two nonlinear functions $h_t(\alpha_t \epsilon_t)$ and $g_t(\alpha_{t-1}, \eta_t)$ by Taylor series expansions has been used to obtain the estimates $a_{t|t}$ and $a_{t|t-1}$ to some extent. Kitagawa (1987) and Kramer and Sorenson (1988)) proposed the numerical integration filter (NIF), where each density function is approximated by a piecewise linear (first-order spline) function and numerical integration is used for the density evaluation. Each density is specified by number of segments, location of nodes, and the value at each node. In Kitagawa (1987), the nodes are chosen as:

$$(\alpha_{1,t}, \alpha_{2,t}, \cdots, \alpha_{n,t}) = (\alpha_{0,t} + h, \alpha_{0,t} + 2h, \cdots, \alpha_{0,t} + nh),$$

where

$$h = \frac{\alpha_{n,t} - \alpha_{0,t}}{n},$$

and $\alpha_{0,t}$ is an initial node for t. $\alpha_{0,t}$ and $\alpha_{n,t}$ are given and constant for all t. That is, the values of nodes are fixed for any time t. According to this approach, in the case where the range of the distributions shifts over time, approximation of the density is quite imprecise, and therefore the filtering estimates $a_{t|t-1}$, $\Sigma_{t|t-1}$, $a_{t|t}$ and $\Sigma_{t|t}$ are biased. To reduce this imprecise approximation of the density and the biased parameter estimates, an extremely large number of nodes is necessary, which implies an extremely wide range of nodes. This extremely large number of nodes is a burden from a computational point of view.

The nodes are given by $\alpha_{i,t}$ for $i = 1, \cdots, n$ and $t = 1, \cdots, T$. Hence, taking into account the shift of the densities of the state-variable over time, we can utilize the estimated filtering mean and variance from the extended Kalman filter (EKF).

First of all, let $P_\epsilon(\epsilon_t)$ and $P_\eta(\eta_t)$ be the probability density functions of ϵ_t and η_t, where $\epsilon_t = q_t(y_t, \alpha_t)$ and $\eta_t = f_t(\alpha_t, \alpha_{t-1})$. Then, by transforming the variables, the density functions of α_t given α_{t-1} in equation (2.16) and y_t given α_t in equation (2.17) are easily obtained as follows:

$$P(y_t|\alpha_t) = \left|\frac{\partial q_t(y_t, \alpha_t)}{\partial y_t'}\right| P_\epsilon\big(q_t(y_t, \alpha_t)\big), \tag{4.18}$$

$$P(\alpha_t|\alpha_{t-1}) = \left|\frac{\partial f_t(\alpha_t, \alpha_{t-1})}{\partial \alpha_t'}\right| P_\eta\big(f_t(\alpha_t, \alpha_{t-1})\big). \tag{4.19}$$

Thus, using (4.18) and (4.19), we can obtain the functional forms of $P(y_t|\alpha_t)$ and $P(\alpha_t|\alpha_{t-1})$ explicitly without any approximation about these densities. Now, we consider evaluating $P(\alpha_t|Y_t)$ and $P(\alpha_t|Y_{t-1})$.

For simplicity of discussion, we consider a scalar case of the state-variable. It can be easily extended to the higher-dimensional cases.

If we assume that the initial value α_0 is nonstochastic, then we have:

$$P(\alpha_1|Y_0) = P(\alpha_1|\alpha_0),$$

where $P(\alpha_1|\alpha_0)$ can be explicitly derived from the transition equation $\alpha_1 = g_1(\alpha_0, \eta_1)$. Let $\alpha_{i,1}$, $i = 1, \cdots, n$, be nodes of α_1, where $\alpha_{i,1}$, $i = 1, \cdots, n$, are assumed to be already sorted in order of size for α_1. That is, for all $t = 1, \cdots, T$, $\alpha_{1,t}$ is the smallest value and $\alpha_{n,t}$ the largest one, where $\alpha_{i,t}$ denotes the i-th smallest value of α_t. Then, using the relationship (4.19), the initial one-step ahead prediction density $P(\alpha_{i,1}|Y_0)$ is represented as:

$$P(\alpha_{i,1}|Y_0) = P(\alpha_{i,1}|\alpha_0),$$

for fixed α_0.

If α_0 is assumed to be distributed as some random vector, e.g., typically normal, then $P(\alpha_1|Y_0)$ is obtained as:

$$\begin{aligned} P(\alpha_1|Y_0) &= \int P(\alpha_1|\alpha_0)P(\alpha_0)\mathrm{d}\alpha_0 \\ &\approx \sum_{j=1}^{n} P(\alpha_1|\alpha_{j,0})P(\alpha_{j,0})(\alpha_{j,0} - \alpha_{j-1,0}), \end{aligned}$$

where $P(\alpha_0)$ is a distribution function of α_0. Let them be $\alpha_{i,0}$, $i = 1, \cdots, n$, which are sorted in order of size for the initial vector α_0. Using numerical integration, the initial one-step ahead prediction density $P(\alpha_1|Y_0)$ is evaluated as follows:

$$P(\alpha_{i,1}|Y_0) = \sum_{j=1}^{n} P(\alpha_{i,1}|\alpha_{j,0})P(\alpha_{j,0})(\alpha_{j,0} - \alpha_{j-1,0}).$$

To evaluate the density $P(\alpha_1|Y_1)$, we can utilize $P(y_1|\alpha_1)$ derived from the measurement equation (i.e., equation (4.18)) and the one-setp ahead prediction density $P(\alpha_{i,1}|Y_0)$ obtained above. That is, evaluation of $P(\alpha_{i,1}|Y_0)$

for each i is performed through equation (4.19), and similarly we can compute the value of $P(y_1|\alpha_{i,1})$ for each i from equation (4.18). Finally, noting that y_1 is available and approximating the integration in equation (2.17), the filtering density at time $t = 1$, i.e., $P(\alpha_{i,1}|Y_1)$, is evaluated as follows:

$$P(\alpha_1|Y_1) = \frac{P(y_1|\alpha_1)P(\alpha_1|Y_0)}{\int P(y_1|\alpha_1)P(\alpha_1|Y_0)\mathrm{d}\alpha_1}$$

$$\approx \frac{P(y_1|\alpha_1)P(\alpha_1|Y_0)}{\sum_{j=1}^{n} P(y_1|\alpha_{j,1})P(\alpha_{j,1}|Y_0)(\alpha_{j,1} - \alpha_{j-1,1})},$$

which implies:

$$P(\alpha_{i,1}|Y_1) = \frac{P(y_1|\alpha_{i,1})P(\alpha_{i,1}|Y_0)}{\int P(y_1|\alpha_1)P(\alpha_1|Y_0)\mathrm{d}\alpha_1},$$

for $\alpha_1 = \alpha_{i,1}$, $i = 1, \cdots, n$.

Similarly, using the nodes of α_2 (i.e., $\alpha_{i,2}$), one-step ahead prediction density at time 2 , i.e., $P(\alpha_{i,2}|Y_1)$, is given by:

$$P(\alpha_{i,2}|Y_1) = \sum_{j=1}^{n} P(\alpha_{i,2}|\alpha_{j,1})P(\alpha_{j,1}|Y_1)(\alpha_{j,1} - \alpha_{j-1,1}).$$

By repeating the same procedure for time t, we can obtain numerical approximation of $P(\alpha_t|Y_{t-1})$ and $P(\alpha_t|Y_t)$ for $t = 1, \cdots, T$, recursively. Therefore, the filtering algorithm to this problem is represented by the following two equations:

$$P(\alpha_t|Y_{t-1}) = \int P(\alpha_t|\alpha_{t-1})P(\alpha_{t-1}|Y_{t-1})\mathrm{d}\alpha_{t-1}$$

$$\approx \sum_{j=1}^{n} P(\alpha_t|\alpha_{j,t-1})P(\alpha_{j,t-1}|Y_{t-1})(\alpha_{j,t-1} - \alpha_{j-1,t-1}),$$

$$P(\alpha_t|Y_t) = \frac{P(y_t|\alpha_t)P(\alpha_t|Y_{t-1})}{\int P(y_t|\alpha_t)P(\alpha_t|Y_{t-1})\mathrm{d}\alpha_t}$$

$$\approx \frac{P(y_t|\alpha_t)P(\alpha_t|Y_{t-1})}{\sum_{j=1}^{n} P(y_t|\alpha_{j,t})P(\alpha_{j,t}|Y_{t-1})(\alpha_{j,t} - \alpha_{j-1,t})},$$

which yield the following prediction equation and updating equation:

$$P(\alpha_{i,t}|Y_{t-1}) = \sum_{j=1}^{n} P(\alpha_{i,t}|\alpha_{j,t-1})P(\alpha_{j,t-1}|Y_{t-1})(\alpha_{j,t-1} - \alpha_{j-1,t-1}), \tag{4.20}$$

$$P(\alpha_{i,t}|Y_t) = \frac{P(y_t|\alpha_{i,t})P(\alpha_{i,t}|Y_{t-1})}{\sum_{j=1}^{n} P(y_t|\alpha_{j,t})P(\alpha_{j,t}|Y_{t-1})(\alpha_{j,t} - \alpha_{j-1,t})}, \qquad (4.21)$$

for $t = 1, \cdots, T$, where the initial distribution is given by:

$$P(\alpha_{i,1}|Y_0) = P(\alpha_{i,1}|\alpha_0),$$

if α_0 is nonstochastic, and

$$P(\alpha_{i,1}|Y_0) = \sum_{j=1}^{n} P(\alpha_{i,1}|\alpha_{j,0})P(\alpha_{j,0})(\alpha_{j,0} - \alpha_{j-1,0}),$$

if the density function of α_0 is expressed as $P(\alpha_0)$, i.e., if α_0 is stochastic.

For more precise evaluation of the integration in equations (4.20) and (4.21), it is possible to replace

$$P(\alpha_{i,t}|\alpha_{j,t-1})P(\alpha_{j,t-1}|Y_{t-1}),$$

and

$$P(y_t|\alpha_{j,t})P(\alpha_{j,t}|Y_{t-1}),$$

by

$$\frac{1}{2}\big(P(\alpha_{i,t}|\alpha_{j,t-1})P(\alpha_{j,t-1}|Y_{t-1}) + P(\alpha_{i,t}|\alpha_{j-1,t-1})P(\alpha_{j-1,t-1}|Y_{t-1})\big),$$

and

$$\frac{1}{2}\big(P(y_t|\alpha_{j,t})P(\alpha_{j,t}|Y_{t-1}) + P(y_t|\alpha_{j-1,t})P(\alpha_{j-1,t}|Y_{t-1})\big),$$

respectively. In equations (4.20) and (4.21), each density is approximated as a sum of rectangles, but it can be approximated as a sum of trapezoids. In the proceeding chapters (i.e., Chapters 5 and 6), each density is approximated by a sum of trapezoids.

In the algorithm above, the problem is how to choose $\alpha_{i,t}$ for $i = 1, \cdots, n$ and $t = 1, \cdots, T$. According to the numerical integration procedure, the nodes $\alpha_{i,t}$ are fixed for all t. Taking into account the shift of the densities over time, we choose n nodes based on the filtering estimates from the extended Kalman filter (EKF) or the second-order nonlinear filter (SNF). We will mention later in more detail how to determine the nodes $\alpha_{i,t}$ for $i = 1, \cdots, n$ and $t = 1, \cdots, T$. For now, we consider them as given. First, we consider evaluating the estimates $a_{t|t-1}$, $\Sigma_{t|t-1}$, $a_{t|t}$ and $\Sigma_{t|t}$.

We utilize the densities given by equations (4.20) and (4.21) to obtain the conditional expectations, i.e., $a_{t|t-1}$ and $a_{t|t}$, and the variances, i.e., $\Sigma_{t|t-1}$ and $\Sigma_{t|t}$. Based on $\alpha_{i,t}$, $i = 1, \cdots, n$, they can be approximated as:

$$a_{t|t-1} = \int \alpha_t P(\alpha_t|Y_{t-1}) \mathrm{d}\alpha_t$$
$$= \sum_{i=1}^{n} \alpha_{i,t} P(\alpha_{i,t}|Y_{t-1})(\alpha_{i,t} - \alpha_{i-1,t}), \tag{4.22}$$

$$\Sigma_{t|t-1} = \int (\alpha_t - a_{t|t-1})(\alpha_t - a_{t|t-1})' P(\alpha_t|Y_{t-1}) \mathrm{d}\alpha_t$$
$$= \sum_{i=1}^{n} (\alpha_{i,t} - a_{t|t-1})(\alpha_{i,t} - a_{t|t-1})' P(\alpha_{i,t}|Y_{t-1})(\alpha_{i,t} - \alpha_{i-1,t}), \tag{4.23}$$

$$a_{t|t} = \sum_{i=1}^{n} \alpha_{i,t} P(\alpha_{i,t}|Y_t)(\alpha_{i,t} - \alpha_{i-1,t}), \tag{4.24}$$

$$\Sigma_{t|t} = \sum_{i=1}^{n} (\alpha_{i,t} - a_{t|t})(\alpha_{i,t} - a_{t|t})' P(\alpha_{i,t}|Y_t)(\alpha_{i,t} - \alpha_{i-1,t}) \tag{4.25}$$

Thus, the prediction and filtering estimates can be computed. Each integration in equations (4.22) – (4.25) is approximated as a sum of rectangles. It is also possible to approximate each integration by a sum of trapezoids. That is, we can approximate equation (4.22) as:

$$a_{t|t-1} = \sum_{i=1}^{n} \frac{1}{2} \Big(\alpha_{i,t} P(\alpha_{i,t}|Y_{t-1}) + \alpha_{i-1,t} P(\alpha_{i-1,t}|Y_{t-1}) \Big)(\alpha_{i,t} - \alpha_{i-1,t})$$

However, for simplicity of discussion, we have utilized the rectangle rule in this section.

Now, we consider choice of the nodes. If the nodes $\alpha_{i,t}$ for $i = 1, \cdots, n$ is distributed away from the true distribution of α_t given Y_{t-1} or Y_t, the filtering estimates obtained from equations (4.22) – (4.25) become unrealistic. Therefore, the nodes $\alpha_{i,t}$ for $i = 1, \cdots, n$ have to be overlapped with the true distributions of α_t given Y_{t-1} and Y_t. Note that the true distribution of α_t given Y_t or Y_{t-1} is unknown in practice.

For the nodes $\alpha_{i,t}$, we can take half of the nodes from the interval:

$$[a^*_{t|t-1} - \sqrt{c\Sigma^*_{t|t-1}}, a^*_{t|t-1} + \sqrt{c\Sigma^*_{t|t-1}}]$$

and half of the nodes from the interval:

$$[a^*_{t|t} - \sqrt{c\Sigma^*_{t|t}}, a^*_{t|t} + \sqrt{c\Sigma^*_{t|t}}],$$

where $a^*_{t|t}$ and $\Sigma^*_{t|t}$ are the filtering estimate and its variance from either the extended Kalman filter (EKF) or the second-order nonlinear filter (SNF), and also $a^*_{t|t-1}$ and $\Sigma^*_{t|t-1}$ are the one-step ahead prediction estimate and its variance from the extended Kalman filter (EKF) or the second-order nonlinear filter (SNF). c is a fixed value and $c \geq 1$ is preferable (remember that the probability is 99% when $c = 9$ in the case of normal distribution). It might be natural to assume that the true distributions of α_t given Y_{t-1} and Y_t are close to the extended Kalman filter (EKF) estimate or the second-order nonlinear filter (SNF) estimate. Thus, the filtering estimates by numerical integration are obtained modifying the nonlinear filtering estimates based on the Taylor series expansions.

Summarizing the numerical integration approach discussed in this section, the estimation procedure is represented as follows:

(i) Given the prediction and filtering estimates of the extended Kalman filter or the second-order nonlinear filter (i.e., $a^*_{t|t-1}$, $\Sigma^*_{t|t-1}$, $a^*_{t|t}$ and $\Sigma^*_{t|t}$), we choose n nodes of α_t from the two intervals:

$$[a^*_{t|s} - \sqrt{c\Sigma^*_{t|s}}, a^*_{t|s} + \sqrt{c\Sigma^*_{t|s}}],$$

for $s = t-1, t$. where $c = 1, 4, 9, 16, 25$ is taken in Chapter 5 and $c = 25$ is taken in Chapter 6. Let the nodes be $\alpha_{i,t}$, $i = 1, \cdots, n$, which should be already sorted in order of size for $i = 1, \cdots, n$.

(ii) Based on the nodes $\alpha_{i,t}$, $i = 1, \cdots, n$, and $\alpha_{j,t-1}$, $j = 1, \cdots, n$, we can numerically compute the density function $P(\alpha_{i,t}|Y_{t-1})$ given by equation (4.20).

(iii) Given the same nodes $\alpha_{i,t}$, $i = 1, \cdots, n$, used in the procedure (ii), the density function $P(\alpha_{i,t}|Y_t)$ can be obtained by equation (4.21).

(iv) Using equations (4.22) – (4.25), the filtering estimates $a_{t|t-1}$, $\Sigma_{t|t-1}$, $a_{t|t}$ and $\Sigma_{t|t}$ are computed.

(v) Repeat the procedures (i) – (iv) for $t = 1, \cdots, T$.

In the procedure (i), it is possible that the nodes $\alpha_{i,t}$, $i = 1, \cdots, n$, are obtained by the other way but $\alpha_{i,t}$ should be close to range of the true densities of α_t given Y_{t-1} and Y_t.

In order to estimate the unknown parameters included in the state-space model, we need to construct the likelihood function corresponding to equation (4.2). In this case, the conditional density function of y_t given Y_{t-1} is written as:

$$\begin{aligned} P(y_t|Y_{t-1}) &= \int P(y_t|\alpha_t)P(\alpha_t|Y_{t-1})\mathrm{d}\alpha_t \\ &\approx \sum_{j=1}^{n} P(y_t|\alpha_{j,t})P(\alpha_{j,t}|Y_{t-1})(\alpha_{j,t} - \alpha_{j-1,t}), \end{aligned}$$

which is equivalent to the denominator of equation (4.21). Therefore, the likelihood function is given by:

$$\begin{aligned} P(y_T, \cdots, y_1) &= \prod_{t=1}^{T} P(y_t|Y_{t-1}) \\ &\approx \prod_{t=1}^{T} \left(\sum_{j=1}^{n} P(y_t|\alpha_{j,t}) P(\alpha_{j,t}|Y_{t-1})(\alpha_{j,t} - \alpha_{j-1,t}) \right), \end{aligned} \quad (4.26)$$

which is maximized with respect to the unknown parameters if the state-space model depends on them. Thus, evaluation of the likelihood function is also performed by numerical integration. Precision of the likelihood function depends on number of the nodes (i.e., n) and location of the nodes.

Thus, Kitagawa (1987) and Kramer and Sorenson (1988) proposed a nonlinear and nonnormal filter using numerical integration. Numerical integration requires the nodes, which are denoted by $\alpha_{i,t}$, $i = 1, \cdots, n$. For all t, $\alpha_{i,t}$ are sorted by size with respect to i, i.e., $\alpha_{1,t}$ is the smallest value and $\alpha_{n,t}$ the largest one. There are some methods to evaluate integration numerically; a rectangle rule, a sum of trapezoids, Simpson's formula and so on. In Chapters 5 and 6, we take the numerical integration method by a sum of trapezoids.

According to the numerical integration approach, clearly we can obtain asymptotically unbiased estimates of the filtering estimates $a_{t|t-1}$, $\Sigma_{t|t-1}$, $a_{t|t}$ and $\Sigma_{t|t}$ by approximating the densities through numerical integration. Note that an asymptotically unbiased estimate in this case implies that the estimate goes to the true state-variable as n increases for all t.

However, some problems are encountered. First, when we compute the densities numerically, computation errors are accumulated and sometimes integration of $P(\alpha_t|Y_{t-1})$ or $P(\alpha_t|Y_t)$ with respect to α_t is not equal to one in practice. We need *ad hoc* modification which satisfies the following two conditions:

$$\sum_{i=1}^{n} P(\alpha_{i,t}|Y_{t-1})(\alpha_{i,t} - \alpha_{i-1,t}) = 1,$$

and

$$\sum_{i=1}^{n} P(\alpha_{i,t}|Y_t)(\alpha_{i,t} - \alpha_{i-1,t}) = 1,$$

which come from

$$\int P(\alpha_t|Y_{t-1}) \mathrm{d}\alpha_t = 1,$$

and

$$\int P(\alpha_t|Y_t) \mathrm{d}\alpha_t = 1.$$

Thus, $P(\alpha_t|Y_{t-1})$ and $P(\alpha_t|Y_t)$ should be re-computed so as to satisfy the condition that integration is equal to one.

Another problem is that density approximation is imprecise when number of nodes is small and/or location of the nodes is not appropriate, because each density function is approximated by a piecewise linear function. In general, $P(\alpha_t|Y_{t-1})$ and $P(\alpha_t|Y_t)$ are not known explicitly. We need a large number of nodes in this approach. Also, a large number of nodes leads to the increase in the computational burden.

Moreover, when the numerical integration approach is applied to the higher-dimensional cases, it takes an extraordinarily long time and also results in tedious programing. The approaches shown in the proceeding sections can be easily extended to the higher-dimensional cases in the sense of both computational time and programing.

4.4 Importance Sampling Filter

According to numerical integration, each density function is approximated as a collection of lines. The precision of approximation depends on number of nodes. For improvement of precision, a large number of nodes are essential. Also, a large number of nodes causes a greater computational burden.

In this section, a recursive algorithm on the weight functions, rather than the density functions, are derived. The weight function is represented as a ratio of two density functions. According to the numerical integration procedure, the nodes have to be sorted by size to perform numerical integration. In this section we introduce an estimation method that does not need numerical integration nor sorting of the nodes. Therefore, we expect less of a computational burden.

The approach discussed here is based on Monte-Carlo density approximation in Bayesian framework. See Geweke (1988, 1989a, 1989b) and Shao (1989) for the outline. There, they approximate the posterior mean by generating random numbers for the nuisance parameters based on the importance density function, which is called the importance sampling theory (see Appendix A4.4 for the detail).

Tanizaki (1991), Tanizaki and Mariano (1994) and Mariano and Tanizaki (1995) developed a nolinear and nonnormal filter with Monte-Carlo integration with importance sampling, where a recursive algorithm of the density functions is converted to that of the weight functions. By Monte-Carlo integration with importance sampling, a recursive filtering algorithm of the weight functions is derived, which is called the importance sampling filter (ISF) in this book.

Let $P_\alpha(\alpha)$ be a probability density function of α_t, which is called the importance density and is appropriately chosen by a researcher. Then, utilizing the appropriately chosen density function $P_\alpha(\alpha_t)$, equation (2.16) can be rewritten as:

$$
\begin{aligned}
\omega_{t|t-1} &\equiv \frac{P(\alpha_t|Y_{t-1})}{P_\alpha(\alpha_t)} \\
&= \int \frac{P(\alpha_t|\alpha_{t-1})}{P_\alpha(\alpha_t)} \frac{P(\alpha_{t-1}|Y_{t-1})}{P_\alpha(\alpha_{t-1})} P_\alpha(\alpha_{t-1}) \mathrm{d}\alpha_{t-1} \\
&= \int \frac{P(\alpha_t|\alpha_{t-1})}{P_\alpha(\alpha_t)} \omega_{t-1|t-1} P_\alpha(\alpha_{t-1}) \mathrm{d}\alpha_{t-1} \\
&\approx \frac{1}{n} \sum_{j=1}^{n} \frac{P(\alpha_t|\alpha_{j,t-1})}{P_\alpha(\alpha_t)} \omega_{j,t-1|t-1},
\end{aligned}
$$

where we define the weight functions $\omega_{t|s}$ and $\omega_{j,t|s}$ as:

$$
\begin{aligned}
\omega_{t|s} &\equiv \frac{P(\alpha_t|Y_s)}{P_\alpha(\alpha_t)}, \\
\omega_{j,t|s} &\equiv \frac{P(\alpha_{j,t}|Y_s)}{P_\alpha(\alpha_{j,t})},
\end{aligned}
$$

for $s = t-1, t$.

Thus, using the weight functions $\omega_{t|s}$ for $s = t-1, t$, equation (2.16) is represented as:

$$
\omega_{t|t-1} = \frac{1}{n} \sum_{j=1}^{n} \frac{P(\alpha_t|\alpha_{j,t-1})}{P_\alpha(\alpha_t)} \omega_{j,t-1|t-1}. \tag{4.27}
$$

Similarly, using the weight function, equation (2.17) is rewritten as:

$$
\begin{aligned}
\omega_{t|t} &\equiv \frac{P(\alpha_t|Y_t)}{P_\alpha(\alpha_t)} \\
&= \frac{P(y_t|\alpha_t) \dfrac{P(\alpha_t|Y_{t-1})}{P_\alpha(\alpha_t)}}{\displaystyle\int P(y_t|\alpha_t) \frac{P(\alpha_t|Y_{t-1})}{P_\alpha(\alpha_t)} P_\alpha(\alpha_t) \mathrm{d}\alpha_t} \\
&= \frac{P(y_t|\alpha_t)\omega_{t|t-1}}{\displaystyle\int P(y_t|\alpha_t)\omega_{t|t-1} P_\alpha(\alpha_t) \mathrm{d}\alpha_t} \\
&\approx \frac{P(y_t|\alpha_t)\omega_{t|t-1}}{\displaystyle\frac{1}{n}\sum_{j=1}^{n} P(y_t|\alpha_{j,t})\omega_{j,t|t-1}},
\end{aligned}
$$

which implies:

$$
\omega_{t|t} = \frac{P(y_t|\alpha_t)\omega_{t|t-1}}{\displaystyle\frac{1}{n}\sum_{j=1}^{n} P(y_t|\alpha_{j,t})\omega_{j,t|t-1}}. \tag{4.28}
$$

Here, $\omega_{t|t}$ and $\omega_{t|t-1}$ are regarded as the weight functions. We can choose an appropriate density function randomly for $P_\alpha(\alpha_t)$, which is typically normal.

Let $\alpha_{i,t}$, $i = 1, \cdots, n$, be the random draws based on the appropriately chosen importance density, $P_\alpha(\alpha_t)$. Then, equations (4.27) and (4.28) reduce to the following filtering algorithm:

$$\omega_{i,t|t-1} = \frac{1}{n}\sum_{j=1}^{n} \frac{P(\alpha_{i,t}|\alpha_{j,t-1})}{P_\alpha(\alpha_{i,t})}\omega_{j,t-1|t-1}, \tag{4.29}$$

$$\omega_{i,t|t} = \frac{P(y_t|\alpha_{i,t})\omega_{i,t|t-1}}{\displaystyle\frac{1}{n}\sum_{j=1}^{n} P(y_t|\alpha_{j,t})\omega_{j,t|t-1}}, \tag{4.30}$$

for $i = 1, \cdots, n$ and $t = 1, \cdots, T$, where, as mentioned above, $\omega_{i,t|s}$ is defined as:

$$\omega_{i,t|s} \equiv \frac{P(\alpha_{i,t}|Y_s)}{P_\alpha(\alpha_{i,t})},$$

for $s = t-1, t$. $\omega_{i,t|s}$ is the weight function evaluated at $\alpha_{i,t}$, which is the i-th random draw from the importance density $P_\alpha(\alpha_t)$.

Based on the random numbers from $P_\alpha(\alpha_t)$, the weight functions are obtained recursively, using equations (4.29) and (4.30). Note that we need to give some values to the initial weight function $\omega_{i,0|0}$ for $i = 1, \cdots, n$. In the case where the initial value α_0 is nonstochastic, we take:

$$\omega_{i,0|0} = 1$$

and $\alpha_{i,0} = \alpha_0$ for all i. Otherwise, the following initial weight function should be taken:

$$\omega_{i,0|0} = \frac{P(\alpha_{i,0})}{P_\alpha(\alpha_{i,0})},$$

where $\alpha_{i,0}$ is generated from the importance density $P_\alpha(\alpha_0)$ and $P(\alpha_{i,0})$ denotes the initial density function evaluated at $\alpha_0 = \alpha_{i,0}$.

Evaluation of the density function is performed by:

$$P(\alpha_{i,t}|Y_s) = \omega_{i,t|s}P_\alpha(\alpha_{i,t}),$$

for $s = t, t-1$, and $i = 1, \cdots, n$. $\omega_{i,t|s}$ is computed in the recursive algorithm (4.29) and (4.30) and $P_\alpha(\alpha_t)$ is specified by a researcher. Therefore, we can easily obtain each density function evaluated at $\alpha_{i,t}$, i.e., $P(\alpha_{i,t}|Y_s)$.

Similarly, the filtering estimates, $a_{t|t-1}$, $\Sigma_{t|t-1}$, $a_{t|t}$ and $\Sigma_{t|t}$, are given by:

$$\begin{aligned} a_{t|t-1} &= \int \alpha_t P(\alpha_t|Y_{t-1}) \mathrm{d}\alpha_t \\ &= \int \alpha_t \frac{P(\alpha_t|Y_{t-1})}{P_\alpha(\alpha_t)} P_\alpha(\alpha_t) \mathrm{d}\alpha_t \\ &= \int \alpha_t \omega_{t|t-1} P_\alpha(\alpha_t) \mathrm{d}\alpha_t \\ &\approx \frac{1}{n} \sum_{i=1}^{n} \alpha_{i,t} \omega_{i,t|t-1}, \end{aligned} \tag{4.31}$$

$$\Sigma_{t|t-1} = \frac{1}{n} \sum_{i=1}^{n} (\alpha_{i,t} - a_{t|t-1})(\alpha_{i,t} - a_{t|t-1})' \omega_{i,t|t-1}, \tag{4.32}$$

$$a_{t|t} = \frac{1}{n} \sum_{i=1}^{n} \alpha_{i,t} \omega_{i,t|t}, \tag{4.33}$$

$$\Sigma_{t|t} = \frac{1}{n} \sum_{i=1}^{n} (\alpha_{i,t} - a_{t|t})(\alpha_{i,t} - a_{t|t})' \omega_{i,t|t}, \tag{4.34}$$

for all t.

It is important that the importance density $P_\alpha(\alpha_t)$ be not too different from $P(\alpha_t|Y_{t-1})$ and $P(\alpha_t|Y_t)$ (see Geweke (1988, 1989a, 1989b), Tanizaki and Mariano (1994) and Mariano and Tanizaki (1995)). In the proceeding chapters, $P_\alpha(\alpha_t)$ is chosen as the following bimodal distribution:

$$P_\alpha(\alpha_t) = \frac{1}{2} \Phi(\alpha_t - a^*_{t|t-1}, c\Sigma^*_{t|t-1}) + \frac{1}{2} \Phi(\alpha_t - a^*_{t|t}, c\Sigma^*_{t|t}),$$

which denotes the average of two normal densities, where $a^*_{t|t}$ and $\Sigma^*_{t|t}$ are the filtering estimates obtained from either the extended Kalman filter (EKF) or the second-order nonlinear filter (SNF), and also $a^*_{t|t-1}$ and $\Sigma^*_{t|t-1}$ are the prediction estimates from the extended Kalman filter (EKF) or the second-order nonlinear filter (SNF). c is constant ($c = 1, 9, 16, 25$ is taken in Chapter 5 and $c = 25$ is taken in Chapter 6). Since $P_\alpha(\alpha_t)$ should be broader than $P(\alpha_t|Y_{t-1})$ and $P(\alpha_t|Y_t)$, c greater than one must be chosen. It might be natural to assume that the true distributions of α_t given Y_{t-1} and Y_{t-1} are close to the bimodal distribution. See Appendix A5.3 in the next chapter for choice of the importance density $P_\alpha(\alpha_t)$. The importance density $P_\alpha(\alpha_t)$ needs to cover $P(\alpha_t|Y_s)$ for $s = t-1, t$ over α_t. Usually, the peak and range of the one-step ahead prediction density $P(\alpha_t|Y_{t-1})$ is different from those of the filtering density $P(\alpha_t|Y_t)$. In general, the range of $P(\alpha_t|Y_{t-1})$ is larger than that of $P(\alpha_t|Y_t)$. For the importance sampling filter, the two densities $P(\alpha_t|Y_s)$, $s = t-1, t$, have to be approximated by one importance density $P_\alpha(\alpha_t)$. Therefore, it might be plausible to take the bimodal distribution for the importance density. Also, The peak and range of $P(\alpha_t|Y_s)$ are not known, but mean and variance of the state-variable can be estimated by the extended Kalman filter (EKF) or the second-order nonlinear filter (SNF) even if EKF and SNF are the biased estimators. It is appropriate to consider that the extended Kalman filter (EKF) estimates or the second-order nonlinear filter (SNF) estimates are not too far from the true values. Therefore, the importance sampling filter (ISF) would be improved by utilizing the importance density based on the extended Kalman filter (EKF) or the second-order nonlinear filter (SNF).

The estimation procedure is as follows:

(i) Generate n random draws of α_t from the importance density function $P_\alpha(\alpha_t)$, which are denoted by $\alpha_{i,t}$, $i = 1, \cdots, n$.
(ii) Given the weight function $\omega_{j,t-1|t-1}$ and the random draws $\alpha_{i,t}$, the weight function $\omega_{i,t|t-1}$ for $i = 1, \cdots, n$ is obtained from equation (4.29).
(iii) Given the weight function $\omega_{i,t|t-1}$ and the random numbers $\alpha_{i,t}$, compute $\omega_{i,t|t}$ using equation (4.30).
(iv) Based on the weight functions $\omega_{i,t|t-1}$ and $\omega_{i,t|t}$, each estimate $a_{t|t-1}$, $\Sigma_{t|t-1}$, $a_{t|t}$ and $\Sigma_{t|t}$ can be obtained from equations (4.31) – (4.34).
(v) Repeat the procedure (i) – (iv) for $t = 1, \cdots, T$.

Finally, consider evaluating the innovation form of the likelihood function. Equation (4.1) is transformed into:

$$\begin{aligned} P(y_t|Y_{t-1}) &= \int P(y_t|\alpha_t)P(\alpha_t|Y_{t-1})\mathrm{d}\alpha_t \\ &= \int P(y_t|\alpha_t)\frac{P(\alpha_t|Y_{t-1})}{P_\alpha(\alpha_t)}P_\alpha(\alpha_t)\mathrm{d}\alpha_t \\ &= \int P(y_t|\alpha_t)\omega_{t|t-1}P_\alpha(\alpha_t)\mathrm{d}\alpha_t \\ &\approx \frac{1}{n}\sum_{i=1}^{n} P(y_t|\alpha_{i,t})\omega_{i,t|t-1}, \end{aligned} \tag{4.35}$$

which is equivalent to the denominator of equation (4.30). Therefore, the likelihood function is given by:

$$\begin{aligned} P(y_T, \cdots, y_1) &= \prod_{t=1}^{T} P(y_t|Y_{t-1}) \\ &\approx \prod_{t=1}^{T} \left(\frac{1}{n}\sum_{i=1}^{n} P(y_t|\alpha_{i,t})\omega_{i,t|t-1} \right), \end{aligned} \tag{4.36}$$

which is maximized with respect to the unknown parameters if the state-space model depends on them.

The advantage of this approach is that we do not have to compute each density function numerically. In Section 4.3, each segment of the distribution function has to be approximated as a sum of rectangles or trapezoids through numerical integration. This implies that length of the segment determines precision of density approximation. Also, the nodes have to be sorted in order of size to perform numerical integration. According to the Monte-Carlo integration procedure, it might be expected that we can more easily compute the filtering estimates of the state-variables, because we do not need to perform numerical integration.

On the other hand, the disadvantage of the importance sampling filter (ISF) is that the density approximation depends on how to choose the importance distribution $P_\alpha(\alpha_t)$ for each time t. If the importance density $P_\alpha(\alpha_t)$ is chosen away from $P(\alpha_t|Y_{t-1})$ and $P(\alpha_t|Y_t)$, the weight functions $\omega_{t|t-1}$ and $\omega_{t|t}$ become unrealistic. Accordingly, the filtering estimates based on the weight functions are biased in the case. For choice of $P_\alpha(\alpha_t)$, as discussed above, it might be appropriate to take the bimodal distribution based on the extended Kalman filter (EKF) or the second-order nonlinear filter (SNF). Also, similarly, we need to pay attention to accumulation of the computation errors as discussed in Section 4.3. In order to reduce unreliability of the approximation, we must have the restriction of

$$\frac{1}{n}\sum_{t=1}^{T}\omega_{i,t|s} = 1,$$

which comes from

$$\begin{aligned} 1 &= \int P(\alpha_t|Y_s)\mathrm{d}\alpha_t \\ &= \int \omega_{t|s}P_\alpha(\alpha_t)\mathrm{d}\alpha_t \\ &\approx \frac{1}{n}\sum_{i=1}^{n}\omega_{i,t|s}, \end{aligned}$$

for $s = t-1, t$ and $t = 1, \cdots, T$.

In Sections 4.3 and 4.4, each integration included in the density-based filtering algorithm is evaluated by numerical integration or Monte-Carlo integration with importance sampling, where a researcher has to assume the nodes or the importance density. In the following two sections, we introduce the nonlinear filters which do not require *ad hoc* assumption such as choice of the nodes or choice of the importance density.

4.5 Density-Based Monte-Carlo Filter

In this section, we introduce an alternative solution to a nonlinear and non-normal filter using a Monte-Carlo simulation technique, which is called the density-based Monte-Carlo filter (DMF) and proposed by Tanizaki and Mariano (1995a, 1995b) and Mariano and Tanizaki (1996). For the numerical integration filter (NIF) and the importance sampling filter (ISF), choice of $\alpha_{i,t}$ is one of the critical problems, because precision of the filtering estimates depends on location of the nodes for the numerical integration filter (NIF) and choice of the importance density for the importance sampling filter (ISF). The nonlinear and nonnormal filter discussed in this section does not requires such assumptions to be taken by a researcher because the random numbers of α_t are generated from the transition equation in the system represented by (3.1) and (3.2).

Defining a collection of the state-vector as:

$$A_t = \{\alpha_0, \alpha_1, \cdots, \alpha_t\},$$

first, note that the joint density function of (Y_t, A_t) is written as:

$$P(Y_t, A_t) = P(Y_t|A_t)P(A_t),$$

where $P(Y_t|A_t)$ and $P(A_t)$ are rewritten as:

$$P(Y_t|A_t) = \prod_{s=1}^{t} P(y_s|\alpha_s), \tag{4.37}$$

$$P(A_t) = P(\alpha_0) \prod_{s=1}^{t} P(\alpha_s|\alpha_{s-1}), \tag{4.38}$$

$P(y_s|\alpha_s)$ and $P(\alpha_s|\alpha_{s-1})$ are obtained from the measurement and transition equations. Using equations (4.37) and (4.38), the filtering density is given by:

$$P(\alpha_t|Y_t) = \frac{\displaystyle\int P(Y_t|A_t)P(A_t)\mathrm{d}A_{t-1}}{\displaystyle\int P(Y_t|A_t)P(A_t)\mathrm{d}A_t}.$$

Note mean of the state-vector is represented as:

$$a_{t|t} \equiv \frac{\displaystyle\int \alpha_t P(Y_t|A_t)P(A_t)\mathrm{d}A_t}{\displaystyle\int P(Y_t|A_t)P(A_t)\mathrm{d}A_t},$$

which is equivalent to the definition of the filtering mean, i.e., $a_{t|t} = \mathrm{E}(\alpha_t|Y_t)$. Therefore, generating random draws of A_t from $P(A_t)$, the filtering estimate based on the Monte-Carlo technique is given by:

$$a_{t|t} = \frac{\displaystyle\frac{1}{n}\sum_{i=1}^{n} \alpha_{i,t} P(Y_t|A_{i,t})}{\displaystyle\frac{1}{n}\sum_{i=1}^{n} P(Y_t|A_{i,t})}$$

$$= \frac{\sum_{i=1}^{n} \alpha_{i,t} \prod_{s=1}^{t} P(y_s|\alpha_{i,s})}{\sum_{i=1}^{n} \prod_{s=1}^{t} P(y_s|\alpha_{i,s})},$$

where $a_{t|t}$ denotes the filtering estimate given by the Monte-Carlo technique, and $A_{i,t}$ is a collection of random draws, which is defined as:

$$A_{i,t} = \{\alpha_{i,0}, \alpha_{i,1}, \cdots, \alpha_{i,t}\}.$$

Generating n random numbers of α_0 and η_s for $s = 1, \cdots, t$, a set of random draws (i.e., $A_{i,t}$) can be obtained from the transition equation:

$$\alpha_{i,s} = g_s(\alpha_{i,s-1}, \eta_{i,s}),$$

for $i = 1, \cdots, n$ and $s = 1, \cdots, t$.

The following filtering algorithm is easy and convenient to compute the filtering estimate $a_{t|t}$ of the state-vector α_t.

$$a_{t|t} = \sum_{i=1}^{n} \alpha_{i,t}\omega_{i,t}, \tag{4.39}$$

where the weight function $\omega_{i,t}$ is defined as:

$$\omega_{i,t} \equiv \frac{P(y_t|\alpha_{i,t})\omega_{i,t-1}}{\sum_{j=1}^{n} P(y_t|\alpha_{j,t})\omega_{j,t-1}}. \tag{4.40}$$

The initial weight should be set to:

$$\omega_{i,0} = \frac{1}{n}, \tag{4.41}$$

for $i = 1, \cdots, n$. $\omega_{i,t}$ represents the weight function which satisfies:

$$1 = \sum_{j=1}^{n} \omega_{i,t}.$$

The likelihood function (4.2) is evaluated as:

$$\begin{aligned} P(y_T, \cdots, y_1) &= \frac{1}{n} \sum_{i=1}^{n} P(Y_t|A_{i,t}) \\ &= \prod_{t=1}^{T} \left(\frac{1}{n} \sum_{i=1}^{n} P(y_t|\alpha_{i,t|t-1}) \right). \end{aligned} \tag{4.42}$$

The features of the above filter are that the random numbers of α_t (i.e., $\alpha_{i,t}$, $i = 1, \cdots, n$) are generated from the transition equation (3.2) for all t, and that the algorithm requires the functional form of the density function of y_t given α_t (i.e., $P(y_t|\alpha_t)$) for all t. We do not need *ad hoc* assumptions such as choice of the nodes for numerical integration and that of the importance density for Monte-Carlo integration with importance sampling. g.

The computational procedure of the density-based Monte-Carlo filter (DMF) is represented as follows:

(i) The random draws of the initial state-variable α_0 are generated from the initial density $P(\alpha_0)$, which are denoted by $\alpha_{i,0}$, $i = 1, \cdots, n$.

(ii) Given $\alpha_{i,0}$, $\alpha_{i,t}$ are obtained from the transition equation:

$$\alpha_{i,t} = g_t(\alpha_{i,t-1}, \eta_{i,t}),$$

using the random draws $\eta_{i,t}$ for $i = 1, \cdots, n$ and $t = 1, \cdots, T$. Note that we have to store $n \times T$ random draws of α_t.

(iii) Given the initial weight (4.41), use equation (4.40) to obtain the weight functions $\omega_{i,t}$ for $i = 1, \cdots, n$ and $t = 1, \cdots, T$.

(iv) The filtering estimates $a_{t|t}$, $t = 1, \cdots, T$, can be computed by equation (4.39) for all t.

Thus, the transition equation (3.2) is utilized in order to generate the random numbers of A_t. The density function $P(y_t|\alpha_t)$ comes from the measurement equation (3.1). The density-based Monte-Carlo filter (DMF) requires deriving the density $P(y_t|\alpha_t)$ only while both the numerical integration filter (NIF) and the importance sampling filter (ISF) have to compute $P(\alpha_t|\alpha_{t-1})$ and $P(y_t|\alpha_t)$ by hand, which implies easy programming of the density-based Monte-Carlo filter (DMF) discussed in this section. Moreover, the above filter is more attractive with respect to computational time than the numerical integration filter (NIF) and the importance sampling filter (ISF).

Under some conditions, the theorems proved by Geweke (1988, 1989a, 1989b), which are related to the asymptotic behavior of the Bayes mean using Monte-Carlo integration (see Appendix A4.4), hold without any modification in the case of the density-based Monte-Carlo filter (DMF) discussed above (see Appendix A4.5). The asymptotic properties shown in Appendix A4.5 indicate that the filtering estimate with Monte-Carlo simulation is consistent but convergence is quite slow as $\sqrt{n}$, where n denotes number of random draws.

The final comments about the density-based Monte-Carlo filter (DMF) are made as follows. In the case where the transition equation follows a random walk process, α_t has no mean and no variance as time t goes to infinity. Under this situation, the density-based Monte-Carlo filter (DMF) does not work. If the range of the state variable is restricted (for example, see Section 5.2.2), the density-based Monte-Carlo filter (DMF) performs better. However, in the case where the transition equation is a random walk process, the density-based Monte-Carlo filter (DMF) is extremely poor (see Section 5.2.1).

Since for the density-based Monte-Carlo filter (DMF) the random draws are generated from the transition equation, in the case of random walk, we do not know where the random draws of the state variable go as time goes to infinity. Accordingly, in the empirical example, the random number generation is taken as follows. Let $\alpha_{i,t}$ and $\omega_{i,t}$ be the i-th random draw of the state variable and the corresponding weight for $i = 1, \cdots, n$. Choose $\alpha_{i,t}$ with large $\omega_{i,t}$, i.e., abandon $\alpha_{i,t}$ with small $\omega_{i,t}$. Thus, we can choose m random draws at time t. Of course, we have $m < n$. Pick up one out of the m random draws with probability $1/m$ and repeat this procedure n times. Then, we can re-arrange n random draws at time t. Thus, using the m random draws with more contribution and the corresponding weight, we obtain filtering estimates.

4.6 Rejection Sampling Filter

Also, the rejection sampling filter (RSF) proposed by Tanizaki (1995b), Tanizaki and Mariano (1995b) and Mariano and Tanizaki (1996) does not require *ad hoc* assumptions such as choice of the nodes and that of the importance density. In this sense, the rejection sampling filter (RSF) is similar to the density-based Monte-Carlo filter (DMF). The rejection sampling filter (RSF) shows a good performance even in the case where the transition equation follows a random walk process.

For a solution to nonlinear and nonnormal state-space model, in this section, we use the random draws to obtain the filtering estimates. Let $\alpha_{i,t|s}$ be the i-th random draw from the density function of α_t given Y_s, i.e., $P(\alpha_t|Y_s)$, $s = t-1, t$.

The Monte-Carlo filter with rejection sampling is derived as follows. Suppose that the random draws $\alpha_{i,t-1|t-1}$ are available. We consider generating random numbers of α_t from the filtering density $P(\alpha_t|Y_t)$. That is, we consider generating $\alpha_{j,t|t}$, given $\alpha_{i,t-1|t-1}$. Generating random draws $\alpha_{j,t|t}$ given $\alpha_{i|t-1|t-1}$ implies that the filtering random draws are recursively obtained.

We need to transform the filtering density into the form such that we can generate random draws. By substituting the prediction density $P(\alpha_t|Y_{t-1})$ into equation (2.17), the filtering density $P(\alpha_t|Y_t)$ is represented as:

$$\begin{aligned}
P(\alpha_t|Y_t) &= \frac{P(y_t|\alpha_t)P(\alpha_t|Y_{t-1})}{\displaystyle\int P(y_t|\alpha_t)P(\alpha_t|Y_{t-1})\mathrm{d}\alpha_t} \\
&\propto P(y_t|\alpha_t)P(\alpha_t|Y_{t-1}) \\
&= \int P(y_t|\alpha_t)P(\alpha_t|\alpha_{t-1})P(\alpha_{t-1}|Y_{t-1})\mathrm{d}\alpha_{t-1}
\end{aligned}$$

$$\approx \frac{1}{n}\sum_{i=1}^{n} P(y_t|\alpha_t)P(\alpha_t|\alpha_{i,t-1|t-1})$$

$$\propto \frac{1}{n}\sum_{i=1}^{n} \omega_1(\alpha_t;y_t)P(\alpha_t|\alpha_{i,t-1|t-1}), \tag{4.43}$$

where $\omega_1(\alpha_t;y_t)$ satisfies:

$$\omega_1(\alpha_t;y_t) \propto P(y_t|\alpha_t),$$
$$0 \le \omega_1(\alpha_t;y_t) \le 1.$$

$w_1(\alpha_t;y_t)$ is called the acceptance probability. In the approximation between the third line and the fourth line, n random draws of α_{t-1} are generated from the filtering density at time $t-1$, i.e., $P(\alpha_{t-1}|Y_{t-1})$.

Thus, from equation (4.43), the filtering density $P(\alpha_t|Y_t)$ is approximately proportional to:

$$\frac{1}{n}\sum_{i=1}^{n} \omega_1(\alpha_t;y_t)P(\alpha_t|\alpha_{i,t-1|t-1}),$$

which is a mixture of n distributions.

When we cannot obtain the explicit functional form of the density $P(\alpha_t|Y_t)$, we consider directly generating random draws from the filtering density. In this case, we may use rejection sampling to obtain a random observation from the filtering density $P(\alpha_t|Y_t)$. See Appendix A4.6 for rejection sampling. The distribution which we want to sample is dominated by the $P(\alpha_t|\alpha_{i,t-1|t-1})$ density. Therefore, the filtering algorithm is as follows. For $t = 1, \cdots, T$, we choose $\alpha_{i,t-1|t-1}$ with probability $1/n$ (i.e., we choose i with equal probability), sample the $P(\alpha_t|\alpha_{i,t-1|t-1})$ random variable and accept it with probability $\omega_1(\alpha_t;y_t)$. Thus, the random numbers generated from $P(\alpha_t|Y_t)$ are obtained.

Summarizing the procedure,

(i) Choose one out of n random draws from $P(\alpha_{t-1}|Y_{t-1})$ randomly, i.e., choose $\alpha_{i,t-1|t-1}$, $i = 1, \cdots, n$, with probability $1/n$.

(ii) Generating a random draw of η_t (i.e., $\eta_{m,t}$), compute $\alpha_{m,t}$ from the transition equation:

$$\alpha_{m,t} = g_t(\alpha_{i,t-1|t-1}, \eta_{m,t}),$$

for m and fixed i (i.e., i is chosen in Step (i)).

(iii) Compute the acceptance probability $\omega_1(\alpha_{m,t};y_t)$.

  (a) If $\alpha_{m,t}$ is accepted with probability $\omega_1(\alpha_{m,t};y_t)$, take $\alpha_{m,t}$ as a random draw of α_t from $P(\alpha_t|Y_t)$, which is denoted by $\alpha_{i,t|t}$.

  (b) If $\alpha_{m,t}$ is not accepted with probability $\omega_1(\alpha_{m,t};y_t)$, go back to Step (ii) and generate another random draw of η_t.

(iv) Repeat Steps (ii) and (iii) until accepted.

(v) Repeat Steps (i) – (iv) n times, because we need to generate n random draws of α_t from $P(\alpha_t|Y_t)$, i.e., $\alpha_{i,t|t}$ for $i = 1, \cdots, n$.

Note that in Step (iii) a uniform random number (say, U) has to be generated to check whether $\alpha_{m,t}$ is accepted or not. That is, $\alpha_{m,t}$ is accepted if $U < \omega_1(\alpha_{m,t}; y_t)$ and it is rejected otherwise.

The likelihood function (4.2) is evaluated as:

$$P(y_T, \cdots, y_1) = \prod_{t=1}^{T} \left(\frac{1}{n} \sum_{i=1}^{n} P(y_t|\alpha_{i,t|t-1}) \right). \tag{4.44}$$

Note that we have

$$\int P(y_t|\alpha_t) P(\alpha_t|Y_{t-1}) \mathrm{d}\alpha_t = \frac{1}{n} \sum_{i=1}^{n} P(y_t|\alpha_{i,t|t-1}),$$

where $\alpha_{i,t|t-1}$ is generated from the transition equation:

$$\alpha_{i,t|t-1} = g_t(\alpha_{j,t-1|t-1}, \eta_{i,t}).$$

Note that $\alpha_{j,t-1|t-1}$ in the above equation is picked up randomly and the random draw of η_t (i.e., $\eta_{i,t}$) is generated.

The evaluation procedure of the likelihood function is:

(i) Pick up $\alpha_{j,t-1|t-1}$ out of $\alpha_{1,t-1|t-1}, \cdots, \alpha_{n,t-1|t-1}$ with probability $1/n$.

(ii) Generate a random draw for η_t, i.e, $\eta_{i,t}$.

(iii) From the transition equation, $\alpha_{j,t-1|t-1}$ and $\eta_{i,t}$, we can obtain $\alpha_{i,t|t-1}$.

$\alpha_{i,t|t}$ is a random draw generated from $P(\alpha_t|Y_t)$ while $\alpha_{i,t|t-1}$ is from $P(\alpha_t|Y_{t-1})$. Both $\alpha_{i,t|t}$ and $\alpha_{i,t|t-1}$ are based on $\alpha_{i,t-1|t-1}$ and the transition equation (3.2). However, rejection sampling is utilized for $\alpha_{i,t|t}$ but not for $\alpha_{i,t|t-1}$.

Once the random draws of α_t given Y_{t-1} and Y_t are obtained, the one-step ahead prediction estimates (i.e., $a_{t|t-1}$ and $\Sigma_{t|t-1}$) and the filtering estimates (i.e., $a_{t|t}$ and $\Sigma_{t|t}$) are easily computed as follows.

$$a_{t|t-1} = \frac{1}{n} \sum_{i=1}^{n} \alpha_{i,t|t-1}, \tag{4.45}$$

$$\Sigma_{t|t-1} = \frac{1}{n} \sum_{i=1}^{n} (\alpha_{i,t|t-1} - a_{t|t-1})(\alpha_{i,t|t-1} - a_{t|t-1})', \tag{4.46}$$

$$a_{t|t} = \frac{1}{n} \sum_{i=1}^{n} \alpha_{i,t|t}, \tag{4.47}$$

$$\Sigma_{t|t} = \frac{1}{n} \sum_{i=1}^{n} (\alpha_{i,t|t} - a_{t|t})(\alpha_{i,t|t} - a_{t|t})', \tag{4.48}$$

for $t = 1, \cdots, T$. Thus, based on the random draws from the filtering density, the filtering mean and variance are arithmetically obtained as equations (4.45) – (4.48).

The features of the above procedure are:

(i) We do not need the functional form of $P(y_t|\alpha_t)$, $P(\alpha_t|\alpha_{t-1})$ and $P(\alpha_t|Y_s)$ for $s = t-1, t$.

(ii) The random draws of α_t are directly generated from the posterior density $P(\alpha_t|Y_s)$.

(iii) Programming is very easy, compared with the other nonlinear filters such as the numerical integration filter (Kitagawa (1987) and Kramer and Sorenson (1988)), the importance sampling filter (Tanizaki (1993a), Tanizaki and Mariano (1994) and Mariano and Tanizaki (1995)), and Monte-Carlo integration with Gibbs sampling (Carlin, Polson and Stoffer (1992)).

The above procedure gives us a general solution to the nonlinear and nonnormal state-space model represented by equations (3.1) and (3.2). Two examples are shown in the following to implement the above procedure, which correspond to the special cases where the state-space model is linear in error terms.

Example 1 (Linear Measurement Equation and Normal Errors). In the case where the transition equation is nonlinear in the lagged state-vector α_{t-1} and linear in the error term η_t but the measurement equation is linear in both the state-vector α_t and the error ϵ_t, we have the following state-space model:

$$\begin{aligned} &\text{(Measurement equation)} \qquad h_{1,t}(y_t) = Z_t\alpha_t + \epsilon_t, \\ &\text{(Transition equation)} \qquad \alpha_t = g_{1,t}(\alpha_{t-1}) + \eta_t, \end{aligned} \tag{4.49}$$

$$\begin{pmatrix} \epsilon_t \\ \eta_t \end{pmatrix} \sim N\left(\begin{pmatrix} 0 \\ 0 \end{pmatrix}, \begin{pmatrix} H_t & 0 \\ 0 & Q_t \end{pmatrix} \right),$$

where $h_{1,t}(\cdot)$, Z_t, $g_{1,t}(\cdot)$, H_t and Q_t are assumed to be known. We may specify $h_{1,t}(y_t) = y_t$ and $g_{1,t}(\alpha_{t-1}) = T_t\alpha_{t-1}$.

From $P(y_t|\alpha_t)$ and $P(\alpha_t|\alpha_{t-1})$, the posterior density $P(\alpha_t|y_t, \alpha_{t-1})$ is obtained as follows.

$$P(\alpha_t|y_t, \alpha_{t-1}) = \frac{P(y_t|\alpha_t)P(\alpha_t|\alpha_{t-1})}{\displaystyle\int P(y_t|\alpha_t)P(\alpha_t|\alpha_{t-1})\mathrm{d}\alpha_t}.$$

Therefore, in equation (4.43), we have:

$$P(y_t|\alpha_t)P(\alpha_t|\alpha_{t-1}) \propto P(\alpha_t|y_t, \alpha_{t-1}).$$

In the case of the system (4.49), $P(\alpha_t|y_t, \alpha_{t-1})$ is represented by:

$$P(\alpha_t|y_t,\alpha_{t-1}) = \Phi(\alpha_t - \Sigma_t\mu_t, \Sigma_t),$$

where μ_t and Σ_t are computed as:

$$\mu_t = Q_t^{-1}g_{1,t}(\alpha_{t-1}) + Z_t'H_t^{-1}h_{1,t}(y_t),$$
$$\Sigma_t^{-1} = Z_t'H_t^{-1}Z_t + Q_t^{-1}.$$

Accordingly, the filtering density $P(\alpha_t|Y_t)$ is rewritten as:

$$P(\alpha_t|Y_t) \propto \int \Phi(\alpha_t - \Sigma_t\mu_t, \Sigma_t)P(\alpha_{t-1}|Y_{t-1})\mathrm{d}\alpha_{t-1}.$$

Note that μ_t depends on α_{t-1}. Let $\mu_{i,t}$ be μ_t evaluated at $\alpha_{i,t-1|t-1}$, instead of α_{t-1}.

Picking up $\alpha_{i,t-1|t-1}$ randomly, $P(\alpha_t|Y_t)$ is approximately represented as average of the normal densities with mean $\Sigma_t\mu_{i,t}$ and variance Σ_t, i.e.,

$$\frac{1}{n}\sum_{i=1}^{n}\Phi(\alpha_t - \Sigma_t\mu_{i,t}, \Sigma_t),$$

where

$$\mu_{i,t} = Q_t^{-1}g_{1,t}(\alpha_{i,t-1|t-1}) + Z_t'H_t^{-1}h_{1,t}(y_t).$$

Note that $\alpha_{i,t-1|t-1}$ is chosen at random. Thus, the random number of α_t given Y_t is obtained from

$$\frac{1}{n}\sum_{i=1}^{n}N(\Sigma_t\mu_{i,t}, \Sigma_t).$$

That is, picking up $\alpha_{i,t-1|t-1}$ at random, the random number of α_t given Y_t is obtained from $N(\Sigma_t\mu_{i,t}, \Sigma_t)$. There, we do not need to perform rejection sampling in the case of the system (4.49).

Example 2 (Nonlinear Measurement Equation and Normal Errors). In the case where the transition equation is nonlinear in the lagged state-vector α_{t-1} and the measurement equation is also nonlinear in the state-vector α_t, we have the following state-space model:

$$\text{(Measurement equation)} \qquad h_{1,t}(y_t) = h_{2,t}(\alpha_t) + \epsilon_t,$$
$$\text{(Transition equation)} \qquad \alpha_t = g_{1,t}(\alpha_{t-1}) + \eta_t,$$

$$\begin{pmatrix}\epsilon_t\\ \eta_t\end{pmatrix} \sim N\left(\begin{pmatrix}0\\0\end{pmatrix}, \begin{pmatrix}H_t & 0\\ 0 & Q_t\end{pmatrix}\right),$$

where $h_{1,t}(\cdot)$, $g_{1,t}(\cdot)$, H_t and Q_t are assumed to be known.

From $P(y_t|\alpha_t)$ and $P(\alpha_t|\alpha_{t-1})$, the posterior density $P(\alpha_t|y_t,\alpha_{t-1})$ cannot be obtained explicitly. Therefore, we may apply rejection sampling. We pick up $\alpha_{i,t-1|t-1}$ randomly. Then, we generate a normal random draw from

$N\big(g_{1,t}(\alpha_{t-1}), Q_t\big)$ and accept it with probability $\omega_1(\alpha_t; y_t)$, where we can take the acceptance probability $\omega_1(\alpha_t; y_t)$ as:

$$\omega_1(\alpha_t; y_t) = \exp\left(-\frac{1}{2}\big(h_{1,t}(y_t) - h_{2,t}(\alpha_t)\big)' H_t^{-1}\big(h_{1,t}(y_t) - h_{2,t}(\alpha_t)\big)\right).$$

Thus, the Monte-Carlo procedure with rejection sampling is implemented.

Note the following comments. The random number generator by rejection sampling is known to be inefficient if the acceptance probability $w_1(\alpha_t; y_t)$ is close to zero. That is, for the rejection sampling filter (RSF), it sometimes takes a long time, especially when the acceptance probability $\omega_1(\alpha_t; y_t)$ is small. See, for example, Carlin and Polson (1991) and Carlin, Polson and Stoffer (1992). For simplicity of discussion, consider the linear and normal measurement equation. In this case, the small acceptance probability implies the small variance of the error term. Because in the normal and linear case, the acceptance probability is taken as the exponential part of the normal density. This exponential part is small when variance is small or difference between the random variable and its mean is large. Therefore, in general, the rejection sampling filter (RSF) is useful when variance of the error term in the measurement equation is large. By the following procedure, the rejection sampling filter (RSF) is improved a little bit in a sense of computational time. In the procedure (i) – (v), if the acceptance probability $\omega_1(\alpha_t; y_t)$ is small, Steps (ii) and (iii)(b) are repeated forever. In order to avoid this situation, if the acceptance probability $\omega_1(\alpha_t; y_t)$ is too small, we may go to (i), pick up another j and repeat the procedure (i) – (v).

Even if the rejection sampling filter (RSF) takes a lot of time computationally, it might be expected that the obtained filtering estimates approach the true state-vector values.

4.7 Summary

In this chapter, the filtering algorithms are derived through approximation of the density functions. A natural extension of density approximation is the Gaussian sum filter (GSF), where the probability density function of α_t given Y_{t-1} or Y_t is approximated as a sum of Gaussian distributions. The obtained algorithm represents the weighted sum of the extended Kalman filter (EKF) estimates. Since the extended Kalman filter (EKF) is the biased estimator, the weighted sum of the extended Kalman filter (EKF) estimates is expected to be biased.

As another approach, Kitagawa (1987) and Kramer and Sorenson (1988) proposed the density approximation by numerical integration, given the fixed nodes, where each density is approximated as piecewise linear functions. According to this procedure, the recursions are derived with respect to the density functions of α_t given Y_{t-1} and Y_t.

Furthermore, an attempt was made by Tanizaki (1991), Tanizaki and Mariano (1994) and Mariano and Tanizaki (1995) to approximate the densities by a set of points through the weight functions, where the weight functions can be obtained recursively. According to this procedure, we do not have to evaluate integration numerically and do not have to sort the nodes by size. It is necessary that the appropriately chosen density $P_\alpha(\alpha_t)$ should be close to $P(\alpha_t|Y_{t-1})$ and $P(\alpha_t|Y_t)$. Otherwise, the approximation becomes poor.

Comparing the numerical integration approach and the importance sampling density approximation approach, the former is extremely more computational burden than the latter in the higher-dimensional cases of the state-vector α_t. That is, the nodes have to be sorted for all elements of α_t and numerical integration has to be performed for all elements of α_t, which implies that it takes an extremely longer time in the higher-dimensional cases than in the scalar case. In this sense, the importance sampling filter (ISF) can be more applicable in practice. In Chapters 5 and 6, however, we do not take the higher-dimensional cases as an example.

Moreover, in this chapter, the Monte-Carlo procedure using the simulation technique is introduced, which was proposed by Tanizaki and Mariano (1995a, 1995b) and Mariano and Tanizaki (1996). The density-based Monte-Carlo filter (DMF) is the estimator that utilizes the density function of y_t given α_t and the random draws from the transition equation. The procedure improves the other nonlinear filters such as the numerical integration filter (NIF) and the importance sampling filter (ISF) from the following three points, i.e., computational time, simplicity of computer programming and no *ad hoc* assumptions. The numerical integration filter (NIF) proposed in Kitagawa (1987) and Kramer and Sorenson (1988) has the disadvantages:

- Location of nodes has to be set by a researcher.
- We have to derive the densities $P(\alpha_t|\alpha_{t-1})$ and $P(y_t|\alpha_t)$ by hand.
- Computational time increases more than proportionally as the dimension of the state variable is high.

The problems of the importance sampling filter (ISF) developed by Tanizaki (1991), Tanizaki and Mariano (1994) and Mariano and Tanizaki (1995) are:

- The importance density $P_\alpha(\alpha_t)$ has to be appropriately chosen by a researcher.
- We need to derive the densities $P(\alpha_t|\alpha_{t-1})$ and $P(y_t|\alpha_t)$ by hand, which is similar to the numerical integration filter (NIF).

The density-based Monte-Carlo filter (DMF) improves some of these problems, i.e., we have to compute $P(y_t|\alpha_t)$ by hand but not $P(\alpha_t|\alpha_{t-1})$, and we do not need to choose the nodes for the numerical integration procedure and the importance density for the importance sampling approach. The measurement equation is utilized for deriving the density $P(y_t|\alpha_t)$ while the transition equation is used to generate the random numbers of the state variable α_t. Thus, in addition to simplicity of computer programming and no *ad hoc* assumptions, precision of the filtering estimates is sometimes quite good.

An alternative to nonlinear and nonnormal filtering algorithm was proposed in Tanizaki (1995b), Tanizaki and Mariano (1995b) and Mariano and Tanizaki (1996). Given random draws of the state vector which are directly generated from the filtering density, the filtering estimate is recursively obtained, which is called the rejection sampling filter (RSF). There, we do not evaluate any integration included in the density-based filtering algorithm. Thus, the Monte-Carlo procedure of filtering algorithm using the simulation technique is introduced, where we utilize the random draws only and the recursive algorithm of random draws are derived. The rejection sampling filter (RSF) also improves over the other nonlinear filters developed in the numerical integration filter (NIF) and the importance sampling filter (ISF) from the following three points, i.e., computational time, simplicity of computer programming and no *ad hoc* assumptions.

Now, we have introduced the following nonlinear filters: the extended Kalman filter (EKF) in Section 3.2, the second-order nonlinear filter (SNF) in Section 3.2, the Monte-Carlo simulation filter (MSF) in Section 3.2, the single-stage iteration filter (SIF) in Section 3.4, the Gaussian sum filter (GSF) in Section 4.2, the numerical integration filter (NIF) in Section 4.3, the importance sampling filter (ISF) in Section 4.4, the density-based Monte-Carlo filter (DMF) in Section 4.5, and the rejection sampling filter (RSF) in Section 4.6. In the next chapter, these nonlinear filters are examined through Monte-Carlo experiments, based on the criteria of bias (BIAS) and root mean square error (RMSE).

A4 Appendix

A4.1 Gibbs Sampling

Geman and Geman (1984), Tanner and Wong (1987), Gelfand and Smith (1990), Gelfand, Hills, Racine-Poon and Smith (1990), Zeger and Karim (1991) and so on developed the Gibbs sampling theory. Carlin, Polson and Stoffer (1992) applied the theory to the nonlinear state-space model. The Gibbs sampling theory is concisely described as follows.

Consider two random variables x and y in order to make things easier. $P(x|y)$ and $P(y|x)$ are assumed to be known, which are the conditional distribution function of x given y and that of y given x. Pick up an arbitrary value for x, i.e., $x^{(0)}$. Given $x^{(0)}$, generate a random number of y, i.e., $y^{(1)}$, using the density $P(y|x^{(1)})$. Again, based on the density $P(x|y^{(0)})$, generate a random number of x given $y^{(1)}$, i.e., $x^{(1)}$. Repeating this procedure, we can generate random numbers $y^{(i)}$ from $P(y|x^{(i-1)})$ and $x^{(i)}$ from $P(x|y^{(i)})$ for $i = 1, \cdots, L$. From the convergence theory of the Gibbs sampler, as L goes to infinity, we can regard $x^{(L)}$ and $y^{(L)}$ as random draws from $P(x, y)$, which is a joint density function of x and y. Thus, we have one set of random numbers $(x^{(L)}, y^{(L)})$. Taking the entire process in parallel n times, we can obtain n sets of random numbers of (x, y), i.e., $(x_j^{(L)}, y_j^{(L)})$, $j = 1, \cdots, n$. Hereafter, abbreviating the superscript (L), we take $(x_j^{(L)}, y_j^{(L)})$ as (x_j, y_j), which denotes j-th set of random draws generated from the joint density $P(x, y)$.

Our interest is to obtain the marginal density of x, i.e., $P(x)$. Using the random numbers of (x, y) given above, we have the following density approximation:

$$\begin{aligned} P(x) &= \int P(x, y)\mathrm{d}y \\ &= \int P(x|y)P(y)\mathrm{d}y \\ &\approx \frac{1}{n}\sum_{j=1}^{n} P(x|y_j). \end{aligned}$$

In the third line of the above equation, y_j, $j = 1, \cdots, n$, are generated from the joint density $P(x, y)$. Note that $P(x|y_j)$ is available from the assumption. Thus, the density $P(x)$ is evaluated. This approach is called the Gibbs sampling. We have considered the bivariate case, but it is easily extended to the multivariate case.

The Gibbs sampler is applied to the state-space model by Carlin, Polson and Stoffer (1992). They proposed a solution to multivariate state-space modeling, where they allowed for the possibilities of nonnormal errors and nonlinear functions in the state-space model. They introduced nuisance parameters into the model. Their crucial assumptions are that distribution functions of ϵ_t and η_t depend on nuisance parameters and that the distribution functions of the nuisance parameters are assumed to be known, which are called the prior densities in the Bayesian framework. The posterior densities have to be computed. By the conditional densities of the state-variables and the posterior densities of the nuisance parameters, the Gibbs sampler is implemented.

The nonlinear filter with Gibbs sampling, which was proposed by Carlin, Polson and Stoffer (1992), is also feasible for higher-dimensional cases of the state-variable α_t and approximates each density function by random numbers. However, it is not based on the recursive algorithm represented by equations (2.16) and (2.17) and moreover the nuisance parameters conventionally fixed are taken as random variables in a framework of Gibbs sampling (i.e., the nuisance parameters included in the density functions of the error terms are assumed to be fixed in general). Thus, the nonlinear filters based on the recursive algorithm are quite different from the Gibbs sampling procedure proposed by Carlin, Polson and Stoffer (1992) with respect to the underlying assumptions. Therefore, we do not consider the Gibbs sampling procedure in this book even though it is also useful for the higher-dimensional cases. This approach might take an extremely long time from the computational point of view. The outline is briefly discussed as follows.

Carlin, Polson and Stoffer (1992) proposed a solution to multivariate state-space modeling, where they allowed for the possibilities of nonnormal errors and nonlinear functions in the state-space model. They introduced nuisance parameters γ and λ into the model. Their crucial assumptions are that distribution functions of ϵ_t and η_t depend on nuisance parameters γ_t and λ_t and that the distribution functions of γ_t and λ_t are given by $P_\gamma(\gamma_t)$ and $P_\lambda(\lambda_t)$, which are called the prior densities in the Bayesian framework. From equation (3.1) and the density function of ϵ_t, the density of y_t given α_t and γ_t is represented by $P(y_t|\alpha_t,\gamma_t)$ and, from equation (3.2) and η_t, the density of α_t given α_{t-1} and λ_t is $P(\alpha_t|\alpha_{t-1},\lambda_t)$. Note that the initial density of α_0 given λ_0 is assumed to be $P(\alpha_0|\lambda_0)$.

Under the above setup, the Monte-Carlo integration with Gibbs sampler is performed as follows:

(1) Under the assumption that $P(y_t|\alpha_t,\gamma_t)$, $P_\gamma(\gamma_t)$, $P(\alpha_t|\alpha_{t-1},\lambda_t)$ and $P_\lambda(\lambda_t)$ are known, the following three posterior density functions for $t=1,\cdots,T$ (i.e., $P_{\gamma*}(\gamma_t|\cdot)$, $P_{\lambda*}(\lambda_t|\cdot)$ and $P_{\alpha*}(\alpha_t|\cdot)$) are obtained using Bayes's theorem.

   a) $P_{\gamma*}(\gamma_t|y_t,\alpha_t)$ comes from $P(y_t|\alpha_t,\gamma_t)$ and $P_\gamma(\gamma_t)$.

   b) $P_{\lambda*}(\lambda_t|\alpha_t,\alpha_{t-1})$ is obtained from $P(\alpha_t|\alpha_{t-1},\lambda_t)$ and $P_\lambda(\lambda_t)$.

   c) $P_{\alpha*}(\alpha_t|y_t,\alpha_{t-1},\alpha_{t+1},\gamma_t,\lambda_t,\lambda_{t+1})$ is derived from the three density functions, i.e.,

   $$P(y_t|\alpha_t,\gamma_t),$$
   $$P(\alpha_t|\alpha_{t-1},\lambda_t),$$
   $$P(\alpha_{t+1}|\alpha_t,\lambda_{t+1}).$$

   Taking into account the endpoint cases, the posterior density function $P_{\alpha*}(\alpha_t|\cdot)$ is given by:

   $$\begin{aligned} &P_{\alpha*}(\alpha_t|\alpha_{t+1},\lambda_t,\lambda_{t+1}), && \text{for } t=0,\\ &P_{\alpha*}(\alpha_t|y_t,\alpha_{t-1},\alpha_{t+1},\gamma_t,\lambda_t,\lambda_{t+1}), && \text{for } t=1,\cdots,T-1.\\ &P_{\alpha*}(\alpha_t|y_t,\alpha_{t-1},\gamma_t,\lambda_t), && \text{for } t=T, \end{aligned}$$

(2) Given the values with the superscript (j), generate the random numbers with $(j+1)$ for filtering as follows.

   a) Generate $\gamma_s^{(j+1)}$ from $P_{\gamma*}(\gamma_s|y_s,\alpha_s^{(j)})$ for $s=1,\cdots,t$.

   b) Generate $\lambda_s^{(j+1)}$ from $P_{\lambda*}(\lambda_s|\alpha_s^{(j)},\alpha_{s-1}^{(j)})$ for $s=1,\cdots,t$.

   c) Generate

$$\alpha_s^{(j+1)} \text{ from } P_{\alpha_*}(\alpha_s|\alpha_{s+1}^{(j)}, \lambda_s^{(j)}, \lambda_{s+1}^{(j)}), \text{ for } s = 0,$$

$$\alpha_s^{(j+1)} \text{ from } P_{\alpha_*}(\alpha_s|y_s, \alpha_{s-1}^{(j)}, \alpha_{s+1}^{(j)}, \gamma_s^{(j)}, \lambda_s^{(j)}, \lambda_{s+1}^{(j)}), \quad \text{for } s = 1, \cdots, t-1,$$

$$\alpha_s^{(j+1)} \text{ from } P_{\alpha_*}(\alpha_s|y_s, \alpha_{s-1}^{(j)}, \gamma_s^{(j)}, \lambda_s^{(j)}), \text{ for } s = t,$$

for $j = 1, \cdots, J$ and $t = 1, \cdots, T$.

As J goes to infinity,

$$\{\gamma_s^{(J+1)}, \lambda_s^{(J+1)}, \alpha_s^{(J+1)}\}_{s=1}^t,$$

approaches a set of random draws from the joint distribution of

$$\{\gamma_s, \lambda_s, \alpha_s\}_{s=1}^t,$$

given information set up to time t, i.e., $\{y_s\}_{s=1}^t$. Let us define:

$$\{\gamma_{i,s}, \lambda_{i,s}, \alpha_{i,s}\}_{s=1}^t,$$

as the i-th set of random draws, i.e.,

$$\{\gamma_{i,s}^{(J+1)}, \lambda_{i,s}^{(J+1)}, \alpha_{i,s}^{(J+1)}\}_{s=1}^t,$$

for enough large J. Suppose that n sets of random draws are generated, i.e., $\{\gamma_{i,s}, \lambda_{i,s}, \alpha_{i,s}\}_{s=1}^t$ for $i = 1, \cdots, n$. Then, the filtering density is approximated as:

$$P(\alpha_t|Y_t) = \frac{1}{n} \sum_{i=1}^n P_{\alpha_*}(\alpha_t|y_t, \alpha_{i,t-1}, \gamma_{i,t}, \lambda_{i,t}).$$

Thus, the filtering estimate based on the Gibbs sampler is obtained.

(3) Let us define $a_{t|s}$ as the estimate of α_t. Since $\alpha_{i,t}$ is the random draw generated from $P(\alpha_t|Y_s)$ for $s = t-1, t$, we can obtain the expectation as follows:

$$a_{t|s} = \frac{1}{n} \sum_{i=1}^n \alpha_{i,t},$$

$$\Sigma_{t|s} = \frac{1}{n} \sum_{i=1}^n (\alpha_{i,t} - a_{t|s})(\alpha_{i,t} - a_{t|s})',$$

for $s = t-1, t$ and $t = 1, \cdots, T$.

Thus, for the estimate of α_t, the nonlinear and nonnormal state-space models developed by Carlin, Polson and Stoffer (1992) utilize the random draws only.

The important point is to generate random draws from the three posterior densities.

The likelihood function (2.30) is evaluated as

$$P(y_T, \cdots, y_1) = \prod_{t=1}^{T} \frac{1}{n} \sum_{i=1}^{n} P_y(y_t|\alpha_{i,t}, \gamma_{i,t}),$$

because the joint density of $\{Y_T, A_T, \Gamma_T, \Lambda_T\}$ is given by

$$P(Y_T, A_T, \Gamma_T, \Lambda_T) = P_\alpha(\alpha_0|\lambda_0) \prod_{t=1}^{T} P_y(y_t|\alpha_t, \gamma_t) P_\alpha(\alpha_t|\alpha_{t-1}, \lambda_t) \times P_\gamma(\gamma_t) P_\lambda(\lambda_t),$$

where A_T, Γ_T and Λ_T denote:

$$A_T = \{\alpha_0, \alpha_1, \cdots, \alpha_T\},$$
$$\Gamma_T = \{\gamma_1, \gamma_2, \cdots, \gamma_T\},$$
$$\Lambda_T = \{\lambda_0, \lambda_1, \cdots, \lambda_T\}.$$

For the numerical integration and the importance sampling procedures, we have to compute $P(\alpha_t|\alpha_{t-1})$ and $P(y_t|\alpha_t)$ by hand and choose location of the nodes for numerical integration and the random draws from the importance density for importance sampling. The Gibbs sampling approach requires the random draws only, using the Gibbs sampler, where we do not need *ad hoc* assumptions such as choice of nodes or importance density. However, for the Gibbs sampling filter, we have another *ad hoc* assumption that the density of nuisance parameters have to be introduced.

The computational disadvantage of the above procedure is a great amount of data storage. $\alpha_{i,t}$, $\gamma_{i,t}$ and $\lambda_{i,t}$ for $t = 1, \cdots, T$ and $i = 1, \cdots, n$ have to be stored in order to obtain the unconditional random draws, i.e., $\{\alpha_{i,t}^{(J+1)}, \gamma_{i,t}^{(J+1)}, \lambda_{i,t}^{(J+1)}\}$ for large J. Also, a computational amount of order $J \times T \times n$ is required, which is quite large. When random draws are generated from each conditional density, rejection sampling is sometimes utilized. Especially, the random number generation by rejection sampling sometimes takes a lot of time computationally, which is also one of the disadvantages.

A4.2 Note on Gaussian Sum Filter

We discuss about difference between the Gaussian sum approach introduced here and that in Anderson and Moore (1979). Anderson and Moore (1979) discussed the following model:

$$y_t = h_t(\alpha_t) + \epsilon_t,$$
$$\alpha_{t+1} = g_t(\alpha_t) + R_t(\alpha_t)\eta_t.$$

In this case, both measurement and transition equations are linear in the error terms. Therefore, we can expand functions $h_t(\alpha_t)$ and $g_t(\alpha_t)$ only with respect to α_t, and apply the Gaussian sum approach to obtain the nonlinear filtering algorithm.

However, since our model is different, i.e.,

$$y_t = h_t(\alpha_t, \epsilon_t),$$
$$\alpha_t = g_t(\alpha_{t-1}, \eta_t),$$

which are nonlinear with respect to ϵ_t and η_t, we need to expand ϵ_t and η_t around zero in order to obtain the estimates $a_{t|t-1}$ and $a_{t|t}$. Therefore, here we approximate the functions $h_t(\alpha_t, \epsilon_t)$ and $g_t(\alpha_{t-1}, \eta_t)$ by the first-order Taylor series expansion around $(\alpha_t, \epsilon_t) = (a_{i,t|t-1}, 0)$ and $(\alpha_{t-1}, \eta_t) = (a_{i,t-1|t-1}, 0)$.

If the densities can be approximated adequately by only one Gaussian density, the Gaussian sum filter reduces to the extended Kalman filter. The filtering algorithm by the Gaussian sum approach follows from the weighted average of the extended Kalman filter equations as shown above.

A4.3 Filtering Estimates by Gaussian Sum Filter

Here, we prove that the filtering estimates by Gaussian sum filter (i.e., $a_{t|t}$ and $\Sigma_{t|t}$) are given by equations (4.3) and (4.4), which are written as:

$$a_{t|t} = \sum_{i=1}^{n} \omega_{i,t} a_{i,t|t},$$

$$\Sigma_{t|t} = \sum_{i=1}^{n} \omega_{i,t} \big(\Sigma_{i,t|t} + (a_{t|t} - a_{i,t|t})(a_{t|t} - a_{i,t|t})' \big).$$

First of all, note that the conditional distribution of α_t, given the information available at time t, is represented as a sum of weighted normal densities, i.e.,

$$\begin{aligned} P(\alpha_t|Y_t) &= \sum_{i=1}^{n} \omega_{i,t} \Phi(\alpha_t - a_{i,t|t}, \Sigma_{i,t|t}) \\ &= \sum_{i=1}^{n} \omega_{i,t} (2\pi)^{-k/2} |\Sigma_{i,t|t}|^{-1/2} \\ &\qquad \times \exp\left(-\frac{1}{2} (\alpha_t - a_{i,t|t}) \Sigma_{i,t|t}^{-1} (\alpha_t - a_{i,t|t})' \right). \end{aligned}$$

Therefore, the expectation of α_t and its covariance are calculated as shown below:

$$\begin{aligned} a_{t|t} &= \mathrm{E}(\alpha_t|Y_t) \\ &= \int \alpha_t P(\alpha_t|Y_t) \mathrm{d}\alpha_t \\ &= \int \alpha_t \sum_{i=1}^{n} \omega_{i,t} \Phi(\alpha_t - a_{i,t|t}, \Sigma_{i,t|t}) \mathrm{d}\alpha_t \end{aligned}$$

$$
\begin{aligned}
&= \sum_{i=1}^{n} \omega_{i,t} \int \alpha_t \Phi(\alpha_t - a_{i,t|t}, \Sigma_{i,t|t}) \mathrm{d}\alpha_t \\
&= \sum_{i=1}^{n} \omega_{i,t} a_{i,t|t},
\end{aligned}
$$

and

$$
\begin{aligned}
&\Sigma_{t|t} \\
&= \mathrm{E}\big((\alpha_t - a_{t|t})(\alpha_t - a_{t|t})' | Y_t\big) \\
&= \int (\alpha_t - a_{t|t})(\alpha_t - a_{t|t})' P(\alpha_t | Y_t) \mathrm{d}\alpha_t \\
&= \int (\alpha_t - a_{t|t})(\alpha_t - a_{t|t})' \sum_{i=1}^{n} \omega_{i,t} \Phi(\alpha_t - a_{i,t|t}, \Sigma_{i,t|t}) \mathrm{d}\alpha_t \\
&= \sum_{i=1}^{n} \omega_{i,t} \int (\alpha_t - a_{t|t})(\alpha_t - a_{t|t})' \Phi(\alpha_t - a_{i,t|t}, \Sigma_{i,t|t}) \mathrm{d}\alpha_t \\
&= \sum_{i=1}^{n} \omega_{i,t} \int \big((\alpha_t - a_{i,t|t}) + (a_{i,t|t} - a_{t|t})\big)\big((\alpha_t - a_{i,t|t}) + (a_{i,t|t} - a_{t|t})\big)' \\
&\qquad\qquad \times \Phi(\alpha_t - a_{i,t|t}, \Sigma_{i,t|t}) \mathrm{d}\alpha_t \\
&= \sum_{i=1}^{n} \omega_{i,t} \Big(\int (\alpha_t - a_{i,t|t})(\alpha_t - a_{i,t|t})' \Phi(\alpha_t - a_{i,t|t}, \Sigma_{i,t|t}) \mathrm{d}\alpha_t \\
&\qquad + \int (\alpha_t - a_{i,t|t})(a_{i,t|t} - a_{t|t})' \Phi(\alpha_t - a_{i,t|t}, \Sigma_{i,t|t}) \mathrm{d}\alpha_t \\
&\qquad + \int (a_{i,t|t} - a_{t|t})(\alpha_t - a_{i,t|t})' \Phi(\alpha_t - a_{i,t|t}, \Sigma_{i,t|t}) \mathrm{d}\alpha_t \\
&\qquad + \int (a_{i,t|t} - a_{t|t})(a_{i,t|t} - a_{t|t})' \Phi(\alpha_t - a_{i,t|t}, \Sigma_{i,t|t}) \mathrm{d}\alpha_t \Big) \\
&= \sum_{i=1}^{n} \omega_{i,t} \big(\Sigma_{i,t|t} + (a_{i,t|t}, -a_{t|t})(a_{i,t|t}, -a_{t|t})' \big).
\end{aligned}
$$

Recall that

$$
\int \alpha_t \Phi(\alpha_t - a_{i,t|t}, \Sigma_{i,t|t}) \mathrm{d}\alpha_t = a_{i,t|t},
$$

$$
\int (\alpha_t - a_{i,t|t})(\alpha_t - a_{i,t|t})' \Phi(\alpha_t - a_{i,t|t}, \Sigma_{i,t|t}) \mathrm{d}\alpha_t = \Sigma_{i,t|t},
$$

$$
\begin{aligned}
&\int (\alpha_t - a_{i,t|t})(a_{i,t|t} - a_{t|t})' \Phi(\alpha_t - a_{i,t|t}, \Sigma_{i,t|t}) \mathrm{d}\alpha_t \\
&= \int (a_{i,t|t} - a_{t|t})(\alpha_t - a_{i,t|t})' \Phi(\alpha_t - a_{i,t|t}, \Sigma_{i,t|t}) \mathrm{d}\alpha_t = 0,
\end{aligned}
$$

$$\begin{aligned}
&\int (a_{i,t|t} - a_{t|t})(a_{i,t|t} - a_{t|t})' \Phi(\alpha_t - a_{i,t|t}, \Sigma_{i,t|t}) \mathrm{d}\alpha_t \\
&= (a_{i,t|t} - a_{t|t})(a_{i,t|t} - a_{t|t})'.
\end{aligned}$$

Thus, equations (4.3) and (4.4) can be derived.

A4.4 Monte-Carlo Integration with Importance Sampling

Here we discuss an approach to Bayesian inference developed in Geweke (1988, 1989a, 1989b).

Let $L(\theta)$, $\pi(\theta)$ and $p(\theta)$ be the likelihood function, the prior density function and the posterior density function, respectively, where θ is the vector of parameters. Then, we have the following relationship between them:

$$p(\theta) = \frac{L(\theta)\pi(\theta)}{\displaystyle\int L(\theta)\pi(\theta)\mathrm{d}\theta}. \tag{4.50}$$

From equation (4.50), the posterior expectation of $g(\theta)$, which is a function of θ, is represented as follows:

$$\begin{aligned}
\mathrm{E}\big(g(\theta)\big) &= \int g(\theta)p(\theta)\mathrm{d}\theta \\
&= \frac{\displaystyle\int g(\theta)L(\theta)\pi(\theta)\mathrm{d}\theta}{\displaystyle\int L(\theta)\pi(\theta)\mathrm{d}\theta} \\
&\equiv \bar{g}.
\end{aligned} \tag{4.51}$$

The integrations in equation (4.51) cannot be carried out analytically in the present case. The expectation is evaluated by Monte-Carlo integration with importance sampling. We utilize another density function of θ, say $I(\theta)$, which can be appropriately assumed by a researcher. Let θ_i, $i = 1, \cdots, n$, be the random numbers from the distribution $I(\theta)$. Then, we have the following approximation of the expectation $\mathrm{E}\big(g(\theta)\big)$.

$$\begin{aligned}
\mathrm{E}\big(g(\theta)\big) &= \frac{\displaystyle\int g(\theta)\frac{L(\theta)\pi(\theta)}{I(\theta)} I(\theta)\mathrm{d}\theta}{\displaystyle\int \frac{L(\theta)\pi(\theta)}{I(\theta)} I(\theta)\mathrm{d}\theta} \\
&= \frac{\displaystyle\int g(\theta)\omega(\theta)I(\theta)\mathrm{d}\theta}{\displaystyle\int \omega(\theta)I(\theta)\mathrm{d}\theta}
\end{aligned}$$

$$= \frac{\frac{1}{n}\sum_{i=1}^{n} g(\theta_i)\omega(\theta_i)}{\frac{1}{n}\sum_{i=1}^{n}\omega(\theta_i)}$$

$$= \frac{\sum_{i=1}^{n} g(\theta_i)\omega(\theta_i)}{\sum_{i=1}^{n}\omega(\theta_i)}$$

$$\equiv \overline{g}_n. \tag{4.52}$$

where $\omega(\theta)$ is defined as:

$$\omega(\theta) = \frac{L(\theta)\pi(\theta)}{I(\theta)},$$

which is called the weight function.

The properties of $\overline{g}_n$ are as follows.

(i) So long as $L(\theta)\pi(\theta)$ is proportional to a proper posterior distribution, $\mathrm{E}\big(g(\theta)\big)$ exists, and the support of $L(\theta)\pi(\theta)$ is included in the support of $I(\theta)$; $\overline{g}_n$ converges almost surely to $\overline{g}$.

(ii) If $\omega(\theta)$ is bounded above, and the posterior variance of $g(\theta)$ exists, then

$$\sqrt{n}(\overline{g}_n - \overline{g}) \longrightarrow N(0, \sigma^2),$$

where

$$\sigma^2 \equiv \frac{\mathrm{E}\Big(\big(g(\theta) - \overline{g}\big)^2\omega(\theta)^2\Big)}{\int L(\theta)\pi(\theta)\mathrm{d}\theta},$$

and the arrow denotes convergence in distribution.

(iii) If we define:

$$\widehat{\sigma}_n^2 \equiv \frac{\sum_{i=1}^{n}\big(g(\theta_i) - \overline{g}_n\big)^2\omega(\theta_i)^2}{\left(\sum_{i=1}^{n}\omega(\theta_i)\right)^2},$$

then

$$n\widehat{\sigma}_n^2 \longrightarrow \sigma^2.$$

Geweke (1988, 1989a, 1989b) pointed out the following. It is important that the importance density $I(\theta)$ is not too different from $L(\theta)\pi(\theta)$ and especially important that the weight function not become very large over the support of $L(\theta)\pi(\theta)$.

Finally, the result (ii) indicates that numerical accuracy is adversely affected by large relative values of the weight function; such large values indicate poor approximation of the posterior density. Also, it is shown from the result (ii) that the estimate based on Monte-Carlo integration with importance sampling (i.e., equation (4.52)) is consistent but convergence speed is quite slow as $\sqrt{n}$.

In Section 4.4, the above approach of Monte-Carlo density approximation is applied to the nonlinear filter.

A4.5 Density-Based Monte-Carlo Filter

Consider the asymptotic properties of the density-based Monte-Carlo filter (DMF) introduced in Section 4.5.

Under some conditions, the theorems proved by Geweke (1989a, 1989b), which are related to the asymptotic behavior of the Bayes mean using Monte-Carlo integration, hold without any modification in the case of the density-based Monte-Carlo filter (DMF). See Appendix A4.4 for the theorems on Monte-Carlo integration.

That is, $\Sigma_{t|t}^*$ is defined as:

$$\Sigma_{t|t}^* = \frac{\displaystyle\int (\alpha_t - a_{t|t})(\alpha_t - a_{t|t})' \big(P(Y_t|A_t)\big)^2 P(A_T)\mathrm{d}A_T}{\Big(\displaystyle\int P(Y_t|A_t)P(A_T)\mathrm{d}A_T\Big)^2},$$

and the sample variance of $a_{t|t}$ as $\Sigma_{t|t}$, i.e.,

$$\Sigma_{t|t} = \frac{\displaystyle\sum_{i=1}^{n} (\alpha_{i,t} - a_{t|t})(\alpha_{i,t} - a_{t|t})' \big(P(Y_t|A_{i,t})\big)^2}{\Big(\displaystyle\sum_{i=1}^{n} P(Y_t|A_{i,t})\Big)^2}.$$

Then, as n goes to infinity, the following two theorems can be shown.

(i) $a_{t|t} \xrightarrow{\text{a.s.}} \mathrm{E}(\alpha_t|Y_t)$,

(ii) $\sqrt{n}\big(a_{t|t} - \mathrm{E}(\alpha_t|Y_t)\big) \xrightarrow{\text{d}} N(0, \Sigma_{t|t}^*)$,

$n\Sigma_{t|t} \xrightarrow{\text{a.s.}} \Sigma_{t|t}^*$,

where " $\xrightarrow{\text{a.s.}}$ " and " $\xrightarrow{\text{d}}$ " denote almost sure convergence and convergence in distribution, respectively. The above asymptotic properties indicate that $a_{t|t}$ is consistent but convergence is slow as $\sqrt{n}$.

A4.6 Rejection Sampling

The rejection sampling is as follows: Let x be a random variable from a density function $f(\cdot)$. When we want to generate random draws from $f(\cdot)$, we need to find the density $g(\cdot)$ which satisfies $f(x) \leq cg(x)$ for all x, where c is constant. For $g(\cdot)$ we should choose the distribution function such that we can easily generate random draws. Define the acceptance probability $\omega(x)$, i.e.,

$$\omega(x) \equiv \frac{f(x)}{cg(x)}.$$

Note that $0 \leq \omega(x) \leq 1$. Let u be a uniform random number between zero and one and v be a random number from $g(\cdot)$. Then, we take the following procedures:

(i) Generate u from a uniform distribution between zero and one.
(ii) Generate v from a density $g(\cdot)$.
(iii) Take v as x if $u \leq \omega(v)$, and return to (i) otherwise.

We can prove the above random number generation procedure as follows: Note that

$$\text{Prob}\big(X \leq x | u \leq \omega(v)\big) = \frac{\text{Prob}\big(X \leq x, u \leq \omega(v)\big)}{\text{Prob}\big(u \leq \omega(v)\big)},$$

where the numerator and the denominator are represented as:

$$\begin{aligned} \text{Prob}\big(X \leq x, u \leq \omega(v)\big) &= \int_{-\infty}^{x} \text{Prob}\big(u \leq \omega(v) | v = t\big) g(t) \text{d}t \\ &= \int_{-\infty}^{x} \omega(t) g(t) \text{d}t \\ &= F(x)/c, \end{aligned}$$

$$\begin{aligned} \text{Prob}\big(u \leq \omega(v)\big) &= \text{Prob}\big(X \leq \infty, u \leq \omega(v)\big) \\ &= F(\infty)/c \\ &= 1/c. \end{aligned}$$

Therefore, we have:

$$\text{Prob}\big(X \leq x | u \leq \omega(v)\big) = F(x).$$

See, for example, Knuth (1981), Boswell, Gore, Patil and Taillie (1993) and O'Hagan (1994) for rejection sampling.

5. Monte-Carlo Experiments

5.1 Introduction

In this chapter, based on the criteria of bias (BIAS) and root mean square error (RMSE), we examine the nonlinear filters introduced and developed in the previous chapters. Monte-Carlo experiments are performed in Section 5.2. There, each nonlinear filter is compared using various types of nonlinear functions. One set of data y_t and α_t for $t = 1, \cdots, T$ is artificially simulated and, given y_t , each filtering estimate of α_t is compared with the artificially simulated α_t. This procedure is performed 4,000 times (i.e., 4,000 sets of data are generated) and BIAS and RMSE between the estimated α_t and the simulated one are computed for each time t.

In the first simulation study, a linear and normal case is taken for the state-space model. In this simulation study, it is expected that the standard Kalman filter shows the best estimator. In the second simulation study, a logistic type of nonlinear function is taken for the measurement and transition equations. The third simulation study considers estimating an ARCH(1) effect. Finally, the nonlinear function taken in Kitagawa (1987) and Carlin, Polson and Stoffer (1992) is examined. For all the cases, we obtain the following results. The extended Kalman filter (EKF), the second-order nonlinear filter (SNF) and the Gaussian sum filter (GSF) give us the biased estimators, while the numerical integration filter (NIF) and the importance sampling filter (ISF) are the unbiased estimators. Moreover, the Monte-Carlo simulation filter (MSF) gives us the unbiased estimator but the inefficient one.

5.2 Simulation Studies

In this section, various types of nonlinear function are taken for the measurement and transition equations. A comparison of the nonlinear filters is examined by Monte-Carlo simulations. In this section, 4,000 runs are performed to reduce unreliability of the simulations. Note that in this section we compare the following nonlinear filters:

EKF: Extended Kalman Filter (Section 3.2),
SNF: Second-Order Nonlinear Filter (Section 3.2),
MSF: Monte-Carlo Simulation Filter (Section 3.2),
SIF: Single-Stage Iteration Filter,
  SIFa: Single-Stage Iteration Filter with 1st-Order Approximation,
  SIFb: Single-Stage Iteration Filter with 2nd-Order Approximation,
  SIFc: Single-Stage Iteration Filter with Monte-Carlo Approximation,
GSF: Gaussian Sum Filter (Section 4.2),
NIF: Numerical Integration Filter (Section 4.3),
ISF: Importance Sampling Filter (Section 4.4),
DMF: Density-Based Monte-Carlo Filter (Section 4.5),
RSF: Rejection Sampling Filter (Section 4.6).

In Monte-Carlo experiments, we perform three kinds of SIF, depending on how to approximate the prediction equations (3.9) and (3.10). We call SIFa when the prediction equations are linearized by the first-order Taylor series expansion, SIFb when they are approximated by the second-order Taylor series expansion, and SIFc when they are evaluated by stochastic simulations, respectively.

5.2.1 Simulation I: Linear Model

For the first simulation study, consider the following linear measurement and transition equations:

$$\text{(Measurement equation)} \qquad y_t = \alpha_t + \epsilon_t, \tag{5.1}$$

$$\text{(Transition equation)} \qquad \alpha_t = \alpha_{t-1} + \eta_t, \tag{5.2}$$

where α_t is the state-variable and we assume ϵ_t as a standard normal random variable. ϵ_t and η_t are assumed to be normally distributed as follows:

$$\begin{pmatrix} \epsilon_t \\ \eta_t \end{pmatrix} \sim N\left(\begin{pmatrix} 0 \\ 0 \end{pmatrix}, \begin{pmatrix} 1 & 0 \\ 0 & 1 \end{pmatrix} \right).$$

The initial value α_0 is distributed as a standard normal random variable, i.e.,

$$\alpha_0 \sim N(0,1).$$

In this example, we compare the standard Kalman filter (denoted by EKF in this simulation study) with GSF, NIF, ISF, DMF and RSF. Note that the extended Kalman filter reduces to the standard Kalman filter when the system is linear.

The simulation procedure is as follows.

(i) The random numbers for ϵ_t and η_t are generated from:

$$\begin{pmatrix} \epsilon_t \\ \eta_t \end{pmatrix} \sim N\left(\begin{pmatrix} 0 \\ 0 \end{pmatrix}, \begin{pmatrix} 1 & 0 \\ 0 & 1 \end{pmatrix} \right),$$

which are a bivariate standard normal distribution. The initial density is also assumed to be the standard normal distribution, i.e.,

$$\alpha_0 \sim N(0,1).$$

Thus, we obtain the artificially simulated data for y_t and α_t, $t = 1, \cdots, T$, from the measurement equation (5.1) and the transition equation (5.2), where sample size $T = 40$ is taken.

(ii) Given the variances of the error terms ϵ_t and η_t and the data y_t obtained in Procedure (i), we can compute the filtering estimates using the algorithms introduced in Chapters 3 and 4, i.e, EKF, GSF, NIF, ISF, DMF and RSF. We do not compute SNF, MSF and SIF, because the system we consider is linear and normal.

(iii) Repeat Procedures (i) and (ii) m times. That is, we perform m simulations for each nonlinear filter. Let $\alpha_t^{(i)}$, $t = 1, \cdots, T$, be the i-th set of artificial data (which is taken as the i-th set of true state-variable), and $a_{t|t}^{(i)}$ be the filtering estimate at time t in the i-th simulation run, where $m = 4,000$ is set.

(iv) For each nonlinear filter, compute the bias (BIAS) at time t as:

$$\text{BIAS}_t = \frac{1}{m} \sum_{i=1}^{m} (a_{t|t}^{(i)} - \alpha_t^{(i)}),$$

and the root mean square error (RMSE) at time t as:

$$\text{RMSE}_t = \left(\frac{1}{m} \sum_{i=1}^{m} (a_{t|t}^{(i)} - \alpha_t^{(i)}) \right)^{1/2}.$$

Moreover, BIAS and RMSE are computed as the arithmetic averages of BIAS_t and RMSE_t for $t = 1, \cdots, T$, i.e.,

$$\text{BIAS} = \frac{1}{T} \sum_{t=1}^{T} \text{BIAS}_t,$$

$$\text{RMSE} = \frac{1}{T} \sum_{t=1}^{T} \text{RMSE}_t.$$

BIAS_t, BIAS, RMSE_t and RMSE are compared for evaluation of each nonlinear filters.

It is expected in this simulation study that EKF shows the best performance, because we do not have to approximate any function for EKF (i.e., the system is linear) and accordingly EKF gives us an optimal filtering solution.

The setup of each filter is as follows.

(1) For the extended Kalman filter (EKF) and the Monte-Carlo simulation filter (MSF), we take the initial values as $a_{0|0} = 0$ and $\Sigma_{0|0} = 1$.
(2) For the Gaussian sum filter (GSF), the initial values $a_{i,0|0} \sim N(0,1)$ and $\Sigma_{i,0|0} \sim \chi^2(1)$ for $i = 1, \cdots, n$ are chosen, which are the random draws.
(3) Also, for the numerical integration filter (NIF), we set $P(\alpha_0|Y_0) = \Phi(\alpha_0, 1)$ and we take half of the nodes (i.e., $\alpha_{i,t}$, $i = 1, \cdots, n$) from the interval:

$$[a^*_{t|t-1} - \sqrt{c\Sigma^*_{t|t-1}}, a^*_{t|t-1} + \sqrt{c\Sigma^*_{t|t-1}}],$$

and half of the nodes from the interval:

$$[a^*_{t|t} - \sqrt{c\Sigma^*_{t|t}}, a^*_{t|t} + \sqrt{c\Sigma^*_{t|t}}],$$

for $i = 1, \cdots, n$ and $t = 1 \cdots, T$, where $a^*_{t|s}$ and $\Sigma^*_{t|s}$ for $s = t-1, t$ denote the filtering and one-step ahead prediction estimates from EKF (the standard Kalman filter in this case) and $c = 1, 4, 9, 16, 25$ is taken.
(4) Moreover, for the importance sampling filter (ISF), we take the following importance density:

$$P_\alpha(\alpha_t) = \frac{1}{2}\Phi(\alpha_t - a^*_{t|t-1}, c\Sigma^*_{t|t-1}) + \frac{1}{2}\Phi(\alpha_t - a^*_{t|t}, c\Sigma^*_{t|t}),$$

for $i = 1, \cdots, n$ and $t = 0, \cdots, T$, which is a bimodal distribution consisting of two normal densities. $a^*_{t|s}$ and $\Sigma^*_{t|s}$ for $s = t-1, t$ represent the filtering and one-step ahead prediction estimates from EKF (the standard Kalman filter in this case). $c = 1, 4, 9, 16, 25$ is taken.
(5) For RSF, Example 1 in Section 4.6 is directly applied to this simulation study, where we do not need to perform rejection sampling in order to generate random numbers from the filtering densities.

In the case where the system is linear and normal, EKF implies the exact Kalman filter estimate. Therefore, in this example, it is expected that EKF gives us the optimal solution. Table 5.1 represents BIAS while Table 5.2 shows RMSE.

Each nonlinear filter is compared based on the BIAS and RMSE criteria in Tables 5.1 and 5.2, where we take $n = 80$ for GSF, $n = 80$ and $c = 25$ for NIF and ISF, and $n = 500$ for DMF and RSF, respectively. All the nonlinear filters shown in Table 5.1 indicate the unbiased estimators because all the values in Table 5.1 are very small and very close to zero and the signs in the values are random. According to the RMSE criterion in Table 5.2, all the nonlinear filters except for DMF shows a good performance i.e., small RMSE. ISF has slightly larger RMSE, compared with the other nonlinear filters (i.e., EKF, GSF, NIF and RSF) while GSF, NIF and RSF are close to EKF.

GSF is the exactly same as EKF after time $t = 7$ in Table 5.1 and time $t = 3$ in Table 5.2. It is shown in Appendix A5.1 that the initial values ($a_{0|0}$

and $\Sigma_{0|0}$) do not affect the filtering estimates ($a_{t|t}$, $t = 1, \cdots, T$) as time t is large. GSF is the weighted average of n extended Kalman filters with different initial values. Therefore, GSF approaches EKF as time t increases.

The reason why DMF is the worst estimator is as follows. DMF is the estimator which approximates the filtering densities by generating the unconditional random draws (rather than the conditional random draws) of the state-variable α_t from the transition equation (5.2). In the case where the transition equation is a random walk as in equation (5.2), α_t does not have mean and variance. In such a situation, we do not know where the random draws from the transition equation (5.2) approaches as time t goes to infinity. As a result, we obtained the result that DMF is the worst estimator in this simulation study.

Tables 5.3 – 5.7 represents precision of each nonlinear filter depending on n, which is number of the nodes or the random draws. In Table 5.3, GSF is compared for $n = 5, 20, 80$. After time $t = 4$, the three cases ($n = 5, 20, 80$) are same for BIAS and RMSE because it is shown in Appendix A5.1 that the initial values do not depend on the filtering estimates. In Table 5.4, NIF is shown for the cases $n = 5, 20, 80$ and $c = 25$. From the table, NIF is also insensitive to number of the nodes (i.e., n) because all the values of the three cases ($n = 5, 20, 80$) are very similar. In Table 5.5, ISF is compared for $n = 5, 20, 80$ and $c = 25$, where precision of the filtering estimates is improved as number of the random draws (i.e., n) increases. In Table 5.6, DMF is examined for the cases $n = 5, 20, 80, 500$. There, as n increases, precision becomes better. However, even if we take $n = 500$, DMF shows an extremely poor performance, i.e., large RMSE, compared with the other nonlinear filters. For one reason, as discussed above, the random draws of α_t are generated by the transition equation. In this simulation study, the transition equation follows a random walk process, which has no mean and no variance. From the transition equation, we generate unconditional random draws of α_t as time t increases. To improve precision, therefore, a great amount of random draws is required. In Table 5.7, RSF is compared for $n = 5, 20, 80, 500$. Even in the case of small n, i.e., $n = 20$, RSF performs better, especially, the case $n = 80$ is very close to the case $n = 500$.

Tables 5.8 and 5.9 represents precision of NIF and SIF for $c = 1, 4, 9, 16, 25$ and fixed $n = 80$. In Table 5.8, NIF is compared for $c = 1, 4, 9, 16, 25$ and $n = 80$. It is shown from the table that it is better for numerical integration to take larger range of integration in practice. We have the very similar result for the cases $c = 4, 9, 16, 25$, where RMSE is almost the same. In Table 5.9, ISF is compared for $c = 1, 4, 9, 16, 25$ and $n = 80$. For Monte-Carlo integration, it is better to take a broad range of the importance density, because precision of the filtering estimates becomes better as c increases. For choice of the importance density used in Monte-Carlo integration, there is numerous literature, for example, Geweke (1988, 1989a) and Shao (1989). Precision of Monte-Carlo integration with importance sampling is investigated in Appendix A5.2, Tanizaki and Mariano (1994) and Mariano and Tanizaki (1995), where it is concluded that variance of the importance density should be larger than that of the original density under the condition that mean of the importance density is not too far from that of the original density. Also, ISF is extremely improved by utilizing the fixed nodes based on the importance density, not the random draws from the importance density. See Appendix A5.3 and Tanizaki and Mariano (1994).

Table 5.1. Simulation I: Comparison of BIAS

t	EKF	GSF	NIF	ISF	DMF	RSF
1	.002	-.039	.002	.002	.003	-.023
2	-.001	-.016	-.001	-.000	.000	-.000
3	.004	-.001	.004	.003	.002	.002
4	.019	.017	.019	.023	.019	.038
5	.001	.000	.001	.004	.006	.026
6	.012	.011	.012	.010	.017	-.002
7	-.013	-.013	-.013	-.015	-.011	-.011
8	-.009	-.009	-.009	-.008	-.000	.010
9	-.009	-.009	-.009	-.005	-.010	-.008
10	-.002	-.002	-.002	.001	.012	-.010
11	-.011	-.011	-.011	-.008	-.010	-.007
12	-.018	-.018	-.018	-.019	-.018	-.013
13	.001	.001	.001	-.001	-.006	.000
14	.001	.001	.001	.001	.000	.002
15	-.003	-.003	-.003	-.004	.024	.020
16	-.005	-.005	-.005	-.010	.017	-.008
17	-.015	-.015	-.015	-.016	-.012	-.021
18	-.005	-.005	-.005	-.007	-.007	-.040
19	.026	.026	.026	.025	.037	.024
20	.014	.014	.014	.012	.007	.001
21	.020	.020	.020	.018	.029	.024
22	.002	.002	.002	.002	-.008	.012
23	-.003	-.003	-.003	-.002	-.008	-.006
24	-.014	-.014	-.014	-.013	-.030	-.027
25	-.004	-.004	-.004	-.005	-.038	-.028
26	-.006	-.006	-.006	-.007	-.026	.004
27	.025	.025	.025	.025	.004	.011
28	.009	.009	.009	.011	.000	.031
29	.010	.010	.010	.011	.033	.008
30	-.008	-.008	-.008	-.006	.006	-.008
31	-.009	-.009	-.009	-.003	-.018	-.022
32	-.017	-.017	-.017	-.016	-.026	-.016
33	.010	.010	.010	.010	.004	.019
34	.007	.007	.007	.004	.005	.019
35	.006	.006	.006	.003	.033	.004
36	-.004	-.004	-.004	-.001	.018	-.004
37	.011	.011	.011	.014	.026	.015
38	.009	.009	.009	.010	.011	.009
39	.008	.008	.008	.004	-.021	.040
40	-.004	-.004	-.004	-.006	-.021	-.011
Ave	.000	-.000	.000	.001	.001	.001

Table 5.2. Simulation I: Comparison of RMSE

t	EKF	GSF	NIF	ISF	DMF	RSF
1	0.818	0.828	0.818	0.835	0.820	0.821
2	0.804	0.806	0.804	0.821	0.806	0.812
3	0.795	0.795	0.795	0.810	0.801	0.803
4	0.784	0.784	0.784	0.800	0.796	0.793
5	0.783	0.783	0.783	0.804	0.803	0.795
6	0.787	0.787	0.787	0.808	0.823	0.799
7	0.778	0.778	0.778	0.799	0.839	0.785
8	0.778	0.778	0.778	0.807	0.846	0.786
9	0.768	0.768	0.768	0.794	0.874	0.776
10	0.765	0.765	0.765	0.791	0.899	0.774
11	0.789	0.789	0.789	0.818	0.951	0.797
12	0.788	0.788	0.788	0.814	0.987	0.800
13	0.793	0.793	0.793	0.818	1.040	0.804
14	0.776	0.776	0.776	0.801	1.050	0.786
15	0.780	0.780	0.780	0.806	1.093	0.790
16	0.786	0.786	0.786	0.817	1.141	0.797
17	0.788	0.788	0.788	0.819	1.218	0.799
18	0.789	0.789	0.789	0.820	1.235	0.800
19	0.787	0.787	0.787	0.817	1.264	0.794
20	0.774	0.774	0.774	0.806	1.277	0.786
21	0.790	0.790	0.790	0.829	1.309	0.805
22	0.778	0.778	0.778	0.810	1.336	0.793
23	0.788	0.788	0.788	0.819	1.381	0.800
24	0.782	0.782	0.782	0.812	1.423	0.793
25	0.788	0.788	0.788	0.828	1.487	0.799
26	0.796	0.796	0.796	0.834	1.495	0.806
27	0.791	0.791	0.791	0.832	1.526	0.803
28	0.785	0.785	0.785	0.818	1.544	0.799
29	0.775	0.775	0.775	0.817	1.613	0.782
30	0.776	0.776	0.776	0.809	1.645	0.787
31	0.795	0.795	0.795	0.828	1.672	0.803
32	0.788	0.788	0.788	0.827	1.702	0.793
33	0.785	0.785	0.785	0.822	1.742	0.794
34	0.779	0.779	0.779	0.817	1.779	0.789
35	0.790	0.790	0.790	0.827	1.785	0.799
36	0.775	0.775	0.775	0.816	1.793	0.784
37	0.795	0.795	0.795	0.834	1.851	0.805
38	0.795	0.795	0.795	0.837	1.856	0.803
39	0.779	0.779	0.779	0.824	1.874	0.790
40	0.768	0.768	0.768	0.816	1.890	0.776
Ave	0.785	0.786	0.785	0.817	1.307	0.795

Table 5.3. Simulation I: Gaussian Sum Filter (GSF)

t	$n = 5$		$n = 20$		$n = 80$	
	BIAS	RMSE	BIAS	RMSE	BIAS	RMSE
1	-.016	0.851	-.049	0.832	-.039	0.828
2	-.008	0.810	-.020	0.807	-.016	0.806
3	.001	0.796	-.003	0.795	-.001	0.795
4	.018	0.784	.016	0.784	.017	0.784
5	.001	0.783	.000	0.783	.000	0.783
6	.012	0.787	.011	0.787	.011	0.787
7	-.013	0.778	-.013	0.778	-.013	0.778
8	-.009	0.778	-.009	0.778	-.009	0.778
9	-.009	0.768	-.009	0.768	-.009	0.768
10	-.002	0.765	-.002	0.765	-.002	0.765
11	-.011	0.789	-.011	0.789	-.011	0.789
12	-.018	0.788	-.018	0.788	-.018	0.788
13	.001	0.793	.001	0.793	.001	0.793
14	.001	0.776	.001	0.776	.001	0.776
15	-.003	0.780	-.003	0.780	-.003	0.780
16	-.005	0.786	-.005	0.786	-.005	0.786
17	-.015	0.788	-.015	0.788	-.015	0.788
18	-.005	0.789	-.005	0.789	-.005	0.789
19	.026	0.787	.026	0.787	.026	0.787
20	.014	0.774	.014	0.774	.014	0.774
21	.020	0.790	.020	0.790	.020	0.790
22	.002	0.778	.002	0.778	.002	0.778
23	-.003	0.788	-.003	0.788	-.003	0.788
24	-.014	0.782	-.014	0.782	-.014	0.782
25	-.004	0.788	-.004	0.788	-.004	0.788
26	-.006	0.796	-.006	0.796	-.006	0.796
27	.025	0.791	.025	0.791	.025	0.791
28	.009	0.785	.009	0.785	.009	0.785
29	.010	0.775	.010	0.775	.010	0.775
30	-.008	0.776	-.008	0.776	-.008	0.776
31	-.009	0.795	-.009	0.795	-.009	0.795
32	-.017	0.788	-.017	0.788	-.017	0.788
33	.010	0.785	.010	0.785	.010	0.785
34	.007	0.779	.007	0.779	.007	0.779
35	.006	0.790	.006	0.790	.006	0.790
36	-.004	0.775	-.004	0.775	-.004	0.775
37	.011	0.795	.011	0.795	.011	0.795
38	.009	0.795	.009	0.795	.009	0.795
39	.008	0.779	.008	0.779	.008	0.779
40	-.004	0.768	-.004	0.768	-.004	0.768
Ave	.000	0.786	-.001	0.786	-.000	0.786

Table 5.4. Simulation I: Numerical Integration Filter (NIF)

t	$n = 5$		$n = 20$		$n = 80$	
	BIAS	RMSE	BIAS	RMSE	BIAS	RMSE
1	.002	0.820	.002	0.818	.002	0.818
2	-.001	0.804	-.001	0.805	-.001	0.804
3	.004	0.795	.004	0.795	.004	0.795
4	.019	0.784	.020	0.784	.019	0.784
5	.000	0.783	.001	0.783	.001	0.783
6	.012	0.788	.012	0.787	.012	0.787
7	-.013	0.778	-.013	0.778	-.013	0.778
8	-.009	0.778	-.009	0.779	-.009	0.778
9	-.008	0.768	-.009	0.769	-.009	0.768
10	-.002	0.766	-.002	0.765	-.002	0.765
11	-.011	0.789	-.011	0.789	-.011	0.789
12	-.018	0.789	-.019	0.788	-.018	0.788
13	.001	0.794	.001	0.793	.001	0.793
14	.001	0.777	.001	0.776	.001	0.776
15	-.004	0.780	-.004	0.780	-.003	0.780
16	-.005	0.787	-.006	0.786	-.005	0.786
17	-.015	0.789	-.015	0.788	-.015	0.788
18	-.005	0.790	-.005	0.789	-.005	0.789
19	.026	0.787	.026	0.787	.026	0.787
20	.014	0.775	.014	0.774	.014	0.774
21	.019	0.792	.020	0.791	.020	0.790
22	.001	0.779	.001	0.778	.002	0.778
23	-.003	0.789	-.003	0.788	-.003	0.788
24	-.014	0.782	-.014	0.781	-.014	0.782
25	-.004	0.789	-.004	0.788	-.004	0.788
26	-.007	0.796	-.006	0.796	-.006	0.796
27	.026	0.792	.025	0.792	.025	0.791
28	.008	0.786	.009	0.784	.009	0.785
29	.009	0.775	.009	0.775	.010	0.775
30	-.008	0.778	-.008	0.777	-.008	0.776
31	-.008	0.795	-.009	0.795	-.009	0.795
32	-.017	0.788	-.017	0.789	-.017	0.788
33	.010	0.786	.010	0.785	.010	0.785
34	.006	0.780	.006	0.779	.007	0.779
35	.006	0.791	.006	0.790	.006	0.790
36	-.004	0.775	-.004	0.775	-.004	0.775
37	.011	0.795	.012	0.795	.011	0.795
38	.010	0.795	.010	0.795	.009	0.795
39	.008	0.780	.008	0.780	.008	0.779
40	-.003	0.769	-.003	0.768	-.004	0.768
Ave	.000	0.786	.000	0.785	.000	0.785

Table 5.5. Simulation I: Importance Sampling Filter (ISF)

t	$n=5$		$n=20$		$n=80$	
	BIAS	RMSE	BIAS	RMSE	BIAS	RMSE
1	-.030	1.765	.000	0.906	.002	0.835
2	.016	1.682	.007	0.896	-.000	0.821
3	-.017	1.664	.006	0.890	.003	0.810
4	-.000	1.675	.015	0.869	.023	0.800
5	.017	1.622	-.000	0.871	.004	0.804
6	.052	1.664	.013	0.883	.010	0.808
7	-.022	1.666	-.004	0.855	-.015	0.799
8	-.029	1.661	-.005	0.861	-.008	0.807
9	.029	1.669	-.006	0.856	-.005	0.794
10	.031	1.660	-.001	0.847	.001	0.791
11	.009	1.596	-.010	0.877	-.008	0.818
12	-.030	1.661	-.012	0.882	-.019	0.814
13	.009	1.663	.005	0.873	-.001	0.818
14	.006	1.638	.008	0.865	.001	0.801
15	.006	1.645	-.005	0.872	-.004	0.806
16	.009	1.683	-.013	0.887	-.010	0.817
17	.024	1.701	-.017	0.885	-.016	0.819
18	-.039	1.667	-.006	0.899	-.007	0.820
19	.019	1.668	.023	0.890	.025	0.817
20	-.008	1.642	.015	0.877	.012	0.806
21	.003	1.666	.019	0.890	.018	0.829
22	.005	1.615	.005	0.879	.002	0.810
23	-.039	1.665	-.005	0.881	-.002	0.819
24	-.030	1.689	-.011	0.883	-.013	0.812
25	-.041	1.658	-.008	0.881	-.005	0.828
26	-.005	1.638	-.011	0.898	-.007	0.834
27	.018	1.633	.027	0.895	.025	0.832
28	.018	1.642	.015	0.896	.011	0.818
29	-.004	1.638	.011	0.889	.011	0.817
30	-.026	1.619	-.001	0.878	-.006	0.809
31	-.043	1.682	.003	0.890	-.003	0.828
32	-.012	1.617	-.007	0.885	-.016	0.827
33	-.025	1.656	.008	0.888	.010	0.822
34	-.014	1.669	.006	0.883	.004	0.817
35	-.007	1.653	.011	0.897	.003	0.827
36	-.027	1.662	.004	0.879	-.001	0.816
37	.042	1.688	.014	0.888	.014	0.834
38	-.004	1.672	.009	0.901	.010	0.837
39	.047	1.673	.007	0.884	.004	0.824
40	-.020	1.620	-.002	0.878	-.006	0.816
Ave	-.002	1.659	.002	0.882	.001	0.817

Table 5.6. Simulation I: Density-Based Monte-Carlo Filter (DMF)

t	$n=5$ BIAS	RMSE	$n=20$ BIAS	RMSE	$n=80$ BIAS	RMSE	$n=500$ BIAS	RMSE
1	-.003	1.051	.000	0.860	-.002	0.827	.003	0.820
2	-.032	1.198	-.002	0.892	-.004	0.827	.000	0.806
3	-.013	1.345	-.005	0.935	.007	0.834	.002	0.801
4	.014	1.518	.007	1.024	.020	0.863	.019	0.796
5	.013	1.694	-.020	1.103	.004	0.882	.006	0.803
6	.051	1.859	.000	1.184	.017	0.919	.017	0.823
7	.006	1.996	.004	1.269	-.010	0.974	-.011	0.839
8	.041	2.117	.001	1.327	.006	1.027	-.000	0.846
9	.042	2.204	.005	1.434	-.005	1.098	-.010	0.874
10	.063	2.347	.003	1.503	.016	1.149	.012	0.899
11	.042	2.466	-.009	1.576	-.008	1.208	-.010	0.951
12	.043	2.584	-.021	1.622	-.022	1.254	-.018	0.987
13	.061	2.704	-.004	1.721	.010	1.314	-.006	1.040
14	.097	2.813	-.011	1.814	.002	1.377	.000	1.050
15	.069	2.904	-.008	1.922	.016	1.431	.024	1.093
16	.111	2.987	.025	1.993	.027	1.475	.017	1.141
17	.041	3.075	.017	2.047	-.011	1.527	-.012	1.218
18	-.000	3.108	.018	2.097	.016	1.543	-.007	1.235
19	.045	3.264	.021	2.130	.031	1.623	.037	1.264
20	.029	3.347	.013	2.217	.012	1.685	.007	1.277
21	.066	3.413	.012	2.216	.008	1.705	.029	1.309
22	.042	3.454	.027	2.286	-.027	1.744	-.008	1.336
23	.023	3.522	.025	2.328	-.006	1.802	-.008	1.381
24	-.008	3.595	-.000	2.392	-.032	1.839	-.030	1.423
25	-.041	3.662	-.008	2.458	-.039	1.899	-.038	1.487
26	-.040	3.738	.002	2.565	-.018	1.957	-.026	1.495
27	.011	3.834	.032	2.612	-.004	1.994	.004	1.526
28	.025	3.902	.027	2.635	.028	2.029	.000	1.544
29	.029	4.026	.006	2.678	.049	2.050	.033	1.613
30	-.002	4.058	-.000	2.741	.016	2.101	.006	1.645
31	-.024	4.112	-.030	2.769	.036	2.147	-.018	1.672
32	-.047	4.180	.008	2.800	.043	2.147	-.026	1.702
33	-.002	4.268	.075	2.862	.045	2.209	.004	1.742
34	.007	4.272	.086	2.919	.069	2.225	.005	1.779
35	-.007	4.350	.061	2.955	.044	2.278	.033	1.785
36	.017	4.378	.026	2.996	-.016	2.262	.018	1.793
37	.032	4.447	.010	3.002	.028	2.338	.026	1.851
38	.080	4.455	.017	3.057	.007	2.363	.011	1.856
39	.112	4.579	.058	3.052	.035	2.383	-.021	1.874
40	.066	4.648	.050	3.119	.011	2.429	-.021	1.890
Ave	.026	3.187	.013	2.128	.010	1.643	.001	1.307

Table 5.7. Simulation I: Rejection Sampling Filter (RSF)

t	$n = 5$		$n = 20$		$n = 80$		$n = 500$	
	BIAS	RMSE	BIAS	RMSE	BIAS	RMSE	BIAS	RMSE
1	-.030	0.905	-.020	0.846	-.018	0.825	-.023	0.821
2	-.009	0.877	-.009	0.823	-.002	0.811	-.000	0.812
3	.001	0.872	-.001	0.815	.001	0.802	.002	0.803
4	.021	0.874	.039	0.801	.039	0.792	.038	0.793
5	.018	0.861	.028	0.801	.028	0.797	.026	0.795
6	.012	0.869	-.001	0.807	.004	0.798	-.002	0.799
7	-.010	0.871	-.004	0.791	-.006	0.786	-.011	0.785
8	-.009	0.863	.011	0.797	.011	0.785	.010	0.786
9	-.030	0.841	.005	0.789	-.008	0.776	-.008	0.776
10	-.005	0.843	-.008	0.784	-.009	0.775	-.010	0.774
11	.001	0.872	-.003	0.805	-.005	0.798	-.007	0.797
12	-.014	0.863	-.021	0.811	-.015	0.799	-.013	0.800
13	-.008	0.874	-.001	0.814	.005	0.804	.000	0.804
14	-.001	0.865	-.006	0.796	.005	0.786	.002	0.786
15	.015	0.873	.015	0.798	.017	0.789	.020	0.790
16	-.005	0.869	-.003	0.806	-.010	0.797	-.008	0.797
17	-.006	0.868	-.018	0.804	-.023	0.797	-.021	0.799
18	-.034	0.867	-.044	0.813	-.042	0.799	-.040	0.800
19	.008	0.867	.027	0.802	.024	0.795	.024	0.794
20	.004	0.857	.001	0.798	-.000	0.784	.001	0.786
21	.025	0.869	.030	0.816	.020	0.803	.024	0.805
22	.016	0.862	.021	0.798	.014	0.791	.012	0.793
23	.002	0.869	-.005	0.808	-.009	0.796	-.006	0.800
24	-.029	0.859	-.031	0.797	-.034	0.791	-.027	0.793
25	-.031	0.867	-.027	0.807	-.027	0.797	-.028	0.799
26	-.011	0.881	.009	0.821	.007	0.807	.004	0.806
27	.016	0.872	.010	0.816	.010	0.801	.011	0.803
28	.030	0.851	.036	0.801	.032	0.796	.031	0.799
29	.021	0.853	.004	0.793	.010	0.785	.008	0.782
30	-.022	0.869	-.007	0.797	-.009	0.786	-.008	0.787
31	-.012	0.881	-.016	0.813	-.020	0.800	-.022	0.803
32	-.025	0.875	-.008	0.804	-.010	0.793	-.016	0.793
33	.009	0.867	.012	0.808	.020	0.795	.019	0.794
34	.010	0.864	.019	0.797	.022	0.789	.019	0.789
35	.000	0.872	.007	0.805	.009	0.797	.004	0.799
36	-.005	0.858	-.005	0.800	.000	0.785	-.004	0.784
37	.011	0.871	.009	0.819	.014	0.804	.015	0.805
38	.008	0.870	.009	0.817	.011	0.805	.009	0.803
39	.018	0.858	.045	0.800	.042	0.790	.040	0.790
40	.001	0.855	-.005	0.790	-.007	0.776	-.011	0.776
Ave	-.001	0.867	.002	0.805	.002	0.795	.001	0.795

Table 5.8. Simulation I: BIAS and RMSE of NIF for $c = 1, 4, 9, 16, 25$

t	$c=1$		$c=4$		$c=9$		$c=16$		$c=25$	
	BIAS	RMSE	BIAS	RMSE	BIAS	RMSE	BIAS	RMSE	BIAS	RMSE
1	-.004	0.866	.000	0.822	.002	0.818	.002	0.818	.002	0.818
2	-.004	0.833	-.001	0.806	-.001	0.804	-.001	0.804	-.001	0.804
3	.003	0.828	.003	0.796	.004	0.795	.004	0.795	.004	0.795
4	.022	0.818	.019	0.785	.019	0.784	.019	0.784	.019	0.784
5	.005	0.810	.002	0.783	.001	0.783	.001	0.783	.001	0.783
6	.010	0.819	.012	0.788	.012	0.787	.012	0.787	.012	0.787
7	-.012	0.817	-.013	0.780	-.013	0.778	-.013	0.778	-.013	0.778
8	-.003	0.815	-.007	0.780	-.009	0.778	-.009	0.778	-.009	0.778
9	-.005	0.802	-.008	0.770	-.009	0.768	-.009	0.768	-.009	0.768
10	-.002	0.799	-.002	0.767	-.002	0.765	-.002	0.765	-.002	0.765
11	-.010	0.822	-.011	0.791	-.011	0.789	-.011	0.789	-.011	0.789
12	-.019	0.816	-.018	0.789	-.018	0.788	-.018	0.788	-.018	0.788
13	.001	0.824	.001	0.795	.001	0.793	.001	0.793	.001	0.793
14	.001	0.807	.001	0.777	.001	0.776	.001	0.776	.001	0.776
15	.001	0.813	-.003	0.781	-.003	0.780	-.003	0.780	-.003	0.780
16	-.000	0.812	-.005	0.787	-.005	0.786	-.005	0.786	-.005	0.786
17	-.017	0.818	-.015	0.789	-.015	0.788	-.015	0.788	-.015	0.788
18	-.013	0.818	-.007	0.789	-.005	0.788	-.005	0.789	-.005	0.789
19	.026	0.826	.025	0.790	.026	0.787	.026	0.787	.026	0.787
20	.008	0.805	.013	0.775	.014	0.774	.014	0.774	.014	0.774
21	.018	0.815	.020	0.790	.020	0.790	.020	0.790	.020	0.790
22	.002	0.807	.002	0.778	.002	0.778	.002	0.778	.002	0.778
23	-.003	0.816	-.003	0.788	-.003	0.788	-.003	0.788	-.003	0.788
24	-.018	0.811	-.015	0.783	-.014	0.782	-.014	0.782	-.014	0.782
25	-.011	0.821	-.006	0.790	-.004	0.788	-.004	0.788	-.004	0.788
26	-.007	0.828	-.006	0.797	-.006	0.796	-.006	0.796	-.006	0.796
27	.022	0.824	.024	0.793	.025	0.791	.025	0.791	.025	0.791
28	.011	0.815	.010	0.785	.009	0.785	.009	0.785	.009	0.785
29	.010	0.811	.010	0.778	.010	0.775	.010	0.775	.010	0.775
30	-.008	0.810	-.008	0.778	-.008	0.776	-.008	0.776	-.008	0.776
31	-.012	0.829	-.009	0.796	-.009	0.795	-.009	0.795	-.009	0.795
32	-.017	0.831	-.018	0.791	-.017	0.788	-.017	0.788	-.017	0.788
33	.011	0.822	.010	0.787	.010	0.785	.010	0.785	.010	0.785
34	.008	0.810	.007	0.780	.007	0.779	.007	0.779	.007	0.779
35	.006	0.824	.007	0.792	.006	0.790	.006	0.790	.006	0.790
36	-.003	0.812	-.004	0.777	-.004	0.775	-.004	0.775	-.004	0.775
37	.012	0.826	.011	0.796	.011	0.795	.011	0.795	.011	0.795
38	.010	0.829	.009	0.796	.009	0.795	.009	0.795	.009	0.795
39	.016	0.810	.010	0.779	.008	0.779	.008	0.779	.008	0.779
40	-.002	0.806	-.003	0.770	-.004	0.768	-.004	0.768	-.004	0.768
Ave	.000	0.818	.000	0.787	.000	0.785	.000	0.785	.000	0.785

Table 5.9. Simulation I: BIAS and RMSE of ISF for $c = 1, 4, 9, 16, 25$

t	$c=1$ BIAS	$c=1$ RMSE	$c=4$ BIAS	$c=4$ RMSE	$c=9$ BIAS	$c=9$ RMSE	$c=16$ BIAS	$c=16$ RMSE	$c=25$ BIAS	$c=25$ RMSE
1	.009	0.875	.005	0.838	.004	0.831	.003	0.832	.002	0.835
2	.010	0.972	.002	0.848	.001	0.825	.000	0.821	-.000	0.821
3	.019	1.071	.010	0.863	.006	0.825	.005	0.814	.003	0.810
4	.032	1.141	.022	0.864	.021	0.817	.022	0.804	.023	0.800
5	.003	1.240	-.000	0.885	.001	0.825	.003	0.808	.004	0.804
6	.009	1.324	.011	0.897	.010	0.832	.010	0.813	.010	0.808
7	-.015	1.380	-.013	0.901	-.015	0.827	-.015	0.806	-.015	0.799
8	-.012	1.448	-.011	0.919	-.013	0.839	-.011	0.816	-.008	0.807
9	-.012	1.502	-.011	0.920	-.009	0.829	-.007	0.802	-.005	0.794
10	-.012	1.520	-.005	0.914	-.002	0.824	-.000	0.799	.001	0.791
11	-.017	1.591	-.011	0.953	-.010	0.854	-.009	0.826	-.008	0.818
12	-.032	1.626	-.021	0.960	-.020	0.854	-.021	0.825	-.019	0.814
13	-.012	1.658	-.002	0.968	-.000	0.860	-.001	0.829	-.001	0.818
14	-.006	1.664	-.001	0.956	-.000	0.841	.000	0.811	.001	0.801
15	-.005	1.708	-.009	0.977	-.006	0.853	-.005	0.819	-.004	0.806
16	-.017	1.717	-.012	0.997	-.010	0.866	-.010	0.830	-.010	0.817
17	-.035	1.772	-.023	1.009	-.019	0.873	-.017	0.834	-.016	0.819
18	-.017	1.768	-.010	1.020	-.006	0.874	-.007	0.835	-.007	0.820
19	.020	1.779	.020	1.018	.024	0.869	.024	0.830	.025	0.817
20	.020	1.797	.012	1.016	.013	0.860	.012	0.819	.012	0.806
21	.028	1.824	.019	1.044	.018	0.882	.018	0.841	.018	0.829
22	.034	1.822	.004	1.036	.002	0.866	.001	0.824	.002	0.810
23	.016	1.827	-.001	1.052	-.003	0.877	-.003	0.834	-.002	0.819
24	-.011	1.843	-.011	1.051	-.013	0.871	-.013	0.826	-.013	0.812
25	-.009	1.864	-.000	1.071	-.005	0.891	-.006	0.844	-.005	0.828
26	.002	1.855	-.003	1.085	-.004	0.895	-.007	0.849	-.007	0.834
27	.026	1.865	.028	1.092	.029	0.897	.028	0.849	.025	0.832
28	.017	1.869	.013	1.090	.012	0.889	.012	0.836	.011	0.818
29	.028	1.888	.011	1.109	.013	0.893	.012	0.837	.011	0.817
30	.004	1.864	-.009	1.091	-.007	0.877	-.006	0.825	-.006	0.809
31	.010	1.866	-.007	1.112	-.005	0.898	-.004	0.845	-.003	0.828
32	-.001	1.872	-.013	1.130	-.016	0.903	-.017	0.846	-.016	0.827
33	.036	1.871	.013	1.129	.011	0.893	.010	0.838	.010	0.822
34	.029	1.868	.007	1.136	.009	0.895	.006	0.837	.004	0.817
35	.013	1.871	.007	1.161	.007	0.910	.005	0.847	.003	0.827
36	.000	1.861	-.004	1.162	-.001	0.901	-.001	0.836	-.001	0.816
37	.020	1.854	.011	1.177	.013	0.916	.014	0.855	.014	0.834
38	.024	1.858	.012	1.187	.012	0.919	.011	0.857	.010	0.837
39	.043	1.859	.006	1.179	.008	0.909	.007	0.845	.004	0.824
40	.017	1.866	-.006	1.196	-.004	0.908	-.004	0.838	-.006	0.816
Ave	.006	1.668	.000	1.025	.001	0.869	.001	0.830	.001	0.817

5.2.2 Simulation II: Logistic Model

For the second simulation study, consider the following logistic type of measurement and transition equations:

$$\text{(Measurement equation)} \qquad y_t = \frac{\exp(\alpha_t)}{\exp(\alpha_t) + \exp(\epsilon_t)}, \tag{5.3}$$

$$\text{(Transition equation)} \qquad \alpha_t = \frac{\exp(\alpha_{t-1})}{\exp(\alpha_{t-1}) + \exp(\eta_t)}, \tag{5.4}$$

where ϵ_t and η_t are assumed to be normally distributed with:

$$\mathrm{E}\begin{pmatrix} \epsilon_t \\ \eta_t \end{pmatrix} = \begin{pmatrix} 0 \\ 0 \end{pmatrix},$$

and

$$\mathrm{Var}\begin{pmatrix} \epsilon_t \\ \eta_t \end{pmatrix} = \begin{pmatrix} 1 & 0 \\ 0 & 1 \end{pmatrix}.$$

Note that both y_t and α_t lie on the interval between zero and one and that there is no unknown parameter in the system. For the single-stage iteration filter (SIF), the measurement equation (5.3) is equivalent to taking the following equation:

$$\log\left(\frac{1}{y_t} - 1\right) = -\alpha_t + \epsilon_t. \tag{5.5}$$

Thus, for SIF, the bias of the filtering estimates arises from approximation of the transition equation (5.4). EKF, SNF, MSF and GSF are based on equation (5.3), not equation (5.5).

As an example, a set of the data series $\{y_t\}$ and $\{\alpha_t\}$ is displayed in Figure 5.1. For $t = 1, \cdots, T$, both (t, y_t) and (t, α_t) are drawn in the same figure. Given y_t, we consider estimating α_t.

The simulation procedure is as follows.

(i) The normal random numbers for ϵ_t and η_t are generated from:

$$\begin{pmatrix} \epsilon_t \\ \eta_t \end{pmatrix} \sim N\left(\begin{pmatrix} 0 \\ 0 \end{pmatrix}, \begin{pmatrix} 1 & 0 \\ 0 & 1 \end{pmatrix}\right).$$

The initial density $P(\alpha_0)$ is assumed to be a uniform distribution between zero and one, i.e.,

$$P(\alpha_0) = 1, \qquad \text{for } 0 \le \alpha_0 \le 1.$$

Thus, we obtain the artificial data for y_t and α_t, $t = 1, \cdots, T$, from equations (5.3) and (5.4), where sample size $T = 40$ is taken.

Figure 5.1. An Example of y_t and α_t (Logistic Model)

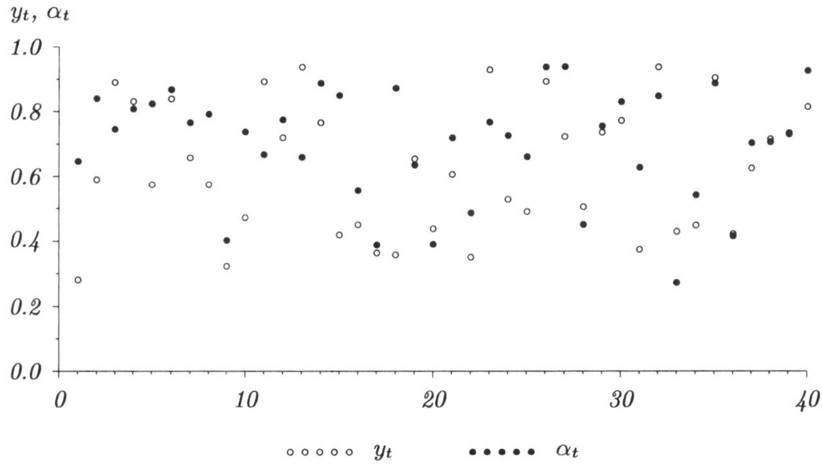

(ii) Given the variances of ϵ_t and η_t, and the data y_t obtained in procedure (i), we can compute the filtered estimates using the algorithms introduced in Chapters 3 and 4. Here, the artificially generated state-variable at time $t = 0$ (i.e., α_0) is taken as the initial value for each filtering algorithm.

(iii) Repeat the procedures (i) – (ii). We perform m simulations for each nonlinear filter. Let $\alpha_t^{(i)}$, $t = 1, \cdots, T$, be the i-th set of artificial data (the i-th true state-variable), and $a_{t|t}^{(i)}$ be the i-th simulation run (the i-th filtered estimate) in a series of m runs, where $m = 4,000$ is set.

(iv) For each nonlinear filter, compute BIAS_t and RMSE_t. See Table 5.10 for BIAS_t and Table 5.11 for RMSE_t, respectively.

The setup of each filter is as follows.

(1) For the extended Kalman filter (EKF), the second-order nonlinear filter (SNF), the Monte-Carlo simulation filter (MSF), the single-stage iteration filter (SIFa, SIFb, SIFc) and the density-based Monte-Carlo filter (DMF), we take $a_{0|0} = 1/2$ and $\Sigma_{0|0} = 1/12$.

(2) For the single-stage iteration filter (SIF), the transition equation is approximated by the first-order Taylor series expansion (SIFa), the second-order Taylor series expansion (SIFb) and the Monte-Carlo stochastic simulations (SIFc).

(3) For the Monte-Carlo simulation filter (MSF), $n = 5, 20, 80, 500$ are chosen, i.e., n normal random numbers are generated for ϵ_t and η_t.

(4) For the Gaussian sum filter (GSF), $a_{i,0|0}$ is generated from a uniform random number between zero and one for $i = 1, \cdots, n$ and $\Sigma_{i,0|0} = a_{i,0|0}(1 - a_{i,0|0})$ is taken for all i are chosen.

(5) Also, for the numerical integration filter (NIF), we set $P(\alpha_0|Y_0) = 1$ and

$$a_{i,0|0} = \frac{i - 0.5}{n},$$

for $i = 0, \cdots, n$ and $t = 1 \cdots, T$.

(6) Moreover, for the importance sampling filter (ISF), we take $P_\alpha(\alpha_t) = 1$ for $i = 1, \cdots, n$ and $t = 0, \cdots, T$. The random draws $\alpha_{i,t}$, $i = 1, \cdots, n$, of α_t are generated based on the uniform density between zero and one.

(7) For the rejection sampling filter (RSF), the acceptance probability is given by:

$$\omega_1(\alpha_t; y_t) = \exp\left(-\frac{1}{2}\left(\log\left(\frac{1}{y_t} - 1\right) + \alpha_t\right)^2\right).$$

The initial density is assumed to be a uniform distribution between zero and one.

In Tables 5.10 and 5.11 where each filter is compared by the bias (BIAS) and the root mean square error (RMSE) for each time t, number of the random draws or the nodes is given by $n = 80$ for GSF, NIF and SIF and $n = 500$ for MSF, SIFc, DMF and RSF. For Tables 5.12 – 5.18, precision of each filtering estimates is examined taking the different number of the random draws or the nodes, i.e., $n = 5, 20, 80$ in Tables 5.12 – 5.16 and $n = 5, 20, 80, 500$ in Tables 5.17 and 5.18. In Tables 5.10, 5.11, 5.15 and 5.16, $c = 25$ is taken. The findings are as follows.

In Table 5.10, each nonlinear filter is compared by the criterion of whether the filtering estimate is biased. For EKF, SIFa and GSF, the estimates for all t are positive, which implies that the filtering estimates based on EKF, SIFa and GSF are clearly upward-biased. Also, SNF and SIFb give us downward-biased filtering estimates, since all of the values are negative. For the other filters, i.e., MSF, SIFc, NIF, ISF, DMF and RSF, we might conclude that their filtering estimates are unbiased.

In Table 5.11, the criterion of the root mean square error (RMSE) is taken. SNF, SIFb, SIFc, NIF, ISF, DMF and RSF indicate a good performance. Accordingly, it might be concluded that the filtering estimates based on EKF, MSF, SIFa and GSF have large RMSE. SIFa gives the largest RMSE of all the filters.

Through Tables 5.10 and 5.11, we might conclude as follows. For EKF and GSF, the estimates after period $t = 4$ are exactly same, which implies that GSF reduces to EKF as time t passes. See Appendix A5.1 for relationship between initial values and filtering estimates. The good estimators are SIFc, NIF, ISF DMF and RSF, which are almost the same values in both Tables 5.10 and 5.11 for all the periods. In the nonlinear filters, shown in Chapter 3, which utilize the Taylor series expansions, MSF shows a good performance in the BIAS criterion (Table 5.10) while MSF is a poor estimator in the

RMSE criterion. SIFc indicates the best estimator in the class of Taylor series expansion approaches. The bias in SIF arises from approximation of the transition equation, i.e., the one-step ahead prediction equations.

In Table 5.12, MSF is compared by taking different number of the random draws, i.e., $n = 5, 20, 80, 500$. There, no improvement in BIAS and certain improvement in RMSE can be found as number of the random numbers (i.e., n) increases. However, RMSE of MSF is still large, compared with SIFc, NIF, ISF DMF and RSF.

In Table 5.13, SIFc is compared by taking different number of the random draws, i.e., $n = 5, 20, 80, 500$. There, the random draws are utilized to approximate the prediction equations (3.9) and (3.10) by Monte-Carlo stochastic simulations. From Table 5.13, no improvement in BIAS and certain improvement in RMSE can be found as number of the random numbers (i.e., n) increases. It might be concluded that in this simulation study the filtering estimates approach the true values as approximation of the prediction equations is improved.

Table 5.14 is related to GSF. As number of the initial values increases, there is basically no improvement in BIAS and RMSE. GSF converges to EKF in spite of number of the initial state-variables (i.e., n) as time passes. See Appendix A5.1 for the relationship between GSF and EKF.

The results of NIF are shown in Table 5.15, where much improvement in BIAS and RMSE is obtained. Also, for small n (i.e., $n = 5$), the estimates are clearly biased because of negative BIAS for all time periods.

The same results as NIF are obtained for ISF in Table 5.16, i.e., certain improvement in BIAS and RMSE. ISF needs larger n than NIF to obtain the similar RMSE (i.e., ISF in the case of $n = 80$ is similar to NIF in the case of $n = 20$), since the simulation errors are included in ISF. Note as follows. Tanizaki and Mariano (1994) examined that ISF is extremely improved by utilizing the fixed nodes based on the importance density, not the random draws from the importance density, and that ISF with the fixed nodes shows a better performance even when number of the random draws or the nodes is small than NIF. See Appendix A5.3 for the fixed nodes of Monte-Carlo integration with importance sampling.

In Table 5.17, DMF is compared for different number of the random draws, i.e, $n = 5, 20, 80, 500$. In the case where range of the sate-variable lies on a certain restricted interval, DMF gives us a good performance as n increases. In the transition equation (5.2) of Simulation I (i.e., a random walk process), the unconditional mean and variance do not exist and accordingly we had the poor results. However, in this simulation study, the better results are obtained since α_t is between zero and one.

RSF is shown in Table 5.18. Even if n is small ($= 20$), RSF gives us the unbiased estimator from the BIAS criterion. Also, it is shown that we have small RMSE as n is large.

Thus, for the estimators which require the random draws or the nodes (i.e., MSF, SIFc, GSF, NIF, ISF, DMF and RSF), we have examined improvement of precision by taking different number of the random draws or the nodes, i.e., $n = 5, 20, 80$ or $n = 5, 20, 80, 500$. We can see that, as number of the random draws or the nodes increases, precision of the filtered estimates improves except for GSF.

Note that the estimators based on density approximation possibly get better when we take more nodes or more random numbers for the state-variable α_t. Generally, an increase in number of the random draws or the nodes improves precision of the filtering estimates. However, we should keep in mind that more random draws or more nodes increase a computational burden.

Thus, summarizing the above results, it might be concluded as follows.

(i) EKF, SNF, SIFa, SIFb and GSF are biased estimators,

(ii) NIF and ISF are better estimators than the other filters in the sense of BIAS and RMSE,

(iii) For MSF, SIFc, NIF, ISF, DMF and RSF, precision of the filtering estimates is certainly improved as number of the random draws or the nodes increases,

(iv) GSF does not improve precision of the filtering estimates even if number of the initial state-variables (i.e., n) increases,

(v) MSF gives us the unbiased filtering estimates but the large root mean square errors (RMSE).

(vi) In the case where α_t lies on a certain restricted interval, DMF performs better.

(vii) Even if n is small, RSF shows a good result.

Table 5.10. Simulation II: Comparison of BIAS

t	EKF	SNF	MSF	SIFa	SIFb	SIFc	GSF	NIF	ISF	DMF	RSF
1	.006	-.005	.001	.006	-.007	.001	.033	.021	.021	.000	.001
2	.013	-.002	.005	.060	-.004	.006	.021	.010	.009	.000	.006
3	.005	-.009	.001	.031	-.012	.000	.007	.001	.000	-.001	.001
4	.005	-.011	-.000	.039	-.013	-.001	.005	-.001	-.001	-.006	-.001
5	.014	-.003	.005	.057	-.006	.005	.014	.005	.004	-.002	.005
6	.002	-.014	-.004	.039	-.017	-.005	.002	-.005	-.006	-.001	-.004
7	.008	-.010	-.001	.051	-.013	-.001	.008	-.001	-.001	.004	-.001
8	.011	-.006	.002	.055	-.009	.002	.011	.002	.001	-.002	.002
9	.013	-.003	.005	.048	-.006	.006	.013	.005	.005	-.000	.005
10	.007	-.010	-.001	.039	-.013	-.000	.007	-.001	-.001	-.000	-.001
11	.008	-.010	-.001	.050	-.013	-.000	.008	-.000	-.002	.002	-.001
12	.004	-.009	.001	.022	-.013	.000	.004	.000	-.001	.000	.001
13	.010	-.007	.001	.048	-.010	.001	.010	.001	.000	-.003	.001
14	.009	-.007	.003	.049	-.010	.002	.009	.002	.001	-.001	.002
15	.006	-.012	-.003	.054	-.015	-.003	.006	-.003	-.004	-.003	-.003
16	.010	-.008	.000	.051	-.011	.000	.010	.000	-.000	.000	-.000
17	.008	-.009	-.000	.054	-.012	-.000	.008	.000	-.000	.003	.000
18	.006	-.010	-.000	.035	-.013	-.001	.006	-.001	-.002	.004	-.000
19	.010	-.006	.002	.050	-.009	.002	.010	.002	.002	-.004	.002
20	.006	-.010	-.000	.043	-.013	-.000	.006	-.000	-.001	.002	-.000
21	.004	-.013	-.004	.048	-.016	-.003	.004	-.003	-.004	-.002	-.004
22	.008	-.008	.001	.034	-.011	.001	.008	.001	.000	.001	.001
23	.005	-.014	-.005	.057	-.016	-.004	.005	-.004	-.005	.001	-.005
24	.010	-.007	.001	.047	-.010	.002	.010	.001	.000	.005	.001
25	.006	-.011	-.001	.043	-.014	-.001	.006	-.001	-.002	.003	-.001
26	.008	-.008	.001	.040	-.011	.001	.008	.000	-.000	.000	.001
27	.007	-.010	-.000	.033	-.013	-.000	.007	-.000	-.001	-.004	-.000
28	.005	-.012	-.002	.042	-.015	-.002	.005	-.002	-.003	-.002	-.002
29	.008	-.009	.000	.046	-.011	.000	.008	.000	-.000	-.001	.000
30	.006	-.010	-.000	.046	-.013	-.001	.006	-.000	-.001	.002	-.000
31	.004	-.014	-.006	.059	-.017	-.005	.004	-.005	-.006	.002	-.006
32	.007	-.010	-.001	.049	-.013	-.001	.007	-.001	-.002	.003	-.001
33	.008	-.011	-.002	.062	-.013	-.001	.008	-.001	-.003	-.005	-.002
34	.008	-.011	-.002	.052	-.013	-.001	.008	-.001	-.002	-.003	-.002
35	.010	-.008	-.000	.055	-.011	.000	.010	.000	.000	-.000	.000
36	.007	-.010	-.001	.036	-.013	-.001	.007	-.001	-.002	.002	-.001
37	.008	-.010	-.001	.050	-.013	-.001	.008	-.001	-.002	-.003	-.001
38	.008	-.010	-.001	.045	-.013	-.001	.008	-.001	-.001	-.001	-.001
39	.008	-.009	.000	.037	-.012	-.000	.008	.000	-.000	-.005	.000
40	.013	-.004	.005	.042	-.007	.005	.013	.005	.005	.002	.005
Ave	.008	-.009	-.000	.045	-.012	-.000	.009	.000	-.000	-.000	-.000

Table 5.11. Simulation II: Comparison of RMSE

t	EKF	SNF	MSF	SIFa	SIFb	SIFc	GSF	NIF	ISF	DMF	RSF
1	.220	.207	.211	.740	.208	.207	.219	.208	.209	.200	.207
2	.218	.200	.205	.753	.201	.201	.217	.201	.202	.198	.201
3	.212	.195	.200	.684	.196	.195	.212	.195	.196	.197	.195
4	.214	.196	.200	.703	.197	.196	.214	.196	.198	.196	.196
5	.215	.199	.204	.684	.199	.199	.215	.199	.200	.194	.199
6	.212	.198	.203	.675	.198	.198	.212	.198	.199	.196	.198
7	.216	.199	.203	.706	.200	.199	.216	.198	.200	.197	.199
8	.214	.199	.204	.687	.199	.199	.214	.199	.199	.196	.199
9	.214	.197	.203	.688	.198	.197	.214	.197	.198	.197	.198
10	.215	.197	.201	.687	.198	.197	.215	.197	.197	.196	.197
11	.211	.193	.197	.695	.194	.193	.211	.193	.194	.203	.193
12	.214	.198	.202	.699	.198	.198	.214	.197	.198	.196	.198
13	.214	.197	.201	.692	.198	.197	.214	.197	.199	.194	.198
14	.215	.197	.201	.689	.198	.197	.215	.197	.198	.195	.197
15	.212	.196	.200	.685	.197	.196	.212	.195	.197	.196	.196
16	.214	.196	.200	.696	.197	.196	.214	.196	.197	.195	.196
17	.213	.198	.203	.677	.198	.197	.213	.197	.199	.197	.198
18	.209	.193	.198	.682	.194	.193	.209	.193	.194	.199	.193
19	.213	.198	.203	.694	.198	.197	.213	.197	.198	.201	.198
20	.211	.197	.202	.696	.197	.197	.211	.196	.198	.194	.197
21	.213	.195	.198	.688	.196	.194	.213	.194	.195	.195	.194
22	.210	.195	.201	.680	.195	.195	.210	.195	.196	.196	.195
23	.212	.196	.200	.682	.197	.196	.212	.196	.197	.197	.196
24	.214	.196	.200	.680	.197	.196	.214	.196	.197	.195	.197
25	.216	.198	.200	.702	.199	.197	.216	.197	.198	.197	.197
26	.210	.195	.200	.673	.196	.195	.210	.195	.196	.201	.195
27	.211	.196	.201	.687	.197	.196	.211	.196	.197	.196	.196
28	.209	.195	.200	.674	.196	.195	.209	.195	.196	.196	.195
29	.214	.197	.201	.689	.198	.197	.214	.197	.198	.197	.197
30	.214	.197	.201	.691	.198	.197	.214	.196	.198	.198	.197
31	.210	.194	.198	.677	.195	.194	.210	.193	.195	.198	.194
32	.212	.197	.202	.682	.197	.196	.212	.196	.198	.201	.196
33	.216	.198	.201	.680	.199	.198	.216	.197	.199	.197	.198
34	.213	.198	.203	.686	.199	.198	.213	.198	.198	.197	.198
35	.213	.197	.202	.688	.198	.197	.213	.197	.198	.198	.198
36	.210	.193	.198	.668	.194	.193	.210	.193	.195	.199	.193
37	.214	.197	.200	.698	.197	.196	.214	.196	.197	.197	.196
38	.211	.195	.200	.692	.196	.195	.211	.195	.196	.198	.195
39	.213	.196	.201	.695	.197	.196	.213	.196	.197	.196	.196
40	.213	.196	.201	.682	.197	.196	.213	.196	.198	.199	.196
Ave	.213	.197	.201	.690	.197	.197	.213	.197	.198	.197	.197

Table 5.12. Simulation II: Monte-Carlo Simulation Filter (MSF)

t	$n = 5$		$n = 20$		$n = 80$		$n = 500$	
	BIAS	RMSE	BIAS	RMSE	BIAS	RMSE	BIAS	RMSE
1	.001	.284	.001	.220	.002	.213	.001	.211
2	.007	.275	.005	.214	.006	.207	.005	.205
3	-.001	.270	.002	.209	.000	.201	.001	.200
4	-.001	.268	-.000	.211	-.001	.202	-.000	.200
5	.003	.263	.004	.214	.005	.206	.005	.204
6	-.007	.269	-.005	.214	-.004	.204	-.004	.203
7	.001	.271	-.002	.214	-.001	.205	-.001	.203
8	.002	.269	.002	.213	.003	.206	.002	.204
9	.004	.273	.004	.211	.005	.204	.005	.203
10	-.002	.280	-.003	.210	-.001	.204	-.001	.201
11	-.000	.264	-.002	.207	-.001	.199	-.001	.197
12	.000	.270	.003	.213	.001	.205	.001	.202
13	.001	.269	.001	.212	.001	.204	.001	.201
14	.002	.269	.003	.211	.002	.203	.003	.201
15	-.006	.270	-.005	.211	-.003	.202	-.003	.200
16	-.004	.269	.000	.213	-.000	.201	.000	.200
17	-.002	.267	.000	.213	-.000	.205	-.000	.203
18	-.006	.280	-.001	.208	.000	.199	-.000	.198
19	-.002	.269	.003	.214	.002	.205	.002	.203
20	-.000	.268	-.003	.211	-.001	.205	-.000	.202
21	-.005	.269	-.005	.209	-.004	.200	-.004	.198
22	-.003	.266	.000	.210	.002	.203	.001	.201
23	-.011	.270	-.005	.211	-.004	.203	-.005	.200
24	.002	.266	.001	.211	.002	.202	.001	.200
25	-.001	.269	-.002	.210	-.001	.203	-.001	.200
26	.006	.273	-.000	.210	.001	.202	.001	.200
27	.005	.261	-.000	.210	.000	.205	-.000	.201
28	.001	.276	-.002	.210	-.002	.202	-.002	.200
29	.003	.276	.002	.211	.000	.204	.000	.201
30	.001	.274	.001	.213	-.000	.203	-.000	.201
31	-.006	.262	-.006	.209	-.006	.201	-.006	.198
32	-.005	.267	-.003	.212	-.001	.204	-.001	.202
33	-.008	.266	-.004	.211	-.002	.203	-.002	.201
34	-.004	.266	-.004	.214	-.002	.206	-.002	.203
35	-.001	.280	-.000	.213	-.000	.204	-.000	.202
36	-.004	.272	-.001	.208	-.001	.199	-.001	.198
37	.001	.269	.000	.212	-.002	.203	-.001	.200
38	-.002	.268	-.002	.211	-.001	.203	-.001	.200
39	-.005	.273	.000	.211	.000	.202	.000	.201
40	.008	.272	.005	.213	.005	.203	.005	.201
Ave	-.001	.270	-.000	.212	.000	.203	-.000	.201

Table 5.13. Simulation II: Single-Stage Iteration Filter (SIFc)

t	$n = 5$		$n = 20$		$n = 80$		$n = 500$	
	BIAS	RMSE	BIAS	RMSE	BIAS	RMSE	BIAS	RMSE
1	.004	.228	.002	.212	.001	.208	.001	.207
2	.005	.221	.006	.205	.006	.202	.006	.201
3	.000	.215	-.001	.201	.000	.197	.000	.195
4	-.002	.215	-.000	.200	-.000	.197	-.001	.196
5	.004	.221	.005	.202	.005	.200	.005	.199
6	-.007	.216	-.006	.202	-.005	.199	-.005	.198
7	-.000	.222	-.001	.202	-.001	.199	-.001	.199
8	.005	.219	.003	.204	.002	.200	.002	.199
9	.006	.217	.005	.202	.005	.198	.006	.197
10	-.000	.217	-.000	.202	-.001	.198	-.000	.197
11	-.001	.213	.000	.198	-.000	.195	-.000	.193
12	-.000	.217	.000	.201	.000	.198	.000	.198
13	.002	.217	.001	.202	.002	.199	.001	.197
14	.001	.218	.003	.202	.002	.198	.002	.197
15	-.002	.219	-.003	.201	-.003	.197	-.003	.196
16	.001	.214	.001	.201	.001	.197	.000	.196
17	-.001	.219	.000	.203	-.000	.199	-.000	.197
18	-.001	.213	-.000	.198	-.001	.195	-.001	.193
19	.005	.217	.003	.204	.002	.199	.002	.197
20	.000	.216	-.000	.200	-.000	.197	-.000	.197
21	-.004	.212	-.004	.199	-.004	.196	-.003	.194
22	.000	.216	.001	.200	.000	.196	.001	.195
23	-.003	.215	-.005	.200	-.004	.197	-.004	.196
24	.003	.217	.001	.202	.002	.197	.002	.196
25	-.001	.216	-.002	.202	-.002	.198	-.001	.197
26	.000	.216	.000	.201	.000	.196	.001	.195
27	-.000	.215	-.000	.201	-.000	.197	-.000	.196
28	-.001	.215	-.002	.200	-.002	.196	-.002	.195
29	.000	.220	.001	.202	.000	.198	.000	.197
30	.000	.217	-.000	.201	-.001	.197	-.001	.197
31	-.003	.216	-.005	.200	-.005	.194	-.005	.194
32	-.001	.215	-.000	.202	-.000	.197	-.001	.196
33	-.001	.214	-.001	.203	-.002	.198	-.001	.198
34	.001	.217	.000	.203	-.001	.199	-.001	.198
35	.002	.217	.002	.202	.001	.198	.000	.197
36	-.002	.214	-.001	.199	-.001	.194	-.001	.193
37	-.004	.215	-.000	.201	-.001	.197	-.001	.196
38	-.001	.216	-.000	.199	-.000	.195	-.001	.195
39	.001	.214	-.000	.201	.000	.197	-.000	.196
40	.004	.218	.004	.201	.005	.197	.005	.196
Ave	.000	.217	.000	.202	.000	.198	-.000	.197

Table 5.14. Simulation II: Gaussian Sum Filter (GSF)

t	$n = 5$ BIAS	$n = 5$ RMSE	$n = 20$ BIAS	$n = 20$ RMSE	$n = 80$ BIAS	$n = 80$ RMSE
1	.026	.219	.021	.219	.033	.219
2	.019	.217	.018	.217	.021	.217
3	.007	.212	.007	.212	.007	.212
4	.005	.214	.005	.214	.005	.214
5	.014	.215	.014	.215	.014	.215
6	.002	.212	.002	.212	.002	.212
7	.008	.216	.008	.216	.008	.216
8	.011	.214	.011	.214	.011	.214
9	.013	.214	.013	.214	.013	.214
10	.007	.215	.007	.215	.007	.215
11	.008	.211	.008	.211	.008	.211
12	.004	.214	.004	.214	.004	.214
13	.010	.214	.010	.214	.010	.214
14	.009	.215	.009	.215	.009	.215
15	.006	.212	.006	.212	.006	.212
16	.010	.214	.010	.214	.010	.214
17	.008	.213	.008	.213	.008	.213
18	.006	.209	.006	.209	.006	.209
19	.010	.213	.010	.213	.010	.213
20	.006	.211	.006	.211	.006	.211
21	.004	.213	.004	.213	.004	.213
22	.008	.210	.008	.210	.008	.210
23	.005	.212	.005	.212	.005	.212
24	.010	.214	.010	.214	.010	.214
25	.006	.216	.006	.216	.006	.216
26	.008	.210	.008	.210	.008	.210
27	.007	.211	.007	.211	.007	.211
28	.005	.209	.005	.209	.005	.209
29	.008	.214	.008	.214	.008	.214
30	.006	.214	.006	.214	.006	.214
31	.004	.210	.004	.210	.004	.210
32	.007	.212	.007	.212	.007	.212
33	.008	.216	.008	.216	.008	.216
34	.008	.213	.008	.213	.008	.213
35	.010	.213	.010	.213	.010	.213
36	.007	.210	.007	.210	.007	.210
37	.008	.214	.008	.214	.008	.214
38	.008	.211	.008	.211	.008	.211
39	.008	.213	.008	.213	.008	.213
40	.013	.213	.013	.213	.013	.213
Ave	.008	.213	.008	.213	.009	.213

Table 5.15. Simulation II: Numerical Integration Filter (NIF)

t	$n=5$ BIAS	$n=5$ RMSE	$n=20$ BIAS	$n=20$ RMSE	$n=80$ BIAS	$n=80$ RMSE
1	.000	.207	.019	.207	.021	.208
2	-.011	.201	.008	.201	.010	.201
3	-.020	.196	-.000	.195	.001	.195
4	-.023	.197	-.003	.196	-.001	.196
5	-.016	.200	.003	.199	.005	.199
6	-.027	.199	-.007	.198	-.005	.198
7	-.023	.200	-.003	.198	-.001	.198
8	-.019	.200	.001	.199	.002	.199
9	-.016	.198	.003	.197	.005	.197
10	-.023	.198	-.002	.197	-.001	.197
11	-.023	.194	-.002	.193	-.000	.193
12	-.021	.199	-.001	.197	.000	.197
13	-.020	.198	-.000	.197	.001	.197
14	-.019	.198	.000	.197	.002	.197
15	-.025	.197	-.005	.195	-.003	.195
16	-.021	.197	-.001	.196	.000	.196
17	-.022	.199	-.001	.197	.000	.197
18	-.023	.194	-.002	.193	-.001	.193
19	-.019	.198	.000	.197	.002	.197
20	-.023	.198	-.002	.196	-.000	.196
21	-.026	.196	-.005	.194	-.003	.194
22	-.020	.196	-.000	.195	.001	.195
23	-.027	.198	-.006	.196	-.004	.196
24	-.020	.197	.000	.196	.001	.196
25	-.023	.198	-.003	.197	-.001	.197
26	-.021	.196	-.001	.195	.000	.195
27	-.022	.197	-.002	.196	-.000	.196
28	-.024	.196	-.004	.195	-.002	.195
29	-.021	.198	-.001	.197	.000	.197
30	-.022	.198	-.002	.196	-.000	.196
31	-.027	.195	-.007	.193	-.005	.193
32	-.023	.198	-.003	.196	-.001	.196
33	-.024	.199	-.003	.197	-.001	.197
34	-.023	.200	-.003	.198	-.001	.198
35	-.021	.198	-.001	.197	.000	.197
36	-.023	.195	-.003	.193	-.001	.193
37	-.023	.197	-.003	.196	-.001	.196
38	-.023	.196	-.002	.195	-.001	.195
39	-.021	.197	-.001	.196	.000	.196
40	-.016	.197	.003	.196	.005	.196
Ave	-.021	.198	-.001	.197	.000	.197

Table 5.16. Simulation II: Importance Sampling Filter (ISF)

t	$n = 5$		$n = 20$		$n = 80$	
	BIAS	RMSE	BIAS	RMSE	BIAS	RMSE
1	.007	.230	.017	.212	.021	.209
2	-.006	.226	.006	.205	.009	.202
3	-.014	.221	-.003	.200	.000	.196
4	-.013	.220	-.004	.201	-.001	.198
5	-.010	.223	.001	.202	.004	.200
6	-.022	.224	-.008	.203	-.006	.199
7	-.021	.224	-.005	.205	-.001	.200
8	-.015	.225	.000	.205	.001	.199
9	-.015	.221	.002	.203	.005	.198
10	-.022	.226	-.005	.202	-.001	.197
11	-.016	.219	-.005	.197	-.002	.194
12	-.020	.224	-.003	.202	-.001	.198
13	-.017	.223	-.001	.203	.000	.199
14	-.016	.224	-.001	.203	.001	.198
15	-.022	.221	-.007	.201	-.004	.197
16	-.018	.223	-.002	.199	-.000	.197
17	-.015	.221	-.003	.203	-.000	.199
18	-.019	.218	-.004	.198	-.002	.194
19	-.015	.223	-.001	.202	.002	.198
20	-.015	.223	-.003	.201	-.001	.198
21	-.021	.221	-.006	.199	-.004	.195
22	-.017	.221	-.003	.201	.000	.196
23	-.023	.224	-.009	.200	-.005	.197
24	-.016	.221	-.001	.201	.000	.197
25	-.018	.226	-.004	.202	-.002	.198
26	-.015	.221	-.003	.200	-.000	.196
27	-.018	.225	-.005	.202	-.001	.197
28	-.021	.220	-.007	.201	-.003	.196
29	-.017	.224	-.002	.202	-.000	.198
30	-.019	.222	-.003	.201	-.001	.198
31	-.023	.221	-.008	.198	-.006	.195
32	-.019	.223	-.006	.202	-.002	.198
33	-.020	.224	-.008	.203	-.003	.199
34	-.018	.222	-.006	.203	-.002	.198
35	-.018	.223	-.003	.202	.000	.198
36	-.020	.220	-.005	.198	-.002	.195
37	-.023	.221	-.005	.202	-.002	.197
38	-.020	.224	-.005	.199	-.001	.196
39	-.016	.221	-.003	.201	-.000	.197
40	-.013	.224	.000	.201	.005	.198
Ave	-.017	.223	-.003	.202	-.000	.198

Table 5.17. Simulation II: Density-Based Monte-Carlo Filter (DMF)

t	$n=5$		$n=20$		$n=80$		$n=500$	
	BIAS	RMSE	BIAS	RMSE	BIAS	RMSE	BIAS	RMSE
1	.001	.223	-.000	.207	-.000	.201	.000	.200
2	.000	.218	-.000	.204	.000	.199	.000	.198
3	.000	.218	-.001	.203	-.001	.198	-.001	.197
4	-.004	.217	-.006	.202	-.006	.198	-.006	.196
5	-.002	.217	-.001	.201	-.003	.196	-.002	.194
6	-.004	.216	-.001	.203	-.001	.198	-.001	.196
7	.000	.216	.003	.203	.004	.199	.004	.197
8	-.004	.219	-.002	.202	-.003	.197	-.002	.196
9	-.000	.220	-.001	.205	-.001	.198	-.000	.197
10	-.001	.221	.000	.202	-.000	.197	-.000	.196
11	.001	.227	.002	.210	.003	.205	.002	.203
12	.001	.223	.001	.202	.000	.198	.000	.196
13	-.003	.219	-.002	.201	-.004	.196	-.003	.194
14	-.002	.223	-.000	.202	-.001	.196	-.001	.195
15	-.003	.224	-.002	.204	-.003	.198	-.003	.196
16	.002	.220	.001	.202	.000	.197	.000	.195
17	.001	.225	.004	.204	.003	.199	.003	.197
18	.005	.225	.005	.207	.005	.201	.004	.199
19	-.006	.230	-.003	.210	-.003	.202	-.004	.201
20	.007	.223	.003	.202	.003	.197	.002	.194
21	-.002	.224	-.001	.205	-.002	.197	-.002	.195
22	.003	.228	.001	.205	.001	.198	.001	.196
23	.004	.228	.002	.207	.001	.199	.001	.197
24	.005	.227	.005	.205	.004	.198	.005	.195
25	.005	.230	.003	.207	.001	.200	.003	.197
26	.004	.230	.004	.210	.000	.204	.000	.201
27	.000	.229	-.003	.205	-.003	.199	-.004	.196
28	-.003	.228	-.002	.205	-.002	.199	-.002	.196
29	-.000	.232	.000	.207	-.000	.200	-.001	.197
30	.002	.232	.001	.208	.004	.200	.002	.198
31	.007	.230	.003	.210	.003	.202	.002	.198
32	.006	.230	.005	.211	.003	.204	.003	.201
33	-.006	.228	-.005	.209	-.005	.201	-.005	.197
34	-.002	.229	-.003	.209	-.002	.200	-.003	.197
35	-.000	.232	.001	.210	.000	.201	-.000	.198
36	.004	.233	.003	.211	.003	.202	.002	.199
37	-.002	.233	-.002	.209	-.003	.200	-.003	.197
38	-.002	.234	-.002	.210	-.001	.200	-.001	.198
39	-.003	.228	-.004	.209	-.005	.199	-.005	.196
40	.001	.235	.003	.210	.001	.203	.002	.199
Ave	.000	.226	.000	.206	-.000	.199	-.000	.197

Table 5.18. Simulation II: Rejection Sampling Filter (RSF)

t	$n = 5$		$n = 20$		$n = 80$		$n = 500$	
	BIAS	RMSE	BIAS	RMSE	BIAS	RMSE	BIAS	RMSE
1	.003	.228	.001	.212	.001	.208	.001	.207
2	.005	.221	.006	.207	.006	.202	.006	.201
3	.001	.215	.000	.201	.002	.196	.001	.195
4	-.000	.215	-.000	.201	-.001	.197	-.001	.196
5	.007	.218	.005	.204	.004	.200	.005	.199
6	-.007	.217	-.004	.202	-.005	.199	-.004	.198
7	-.001	.216	-.001	.203	-.001	.199	-.001	.199
8	.002	.218	.003	.204	.002	.200	.002	.199
9	.005	.218	.005	.203	.005	.199	.005	.198
10	-.002	.214	-.001	.202	-.001	.199	-.001	.197
11	-.001	.211	-.000	.198	-.001	.194	-.001	.193
12	.000	.214	.000	.203	.001	.198	.001	.198
13	.002	.216	.001	.202	.001	.199	.001	.198
14	.001	.216	.003	.203	.003	.198	.002	.197
15	-.004	.214	-.003	.201	-.004	.196	-.003	.196
16	.001	.216	-.000	.200	.000	.197	-.000	.196
17	-.002	.216	.000	.201	-.000	.198	.000	.198
18	-.001	.212	-.001	.199	-.000	.194	-.000	.193
19	.005	.217	.001	.203	.003	.199	.002	.198
20	-.000	.214	.000	.199	-.000	.198	-.000	.197
21	-.004	.212	-.004	.199	-.004	.196	-.004	.194
22	.000	.215	.000	.199	.002	.196	.001	.195
23	-.004	.215	-.007	.201	-.005	.197	-.005	.196
24	.001	.215	.002	.201	.002	.198	.001	.197
25	-.002	.215	-.001	.202	-.002	.198	-.001	.197
26	.000	.214	.002	.201	.001	.196	.001	.195
27	.001	.218	-.000	.201	-.000	.198	-.000	.196
28	-.002	.213	-.003	.200	-.002	.196	-.002	.195
29	-.002	.216	-.001	.201	.000	.198	.000	.197
30	-.000	.216	-.000	.202	-.000	.198	-.000	.197
31	-.004	.214	-.006	.197	-.006	.194	-.006	.194
32	-.001	.216	-.000	.201	-.001	.197	-.001	.196
33	-.001	.216	-.003	.202	-.003	.199	-.002	.198
34	-.000	.217	-.002	.203	-.003	.200	-.002	.198
35	.001	.216	.000	.201	.000	.198	.000	.198
36	-.002	.214	-.000	.199	-.001	.194	-.001	.193
37	-.001	.215	-.002	.201	-.001	.198	-.001	.196
38	-.001	.215	-.002	.200	-.001	.196	-.001	.195
39	-.000	.217	.000	.199	.000	.197	.000	.196
40	.006	.217	.005	.202	.005	.198	.005	.196
Ave	-.000	.216	-.000	.202	-.000	.198	-.000	.197

5.2.3 Simulation III: ARCH(1) Model

In the third example, estimation of an unknown parameter is considered using an ARCH(1) model, which is written as:

$$\alpha_t = \sigma_t \eta_t, \tag{5.6}$$

where σ_t is

$$\sigma_t = (a + b\alpha_{t-1}^2)^{1/2}, \tag{5.7}$$

$a > 0$ and $0 \leq b < 1$. η_t is mutually independently, identically and normally distributed with mean zero and variance one. Eliminating σ_t from equations (5.6) and (5.7), an ARCH(1) process is rewritten as:

$$\alpha_t = (a + b\alpha_{t-1}^2)^{1/2}\eta_t,$$

which corresponds to the transition equation in this simulation study. Also, normalizing the unconditional variance of α_t to be one, $a = 1 - b$ is taken. For the measurement equation, we simply assume that the observed variable y_t consists of a sum of the ARCH(1) process α_t and the random shock ϵ_t. y_t is assumed to be linear in both ϵ_t and η_t. Therefore, we have the following state-space form:

$$\text{(Measurement equation)} \qquad y_t = \alpha_t + \epsilon_t, \tag{5.8}$$

$$\text{(Transition equation)} \qquad \alpha_t = (1 - b + b\alpha_{t-1}^2)^{1/2}\eta_t, \tag{5.9}$$

where α_t is the state-variable. ϵ_t and η_t are assumed to be normally distributed as follows:

$$\begin{pmatrix} \epsilon_t \\ \eta_t \end{pmatrix} \sim N\left(\begin{pmatrix} 0 \\ 0 \end{pmatrix}, \begin{pmatrix} 1 & 0 \\ 0 & 1 \end{pmatrix} \right).$$

The initial value α_0 is also assumed to be a standard normal random variable, i.e.,

$$\alpha_0 \sim N(0, 1).$$

In this simulation study, we examine precision of the filtering estimates as well as the estimate of b. EKF, SNF, SIFc, GSF, NIF, ISF, DMF and RSF are compared for both filtering estimates and parameter estimates.

The simulation procedure is as follows. Given the parameter b, generate normal random draws of ϵ_t and η_t, and compute y_t and α_t for $t = 1, \cdots, T$ ($T = 20, 40$). For the initial value α_0, we take the normal distribution with mean zero and variance one. Then, we perform the following two simulation studies.

(i) Given b and y_t, $t = 1, \cdots, T$, we estimate α_t given Y_s, $s = t - 1, t$.

(ii) Given y_t, $t = 1, \cdots, T$, we estimate α_t given Y_s, $s = t-1, t$, and the unknown parameter b simultaneously using the appropriate likelihood function.

Here, 4,000 sets of data $\{y_t\}_{t=1}^T$ are generated and the corresponding estimates of b are obtained. We choose the true parameter as $b = 0.5, 0.8, 0.9$ and number of the random numbers as $n = 80$ for GSF, NIF and ISF and $n = 500$ for SIFc, DMF and RSF.

For SIFc, each expectation in the prediction equations (3.9) – (3.10) is computed by generating normal random numbers, shown in equations (3.31) and (3.32).

Moreover, for GSF, we take the initial values $a_{i,0|0}$, $i = 1, \cdots, n$, as the random numbers generated from $N(0,1)$ and the initial variances $\Sigma_{i,0|0}$, $i = 1, \cdots, n$, are assumed to be fixed for all i, i.e., $\Sigma_{i,0|0} = 1$ for $i = 1, \cdots, n$.

For NIF, we take half of the nodes (i.e., $\alpha_{i,t}$, $i = 1, \cdots, n$) from the interval:

$$[a^*_{t|t-1} - \sqrt{c\Sigma^*_{t|t-1}}, a^*_{t|t-1} + \sqrt{c\Sigma^*_{t|t-1}}],$$

and half of the nodes from the interval:

$$[a^*_{t|t} - \sqrt{c\Sigma^*_{t|t}}, a^*_{t|t} + \sqrt{c\Sigma^*_{t|t}}],$$

for $i = 1, \cdots, n$ and $t = 1 \cdots, T$, where $a^*_{t|s}$ and $\Sigma^*_{t|s}$ for $s = t-1, t$ denote the filtering and one-step ahead prediction estimates from EKF (the standard Kalman filter in this case) and $c = 25$ is taken.

For ISF, the bimodal distribution consisting of two normal distributions is taken as the appropriately chosen density function $P_\alpha(\alpha_t)$, i.e., importance density:

$$P_\alpha(\alpha_t) = \frac{1}{2}\Phi(\alpha_t - a^*_{t|t-1}, c\Sigma^*_{t|t-1}) + \frac{1}{2}\Phi(\alpha_t - a^*_{t|t}, c\Sigma^*_{t|t}),$$

for $i = 1, \cdots, n$ and $t = 0, \cdots, T$. $a^*_{t|s}$ and $\Sigma^*_{t|s}$ for $s = t-1, t$ represent the filtering and one-step ahead prediction estimates from EKF (the standard Kalman filter in this case). $c = 25$ is taken.

For RSF, Example 1 in Section 4.6 is directly applied to this simulation study, where we do not need to perform rejection sampling in order to generate random numbers from the filtering densities.

(i) First, given $b = 0.5, 0.8, 0.9$ and artificially simulated y_t for $t = 1, \cdots, T$, we estimate α_t given Y_s for $s = t-1, t$.

    The results are in Table 5.19, where the filtering estimates of the state-variable are compared with respect to BIAS and RMSE for each nonlinear filter. Thus, the same procedure as Simulations I and II is taken. From the BIAS criterion, none of the estimators are biased (we had the results that BIAS_t is very small in absolute value for all time t, which is not shown in Table 5.19). SIFc is the worst estimator from RMSE because RMSE is the largest for $b = 0.5, 0.8, 0.9$. Also, DMF shows quite large RMSE. For NIF, ISF and RSF, we have small RMSE and it is shown that RMSE decreases as n increases.

In this simulation study, we do not take MSF for comparison. However, if it is compared with the other nonlinear filters, it might be expected that MSF is worse than SIFc. MSF approximates both the prediction equations and the updating equations by generating random draws, while SIFc approximates the prediction equations only. Therefore, it might be appropriate to consider that MSF has more computational errors than SIFc.

(ii) Next, given y_t, $t = 1, \cdots, T$, we estimate α_t given Y_s, $s = t-1, t$, and the unknown parameter b simultaneously using the appropriate likelihood function.

The maximization of the likelihood function is performed by a simple grid search, in which the function is maximized by changing the parameter value of b by 0.01 in the interval between 0.01 and 0.99.

In Tables 5.20 and 5.21, a comparison between the true parameters and the estimated ones is shown for each procedure. There, b indicates the true value. AVE, SER and RMSE represent arithmetic average, standard error and root mean square error. 10%, 25%, 50%, 75% and 90% give us the 0.10th, 0.25th, 0.50th, 0.75th and 0.90th quantiles of the 4,000 estimates of b. We take $n = 80$ for GSF, NIF and ISF and $n = 500$ for SIFc, DMF and RSF, which is number of the random draws or the nodes.

According to AVE, in the cases 0.5, 0.8, 0.9, the estimates of b based on EKF, SNF and GSF show marked downward bias, while SIFc, NIF and ISF produce estimates that are close to each other and with considerably smaller bias than EKF, SNF and GSF. Also, as the true parameter b goes to one and the sample size is large, SIFc, NIF and ISF perform better than the other filters in terms of AVE, SER and RMSE of the parameter estimates. NIF is the exactly same as ISF for all values.

Thus, we estimated the state-variable and the unknown parameter simultaneously in Simulation III. The estimates of ARCH(1) effect took very low values for EKF, SNF and GSF, while SIFc, NIF and ISF showed better estimates (but still downward-biased).

Table 5.19. Simulation III: Comparison of BIAS and RMSE ($T = 40$)

	n	$b = 0.5$		$b = 0.8$		$b = 0.9$	
		BIAS	RMSE	BIAS	RMSE	BIAS	RMSE
EKF		0.0007	0.7033	0.0002	0.6826	-0.0002	0.6509
SNF		0.0007	0.6995	0.0002	0.6624	-0.0002	0.6236
	5	0.0066	1.4565	0.0043	1.7594	0.0026	1.8909
SIFc	20	0.0009	1.0597	0.0001	1.0831	0.0000	1.0619
	80	-0.0005	1.0186	-0.0018	1.0390	-0.0020	1.0161
	500	0.0003	1.0069	-0.0011	1.0323	-0.0011	1.0184
	5	0.0008	0.7020	0.0004	0.6737	0.0000	0.6322
GSF	20	0.0008	0.7018	0.0003	0.6735	-0.0001	0.6319
	80	0.0008	0.7018	0.0003	0.6736	-0.0001	0.6321
	5	0.0006	0.6905	0.0003	0.6407	-0.0001	0.6064
NIF	20	0.0007	0.6900	0.0003	0.6193	0.0000	0.5745
	80	0.0006	0.6875	0.0003	0.6174	0.0000	0.5697
	5	0.0014	1.1476	-0.0006	0.9207	-0.0010	0.7722
ISF	20	0.0017	0.7321	0.0008	0.6445	0.0004	0.5667
	80	0.0007	0.6961	0.0004	0.6182	0.0002	0.5437
	5	0.0015	1.1219	-0.0002	1.0852	-0.0008	1.0433
DMF	20	0.0027	0.9938	0.0005	0.9909	-0.0001	0.9765
	80	0.0011	0.9125	-0.0004	0.9255	-0.0007	0.9251
	500	0.0002	0.8423	0.0002	0.8517	-0.0001	0.8571
	5	-0.0015	0.7614	-0.0027	0.7025	-0.0025	0.6381
RSF	20	0.0003	0.7105	-0.0011	0.6552	-0.0013	0.5958
	80	0.0003	0.6976	-0.0012	0.6432	-0.0015	0.5848
	500	0.0000	0.6939	-0.0012	0.6396	-0.0015	0.5818

Table 5.20. True Parameter Values and Estimates ($T = 20$)

	AVE	SER	RMSE	10%	25%	50%	75%	90%
				$b = 0.5$				
EKF	0.373	0.346	0.369	0.01	0.01	0.31	0.67	0.92
SNF	0.385	0.352	0.371	0.01	0.01	0.33	0.70	0.94
SIFc	0.481	0.414	0.414	0.01	0.01	0.45	0.97	0.99
GSF	0.385	0.352	0.370	0.01	0.01	0.32	0.70	0.94
NIF	0.424	0.386	0.393	0.01	0.01	0.37	0.82	0.95
ISF	0.434	0.385	0.390	0.01	0.01	0.39	0.83	0.99
DMF	0.590	0.374	0.384	0.01	0.22	0.73	0.94	0.99
RSF	0.485	0.395	0.395	0.01	0.01	0.49	0.91	0.99
				$b = 0.8$				
EKF	0.527	0.357	0.449	0.01	0.20	0.55	0.87	0.99
SNF	0.539	0.362	0.447	0.01	0.19	0.58	0.89	0.99
SIFc	0.675	0.390	0.409	0.01	0.33	0.91	0.99	0.99
GSF	0.548	0.359	0.439	0.01	0.22	0.60	0.90	0.99
NIF	0.553	0.401	0.471	0.01	0.01	0.71	0.93	0.99
ISF	0.601	0.383	0.432	0.01	0.16	0.75	0.96	0.99
DMF	0.751	0.317	0.321	0.07	0.65	0.90	0.99	0.99
RSF	0.655	0.375	0.402	0.01	0.33	0.83	0.99	0.99
				$b = 0.9$				
EKF	0.626	0.344	0.440	0.01	0.36	0.72	0.96	0.99
SNF	0.637	0.348	0.437	0.01	0.38	0.75	0.97	0.99
SIFc	0.776	0.349	0.370	0.01	0.71	0.99	0.99	0.99
GSF	0.655	0.340	0.420	0.01	0.42	0.76	0.99	0.99
NIF	0.645	0.394	0.469	0.01	0.14	0.85	0.99	0.99
ISF	0.703	0.358	0.408	0.01	0.52	0.88	0.99	0.99
DMF	0.835	0.262	0.270	0.45	0.81	0.96	0.99	0.99
RSF	0.756	0.336	0.366	0.01	0.66	0.93	0.99	0.99

Table 5.21. True Parameter Values and Estimates ($T = 40$)

	AVE	SER	RMSE	10%	25%	50%	75%	90%
				$b = 0.5$				
EKF	0.333	0.285	0.330	0.01	0.01	0.30	0.55	0.75
SNF	0.350	0.295	0.331	0.01	0.01	0.33	0.57	0.78
SIFc	0.482	0.383	0.383	0.01	0.01	0.49	0.88	0.99
GSF	0.339	0.289	0.330	0.01	0.01	0.31	0.56	0.76
NIF	0.450	0.349	0.353	0.01	0.03	0.49	0.78	0.90
ISF	0.422	0.346	0.354	0.01	0.01	0.42	0.74	0.90
DMF	0.766	0.284	0.389	0.25	0.72	0.88	0.95	0.99
RSF	0.471	0.352	0.353	0.01	0.08	0.50	0.81	0.94
				$b = 0.8$				
EKF	0.522	0.299	0.408	0.01	0.30	0.54	0.77	0.92
SNF	0.542	0.305	0.399	0.01	0.32	0.58	0.80	0.93
SIFc	0.712	0.347	0.358	0.01	0.51	0.89	0.99	0.99
GSF	0.535	0.302	0.402	0.01	0.32	0.56	0.78	0.94
NIF	0.628	0.347	0.387	0.01	0.38	0.78	0.90	0.96
ISF	0.633	0.338	0.376	0.01	0.40	0.75	0.91	0.99
DMF	0.871	0.183	0.197	0.68	0.85	0.94	0.99	0.99
RSF	0.682	0.323	0.343	0.04	0.52	0.80	0.94	0.99
				$b = 0.9$				
EKF	0.644	0.286	0.384	0.20	0.46	0.69	0.89	0.99
SNF	0.662	0.289	0.374	0.19	0.49	0.72	0.91	0.99
SIFc	0.824	0.287	0.297	0.31	0.81	0.99	0.99	0.99
GSF	0.662	0.284	0.371	0.22	0.48	0.72	0.90	0.99
NIF	0.712	0.337	0.386	0.01	0.64	0.86	0.95	0.99
ISF	0.745	0.303	0.340	0.11	0.67	0.86	0.97	0.99
DMF	0.914	0.142	0.143	0.78	0.90	0.97	0.99	0.99
RSF	0.799	0.267	0.286	0.40	0.74	0.91	0.99	0.99

5.2.4 Simulation IV: Nonstationary Growth Model

The following nonlinear functions are taken by Kitagawa (1987) and Carlin, Polson and Stoffer (1992).

$$\text{(Measurement equation)} \qquad y_t = \frac{\alpha_t^2}{20} + \epsilon_t, \tag{5.10}$$

$$\text{(Transition equation)} \qquad \alpha_t = \frac{1}{2}\alpha_{t-1} + \frac{25\alpha_{t-1}}{1+\alpha_{t-1}^2} + 8\cos(1.2(t-1)) + \eta_t, \tag{5.11}$$

which is called the nonstationary growth model in Kitagawa (1987) and Carlin, Polson and Stoffer (1992). The error terms are mutually independently distributed with the following bivariate normal random variable:

$$\begin{pmatrix} \epsilon_t \\ \eta_t \end{pmatrix} \sim N\left(\begin{pmatrix} 0 \\ 0 \end{pmatrix}, \begin{pmatrix} 1 & 0 \\ 0 & 10 \end{pmatrix} \right).$$

The distribution of the initial value is assumed to be normal, i.e.,

$$\alpha_0 \sim N(0,1).$$

As an example, a set of the data series $\{y_t\}$ and $\{\alpha_t\}$ is displayed in Figure 5.2. For $t = 1, \cdots, T$, both (t, y_t) and (t, α_t) are drawn in the same figure. Given y_t, we consider estimating α_t.

Figure 5.2. An Example of y_t and α_t (Nonstationary Growth Model)

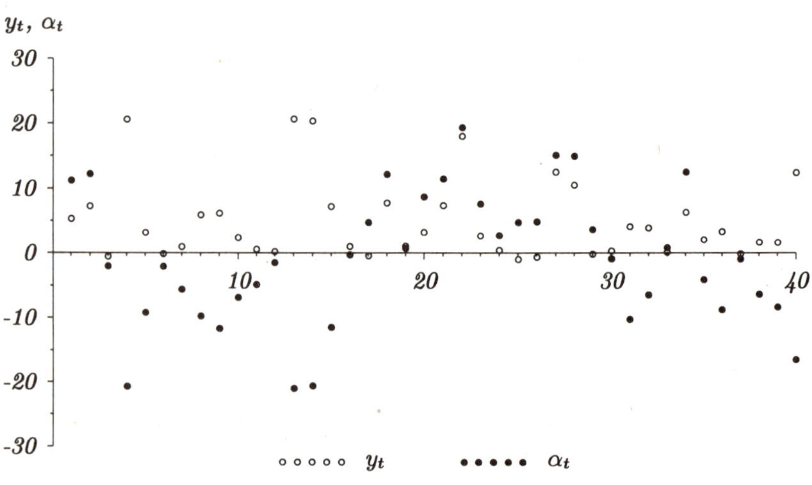

The setup of each filter is as follows.

(1) For EKF, SNF, MSF, SIFa, SIFb and SIFc, we take $a_{0|0} = 0$ and $\Sigma_{0|0} = 1$.
(2) For SIF, the transition equation is approximated by the first-order Taylor series expansion (SIFa), the second-order Taylor series expansion (SIFb) and the Monte-Carlo stochastic simulations (SIFc).
(3) For MSF, n normal random numbers are generated for ϵ_t and η_t to approximate the prediction equations and the updating equations, where $n = 500$ are chosen.
(4) For GSF, the initial values $a_{i,0|0} \sim N(0,1)$, $i = 1, \cdots, n$, are chosen and the initial variances $\Sigma_{i,0|0}$ is taken as: $\Sigma_{i,0|0} = 1$ for $i = 1, \cdots, n$.
(5) Also, for NIF, we set $P(\alpha_0|Y_0) = \Phi(\alpha_0, 1)$ and we take half of the nodes (i.e., $\alpha_{i,t}$, $i = 1, \cdots, n$) from the interval:

$$[a^*_{t|t-1} - \sqrt{c\Sigma^*_{t|t-1}}, a^*_{t|t-1} + \sqrt{c\Sigma^*_{t|t-1}}],$$

and half of the nodes from the interval:

$$[a^*_{t|t} - \sqrt{c\Sigma^*_{t|t}}, a^*_{t|t} + \sqrt{c\Sigma^*_{t|t}}],$$

for $i = 1, \cdots, n$ and $t = 1 \cdots, T$, where $a^*_{t|s}$ and $\Sigma^*_{t|s}$ for $s = t-1, t$ denote the filtering and one-step ahead prediction estimates from EKF and $c = 25$ is taken.
(6) Moreover, for ISF, we take the following importance density:

$$P_\alpha(\alpha_t) = \frac{1}{2}\Phi(\alpha_t - a^*_{t|t-1}, c\Sigma^*_{t|t-1}) + \frac{1}{2}\Phi(\alpha_t - a^*_{t|t}, c\Sigma^*_{t|t}),$$

for $i = 1, \cdots, n$ and $t = 0, \cdots, T$, which is a bimodal distribution consisting of two normal densities. $a^*_{t|s}$ and $\Sigma^*_{t|s}$ for $s = t-1, t$ represent the filtering and one-step ahead prediction estimates from EKF. $c = 25$ is taken.
(7) For the rejection sampling filter, the acceptance probability is given by:

$$\omega_1(\alpha_t; y_t) = \exp\left(-\frac{1}{2}\left(y_t - \frac{\alpha_t^2}{20}\right)^2\right).$$

The initial density is assumed to be a standard normal distribution.

In Tables 5.22 and 5.23, we take $n = 80$ for GSF, NIF and ISF and $n = 500$ for MSF, SIFc, DMF and RSF, respectively.

From the BIAS criterion of Table 5.22, DMF and RSF are the best estimators because of small BIAS in absolute value for all time t. MSF, NIF and ISF are inferior to DMF and RSF but show a quit good performance. According to RMSE in Table 5.23, we can take RSF as the best, DMF as the second best, MSF as the third, ISF as the fourth and moreover SNF as the worst estimator and EKF as the second worst.

Table 5.22. Simulation IV: Comparison of BIAS

t	EKF	SNF	MSF	SIFa	SIFb	SIFc	GSF	NIF	ISF	DMF	RSF
1	7.2	7.2	-0.0	3.1	3.1	3.3	-0.4	-2.1	1.2	0.0	-0.0
2	4.8	4.8	1.4	3.4	2.8	4.0	-1.4	2.5	3.2	-0.1	0.4
3	8.9	9.0	2.5	8.5	7.8	8.7	-0.1	4.1	2.0	-0.0	0.5
4	15.6	13.1	-0.3	13.5	11.7	9.3	1.8	5.3	1.7	0.0	-0.1
5	9.6	2.4	-1.8	8.5	6.3	5.0	1.9	5.5	3.2	0.1	-0.5
6	2.7	-0.0	-1.2	2.4	0.9	0.7	-0.1	2.0	1.6	0.0	-0.6
7	1.7	-1.4	0.1	2.2	0.0	1.3	0.1	1.8	2.1	-0.1	0.4
8	6.9	4.8	1.9	6.5	5.0	5.9	4.0	4.8	3.4	0.0	0.8
9	8.3	7.2	-0.8	5.7	5.4	4.0	4.5	4.2	1.1	0.1	0.4
10	3.2	9.9	-1.4	2.3	4.9	-0.1	0.8	3.4	0.1	0.3	-0.3
11	-2.2	0.2	-1.8	-4.5	-2.3	-6.4	-2.9	-0.8	-0.3	-0.1	-0.3
12	-10.2	-0.8	0.3	-9.4	-4.6	-7.3	-11.1	-1.8	-1.1	-0.3	0.1
13	-2.3	3.2	2.2	-3.2	1.1	-1.1	-3.3	-1.1	0.1	0.0	0.8
14	0.9	4.5	0.7	-0.4	2.0	0.3	-0.3	0.7	-0.5	0.0	0.8
15	-0.4	10.1	0.0	-1.4	3.4	-1.1	-1.5	0.0	-1.7	0.1	-0.1
16	-5.8	-0.3	-1.3	-6.2	-2.7	-6.0	-6.2	-3.3	-2.2	-0.3	-0.3
17	-9.8	-4.5	1.5	-8.8	-5.6	-5.6	-9.9	-3.8	-1.0	-0.4	0.1
18	-3.9	-2.3	2.2	-4.9	-2.9	-1.5	-3.9	-3.2	-0.8	-0.4	0.3
19	0.7	2.4	2.0	0.1	1.6	2.8	0.4	0.2	-0.4	0.0	0.5
20	1.4	5.5	0.1	1.1	3.3	2.4	1.0	0.2	-1.0	0.2	-0.4
21	-3.9	-1.0	-1.6	-3.4	-2.2	-2.9	-4.1	-2.7	-2.8	-0.0	-0.7
22	-5.0	-3.6	0.6	-2.6	-2.7	-1.7	-5.0	-2.8	-0.6	-0.4	-0.6
23	-1.4	-7.1	1.0	0.0	-3.1	1.4	-1.3	-2.0	0.8	0.7	0.3
24	4.1	1.4	1.7	5.9	3.3	6.9	4.1	1.5	0.9	0.1	0.2
25	10.3	5.0	-1.1	11.5	6.5	8.0	10.2	2.7	0.7	0.2	-0.5
26	5.2	-0.1	-2.5	6.8	1.3	2.9	5.2	2.1	0.9	-0.1	-0.8
27	0.6	-2.6	-1.3	1.9	-1.1	0.1	0.7	-0.0	0.9	-0.3	-0.7
28	0.8	-6.4	-0.0	2.0	-2.3	0.8	0.8	0.6	1.8	-0.4	0.3
29	6.5	1.9	1.5	6.5	3.4	5.6	6.5	4.1	3.1	0.2	0.5
30	8.7	5.6	-1.2	6.8	4.9	4.2	8.6	3.9	0.8	0.7	0.2
31	1.8	6.3	-2.0	3.0	3.3	-0.2	1.8	2.9	-0.0	0.5	-0.6
32	-2.3	-0.3	-2.0	-3.3	-3.3	-6.2	-2.3	-1.2	-0.5	-0.1	-0.5
33	-5.7	2.6	0.1	-6.1	-4.6	-6.6	-5.7	-1.6	-0.6	-0.3	0.4
34	0.6	2.2	2.0	-0.4	0.9	-0.1	0.6	0.6	1.2	0.1	0.7
35	2.5	4.3	0.1	0.5	1.7	0.3	2.5	1.5	-0.1	0.4	0.7
36	0.5	7.8	-0.3	-1.0	2.4	-1.4	0.5	0.9	-1.3	0.5	-0.2
37	-5.1	-0.5	-1.3	-6.2	-3.5	-6.4	-5.1	-2.5	-1.8	-0.2	-0.2
38	-9.9	-4.7	1.4	-10.0	-6.7	-6.8	-9.9	-3.0	-0.7	-0.6	0.2
39	-4.5	0.8	2.5	-5.5	-2.8	-2.6	-4.5	-2.4	-0.8	-0.0	0.7
40	0.0	2.6	1.8	-0.7	1.0	0.9	0.0	0.0	-0.4	0.1	0.7
Ave	1.0	2.2	0.1	0.6	0.9	0.3	-0.5	0.5	0.3	-0.0	0.0

Table 5.23. Simulation IV: Comparison of RMSE

t	EKF	SNF	MSF	SIFa	SIFb	SIFc	GSF	NIF	ISF	DMF	RSF
1	9.2	9.2	10.8	6.3	6.3	6.3	4.8	10.6	7.6	4.2	4.7
2	8.2	8.5	9.5	8.2	8.6	7.9	6.6	6.8	7.2	5.3	6.2
3	16.0	16.3	8.8	15.4	15.0	15.4	3.6	11.3	6.5	3.3	4.2
4	30.1	29.1	9.7	21.5	20.3	19.4	12.1	14.3	9.6	4.1	3.9
5	36.9	33.6	9.8	15.5	15.3	15.3	27.5	15.7	14.3	8.9	7.5
6	13.3	14.6	8.4	8.2	9.9	9.8	9.6	15.4	10.0	3.7	3.1
7	17.0	21.7	9.8	9.8	12.1	10.8	25.8	10.0	9.3	6.7	5.9
8	18.3	16.5	8.9	13.7	13.4	13.8	14.2	13.6	10.5	5.9	5.9
9	19.9	19.7	8.8	15.5	15.0	14.5	15.5	13.6	8.1	4.4	3.4
10	27.3	41.4	9.7	14.7	15.3	14.1	26.7	13.6	13.7	10.6	7.9
11	14.4	16.0	8.4	14.3	12.7	15.5	13.7	13.1	9.1	5.1	3.9
12	35.3	26.6	9.9	19.4	16.1	18.4	35.0	13.3	13.8	5.9	4.5
13	20.1	25.8	9.6	13.9	14.0	14.4	18.8	17.8	14.3	8.5	6.8
14	13.7	16.7	8.3	9.9	12.1	10.9	11.9	11.3	6.4	4.8	3.0
15	21.6	35.1	9.5	11.0	14.7	11.4	20.9	11.5	10.8	9.2	6.8
16	16.5	16.6	8.3	14.1	12.8	14.0	16.2	14.0	9.5	6.6	5.1
17	23.7	20.6	9.5	18.3	16.2	16.5	23.5	13.4	8.4	5.9	3.8
18	32.4	31.2	9.9	15.4	14.9	15.1	33.2	15.2	13.6	11.2	7.9
19	14.1	15.0	8.6	10.7	11.9	13.1	13.9	15.3	8.6	6.1	3.3
20	31.7	35.2	9.8	13.0	15.1	14.7	30.9	13.7	11.9	8.5	5.4
21	16.6	16.2	8.9	12.7	13.0	13.3	16.4	17.1	13.2	8.4	6.2
22	16.9	16.2	8.4	12.4	12.6	12.0	16.8	11.9	7.1	5.7	3.1
23	23.5	33.9	9.7	13.1	14.6	12.7	24.9	12.8	11.8	11.3	7.7
24	15.5	15.3	8.4	14.7	13.1	15.2	15.3	12.8	8.6	7.5	4.3
25	28.5	24.8	9.8	20.5	17.5	18.8	27.8	13.0	9.8	7.6	4.2
26	20.5	28.1	9.9	14.5	14.4	14.8	20.4	15.3	12.4	10.6	7.2
27	12.5	15.3	8.4	8.4	11.5	10.4	12.4	10.3	7.1	6.3	2.9
28	19.0	32.6	9.6	9.8	13.9	11.4	19.1	10.6	10.0	10.0	6.1
29	17.9	20.7	8.6	13.8	13.1	13.7	17.9	13.9	10.5	8.6	5.7
30	24.5	19.9	9.0	16.5	15.1	14.8	24.5	13.3	9.4	7.0	3.4
31	29.0	39.4	10.0	15.2	14.8	14.6	29.0	15.4	14.7	12.2	8.1
32	14.8	62.5	8.6	13.7	13.5	15.7	14.8	14.1	9.7	7.8	3.5
33	35.8	33.6	9.9	17.0	16.1	18.1	35.8	14.0	13.7	9.4	5.0
34	17.1	60.1	9.2	13.1	13.3	13.8	17.0	17.7	13.7	9.8	6.5
35	14.6	21.6	8.3	10.7	11.7	10.7	14.6	10.7	6.5	6.7	3.1
36	23.2	35.2	9.5	11.7	13.9	11.6	23.2	12.0	10.4	11.0	7.2
37	16.0	32.9	8.3	14.3	13.1	14.4	16.0	13.7	10.1	8.5	5.0
38	29.2	22.8	9.7	19.3	17.2	17.6	29.2	12.5	9.9	8.3	3.9
39	30.0	28.9	10.1	15.3	14.5	15.1	30.0	15.9	14.9	11.5	7.8
40	13.7	32.5	8.5	9.9	11.3	11.4	13.7	13.9	8.5	7.6	3.2
Ave	21.0	26.0	9.2	13.6	13.7	13.8	19.6	13.4	10.4	7.6	5.2

5.3 Summary

In this chapter, the nonlinear filters introduced in the previous chapters were compared. The various functional forms were taken for the measurement and transition equations in Section 5.2, where 4,000 simulations were performed.

We can conclude in the simulation studies that RSF is better than any other nonlinear filtering estimations, judging from the criteria of BIAS and RMSE.

In Table 5.24, we summarize the results of Simulation I from Tables 5.1 and 5.2, Simulation II from Tables 5.10 and 5.11, Simulation III from Tables 5.19 — 5.21, and Simulation IV from Tables 5.22 and 5.23. There, ○, △ and × denote excellent, fair, poor performances, respectively.

Table 5.24. Summary of Simulations I – IV

	Simulation I	Simulation II	Simulation III	Simulation IV
EKF	○	×	△	×
SNF	—	×	△	×
MSF	—	△	×	○
SIFa	—	×	—	△
SIFb	—	×	—	△
SIFc	—	○	—	△
GSF	○	×	△	×
NIF	○	○	○	△
ISF	△	○	○	○
DMF	×	○	×	○
RSF	○	○	○	○

From Table 5.24, we can observe that precision of the filtering estimates depends on the functional form of the state-space model.

The findings in Simulations II – IV are summarized as follows:

(i) EKF and GSF are clearly biased estimators because of linearization of the nonlinear measurement and transition equations,

(ii) NIF and ISF are better estimators by the criteria of BIAS and RMSE than the other filters,

(iii) For the nonlinear filters that the nodes or random draws are required (except for GSF), precision of the estimates is certainly improved as number of the nodes or the random draws increases,

(iv) For GSF, precision of the filtering estimates does not depend on number of the nodes or random draws,

(v) MSF gives us the unbiased filtering estimates but large RMSE.

(vi) For DMF, precision of the filtering estimates depends on functional form. In Simulations II and IV, DMF indicates quite good results while in Simulations I and III it is the worst estimator.

(vii) Of all the nonlinear filters, RSF performs better for all the simulation studies.

Finally, we set down the following general comments. For three of the nonlinear filters introduced in Chapter 3 (i.e., EKF, SNF and MSF), the disadvantage is that we have to assume the normal distributions for $P(y_t|\alpha_t)$ and $P(\alpha_t|\alpha_{t-1})$ even if they are not normal, because normality assumption is required in the process of derivations of EKF, SNF and MSF to obtain the linear recursive algorithm. Otherwise, we cannot obtain the filtering estimates of the state-variables which are approximately normally distributed. This implies that the nonnormal error terms (i.e., residuals) are approximated as the normal ones. Therefore, it is possible that the extended Kalman filter (EKF) and the second-order nonlinear filter (SNF) are biased. For both MSF and SIF, however, there is a smaller possibility of bias. The reason is that for MSF each expectation in the algorithm (3.9) – (3.16) is evaluated by normal random draws. Also, the filtering estimates obtained from SIF are less biased because the conventional nonlinear least squares estimation procedure such as Newton-Raphson optimization is used.

On the other hand, for the algorithms developed in Chapter 4, we do not need normality approximation for the two distribution functions $P(y_t|\alpha_t)$ and $P(\alpha_t|\alpha_{t-1})$. The exact distributions of $P(y_t|\alpha_t)$ and $P(\alpha_t|\alpha_{t-1})$ are derived from the measurement and transition equations. Normality approximation is not required for these distributions.

The disadvantage of the simulation-based nonlinear filters (i.e., MSF, SIFc, NIF, ISF, DMF and RSF) is that they take a long time computationally. Also, for NIF, ISF and DMF, we have to calculate $P(y_t|\alpha_t)$ and/or $P(\alpha_t|\alpha_{t-1})$ by hand. Sometimes these methods may not be feasible.

Certainly, for the simulation-based procedures (i.e., MSF, SIFc, NIF, ISF, DMF and RSF), precision of the estimates is improved to some extent as number of the nodes or the random draws increases. However, an increase in the nodes or the random draws greatly increases the computational burden, and the estimation may therefore be inapplicable in practice. Approximating the densities yields better filtering estimates than linearizing the nonlinear functions because approximation of nonlinear functions clearly leads to biased estimates, while approximating the densities by nodes or random numbers gives us the asymptotically unbiased and consistent estimates.

A5 Appendix

A5.1 On Initial Value of State-Variable

In this appendix, we consider by Monte-Carlo experiments how the initial value (i.e., $t = 0$) affects the filtering estimates (i.e., $t = 1, \cdots, T$) using the standard Kalman filter algorithm. Tanizaki (1993c) examined the relationship between the initial value and the filtering estimate in the state-space model, where it is concluded that the initial value does not affect the filtering estimate too much under the condition that variance of the initial value is large.

By examining the initial values in this appendix, we can resolve the reason why the Gaussian sum filter (GSF) approaches the extended Kalman filter (EKF) as time t increases.

For simplicity of discussion, we take the following simple linear and normal state-space model:

$$\text{(Measurement equation)} \qquad y_t = \alpha_t + \epsilon_t, \tag{5.1}$$

$$\text{(Transition equation)} \qquad \alpha_t = \alpha_{t-1} + \eta_t, \tag{5.2}$$

where α_t is the state-variable and we assume ϵ_t as a standard normal random variable. ϵ_t and η_t are assumed to be normally distributed as follows:

$$\begin{pmatrix} \epsilon_t \\ \eta_t \end{pmatrix} \sim N\left(\begin{pmatrix} 0 \\ 0 \end{pmatrix}, \begin{pmatrix} 1 & 0 \\ 0 & 1 \end{pmatrix} \right).$$

The initial value α_0 is distributed as a standard normal random variable, i.e.,

$$\alpha_0 \sim N(0, 1).$$

Equations (5.1) and (5.2) are also examined in Section 5.2.1.

The initial values in the true model are given by $a_0 = 0$ and $\sigma_0 = 1$. That is, given $a_0 = 0$ and $\sigma = 1$, both y_t and α_t are artificially generated for $t = 1, \cdots, T$.

We take the filtering initial values as $a_{0|0} = 0.0, 1.0, 2.0, 3.0, 5.0$ and $\Sigma_{0|0} = 0.0, 1.0, 2.0, 3.0, 5.0$. Note that $\Sigma_{0|0} = 0$ implies that the initial value $a_{0|0}$ is nonstochastic. We compare the artificially simulated state-variable (i.e., α_t) with the filtering estimate (i.e., $a_{t|t}$).

Let m be number of the simulation runs. We define $\text{BIAS}_t(\cdot,\cdot)$ and $\text{RMSE}_t(\cdot,\cdot)$ as:

$$\text{BIAS}_t(a_{0|0}, \Sigma_{0|0}) \equiv \frac{1}{m} \sum_{i=1}^{m} (a_{t|t}^{(i)} - \alpha_t^{(i)}),$$

$$\text{RMSE}_t(a_{0|0}, \Sigma_{0|0}) \equiv \left(\frac{1}{m} \sum_{i=1}^{m} (a_{t|t}^{(i)} - \alpha_t^{(i)})^2 \right)^{1/2},$$

for $t = 1, \cdots, T$. $\mathrm{BIAS}_t(\cdot,\cdot)$ and $\mathrm{RMSE}_t(\cdot,\cdot)$ depend on the initial values $a_{0|0}$ and $\Sigma_{0|0}$, because $a_{t|t}^{(i)}$ is a function of the initial values $a_{0|0}$ and $\Sigma_{0|0}$.

$a_{t|t}^{(i)}$ represents the filtering estimate given $a_{0|0}$ and $\Sigma_{0|0}$, and $\alpha_t^{(i)}$ denotes the artificially simulated state-variable given $a_0 = 0$ and $\sigma_0 = 1$. Both $a_{t|t}^{(i)}$ and $\alpha_t^{(i)}$ are obtained in the i-th simulation run. The initial density of the state-variable is normal with mean zero and variance one, i.e., $\alpha_0 \sim N(a_0, \sigma_0^2)$, $a_0 = 0$ and $\sigma_0 = 1$. For number of the simulation runs, $m = 10,000$ is taken.

We compare the filtering estimates given the true initial values and those given the other initial values. The following cases are displayed in each table:

- $a_{0|0} = 0$ and $\Sigma_{0|0} = 0, 1, 2, 3, 5$ in Table 5.25
- $a_{0|0} = 1$ and $\Sigma_{0|0} = 0, 1, 2, 3, 5$ in Table 5.26
- $a_{0|0} = 2$ and $\Sigma_{0|0} = 0, 1, 2, 3, 5$ in Table 5.27
- $a_{0|0} = 3$ and $\Sigma_{0|0} = 0, 1, 2, 3, 5$ in Table 5.28
- $a_{0|0} = 5$ and $\Sigma_{0|0} = 0, 1, 2, 3, 5$ in Table 5.29

In Figures 5.3 – 5.6, the following two differences are drawn:

- $\mathrm{BIAS}_t(a_{0|0}, \Sigma_{0|0}) - \mathrm{BIAS}_t(0, 1)$ in Figures 5.3 and 5.5
- $\mathrm{RMSE}_t(a_{0|0}, \Sigma_{0|0}) - \mathrm{RMSE}_t(0, 1)$ in Figures 5.4 and 5.6

where the initial mean and variance of the filtering estimates are taken as:

$$
\begin{aligned}
(a_{0|0}, \Sigma_{0|0}) &= (1,1), (2,1), (3,1), (5,1), && \text{in Figures 5.3 and 5.4,} \\
&= (5,1), (5,2), (5,3), (5,5), && \text{in Figures 5.5 and 5.6.}
\end{aligned}
$$

Figures 5.3 and 5.4 display BIAS and RMSE given $\Sigma_{0|0} = 1$ (which is the true variance), while Figures 5.5 and 5.6 represent BIAS and RMSE given $a_{0|0} = 5$ (which is the biased initial mean).

We consider these tables and figures from the following points of view:

(i) For BIAS and RMSE, the filtering estimates for $\Sigma_{0|0} = 0, 1, 2, 3, 5$ given unbiased initial mean (i.e., $a_{0|0} = 0$). See Table 5.25.

(ii) For BIAS and RMSE, the filtering estimates for $a_{0|0} = 0, 1, 2, 3, 5$ given unbiased initial variance (i.e., $\Sigma_{0|0} = 1$). See Tables 5.25 – 5.29 and Figures 5.3 and 5.4.

(iii) For BIAS and RMSE, the filtering estimates for $\Sigma_{0|0} = 0, 1, 2, 3, 5$ given the biased initial mean (i.e., $a_{0|0} = 5$). See Table 5.29 and Figures 5.5 and 5.6.

As for (i), consider the BIAS and RMSE criteria in the case where the initial mean of the state-variable is equivalent to the true mean (i.e., $a_{0|0} = 0$). From Table 5.25, Case: $\Sigma_{0|0} = 1$ is very close to the other cases $\Sigma_{0|0} = 0, 2, 3, 5$, especially, exactly equal to all the cases for $t = 4, \cdots, T$. This implies that the filtering estimates $a_{t|t}$, $t = 1, \cdots, T$, are unbiased for any initial variance $\Sigma_{0|0}$ when the initial mean $a_{0|0}$ is true. Note that $\mathrm{BIAS}_t(a_{0|0}, \Sigma_{0|0})$ and $\mathrm{RMSE}_t(a_{0|0}, \Sigma_{0|0})$ when $a_{0|0} = 0$ and $\Sigma_{0|0} = 1$ is equal to the true BIAS and RMSE.

Next, for (ii), we examine BIAS and RMSE of the filtering estimates given the initial variance $\Sigma_{0|0} = 1$, which are shown in Tables 5.25 – 5.29 and Figures 5.3 and 5.4. In the case where the initial mean $a_{0|0}$ is different from the true initial mean (i.e., $a_{0|0} \neq 0$), we have large bias as the initial mean is away from the true mean. Similarly, as $a_{0|0}$ has large bias, the difference between $\mathrm{RMSE}_t(a_{0|0}, 1)$ and $\mathrm{RMSE}_t(0, 1)$ is also large.

For (iii), the filtering estimates given the biased initial mean are shown in Table 5.29 and Figures 5.5 and 5.6. The filtering estimates are close to the true values when the initial variance is large. That is, the case of $\Sigma_{0|0} = 5$ is close to the horizontal line compared with the others. It is concluded that bias of the filtering estimates disappears as $\Sigma_{0|0}$ is large.

Summarizing the results obtained above, we have the followings: First, when the initial mean is true, the filtering estimates are unbiased for any initial variance, where RMSE of such filtering estimates is almost equal to that of the true value. Second, given the initial variance, BIAS and RMSE of $a_{t|t}$ increase when the initial mean $a_{0|0}$ is away from the true value. Third, given the initial mean, BIAS and RMSE of $a_{t|t}$ decrease when the initial variance $\Sigma_{0|0}$ is large.

The best choice is to take the unbiased initial mean and the unbiased initial variance for the filtering estimates. Usually, however, the initial values are not given *a priori*, i.e., they are unknown. Therefore, the second best might be to give a plausible value to the initial mean and a sufficiently large value to the initial variance, which is more realistic in empirical studies. Thus, in this appendix, we have obtained the result that $a_{t|t}$, $t = 1, \cdots, T$, are not biased for any $a_{0|0}$ when $\Sigma_{0|0}$ is sufficiently large.

Table 5.25. BIAS and RMSE: Case $a_{0|0} = 0$

	t \ $\Sigma_{0\|0}$	0	1	2	3	5
	1	-0.004	-0.006	-0.007	-0.008	-0.008
	2	0.007	0.007	0.007	0.007	0.007
B	3	0.018	0.018	0.019	0.019	0.019
I	4	0.017	0.017	0.017	0.017	0.017
A	5	0.005	0.005	0.005	0.005	0.005
S	6	0.008	0.008	0.008	0.008	0.008
	7	0.005	0.005	0.005	0.005	0.005
	8	0.001	0.001	0.001	0.001	0.001
	9	0.001	0.001	0.001	0.001	0.001
	10	0.005	0.005	0.005	0.005	0.005
	1	0.855	0.804	0.856	0.910	0.958
	2	0.802	0.793	0.801	0.809	0.816
R	3	0.790	0.788	0.789	0.790	0.791
M	4	0.791	0.791	0.791	0.791	0.791
S	5	0.788	0.788	0.788	0.788	0.788
E	6	0.785	0.785	0.785	0.785	0.785
	7	0.783	0.783	0.783	0.783	0.783
	8	0.783	0.783	0.783	0.783	0.783
	9	0.781	0.781	0.781	0.781	0.781
	10	0.781	0.781	0.781	0.781	0.781

Table 5.26. BIAS and RMSE: Case $a_{0|0} = 1$

	t \ $\Sigma_{0\|0}$	0	1	2	3	5
	1	0.496	0.328	0.160	0.083	0.029
	2	0.207	0.132	0.065	0.038	0.019
B	3	0.095	0.066	0.041	0.030	0.023
I	4	0.047	0.036	0.026	0.022	0.019
A	5	0.017	0.012	0.009	0.007	0.006
S	6	0.012	0.010	0.009	0.008	0.008
	7	0.007	0.006	0.006	0.005	0.005
	8	0.002	0.001	0.001	0.001	0.001
	9	0.001	0.001	0.001	0.001	0.001
	10	0.005	0.005	0.005	0.005	0.005
	1	0.988	0.868	0.870	0.914	0.958
	2	0.828	0.804	0.804	0.810	0.816
R	3	0.795	0.791	0.790	0.790	0.791
M	4	0.792	0.791	0.791	0.791	0.791
S	5	0.788	0.788	0.788	0.788	0.788
E	6	0.785	0.785	0.785	0.785	0.785
	7	0.783	0.783	0.783	0.783	0.783
	8	0.783	0.783	0.783	0.783	0.783
	9	0.781	0.781	0.781	0.781	0.781
	10	0.781	0.781	0.781	0.781	0.781

Table 5.27. BIAS and RMSE: Case $a_{0|0} = 2$

	t \ $\Sigma_{0\|0}$	0	1	2	3	5
	1	0.996	0.661	0.326	0.174	0.066
	2	0.407	0.257	0.124	0.069	0.032
B	3	0.172	0.114	0.063	0.042	0.028
I	4	0.076	0.054	0.034	0.026	0.021
A	5	0.028	0.019	0.012	0.009	0.007
S	6	0.016	0.013	0.010	0.009	0.008
	7	0.008	0.007	0.006	0.006	0.005
	8	0.002	0.002	0.001	0.001	0.001
	9	0.002	0.001	0.001	0.001	0.001
	10	0.005	0.005	0.005	0.005	0.005
	1	1.312	1.041	0.916	0.926	0.960
	2	0.899	0.834	0.811	0.812	0.816
R	3	0.808	0.796	0.791	0.791	0.791
M	4	0.794	0.792	0.792	0.791	0.791
S	5	0.789	0.788	0.788	0.788	0.788
E	6	0.785	0.785	0.785	0.785	0.785
	7	0.783	0.783	0.783	0.783	0.783
	8	0.783	0.783	0.783	0.783	0.783
	9	0.781	0.781	0.781	0.781	0.781
	10	0.781	0.781	0.781	0.781	0.781

Table 5.28. BIAS and RMSE: Case $a_{0|0} = 3$

	t \ $\Sigma_{0\|0}$	0	1	2	3	5
	1	1.496	0.994	0.493	0.265	0.103
	2	0.607	0.382	0.183	0.100	0.044
B	3	0.249	0.161	0.085	0.054	0.033
I	4	0.106	0.072	0.043	0.031	0.023
A	5	0.039	0.026	0.015	0.011	0.008
S	6	0.021	0.016	0.012	0.010	0.009
	7	0.010	0.008	0.007	0.006	0.005
	8	0.003	0.002	0.002	0.001	0.001
	9	0.002	0.002	0.001	0.001	0.001
	10	0.005	0.005	0.005	0.005	0.005
	1	1.723	1.279	0.987	0.948	0.963
	2	1.006	0.880	0.822	0.815	0.817
R	3	0.828	0.804	0.793	0.791	0.791
M	4	0.798	0.794	0.792	0.792	0.792
S	5	0.789	0.789	0.788	0.788	0.788
E	6	0.786	0.785	0.785	0.785	0.785
	7	0.783	0.783	0.783	0.783	0.783
	8	0.783	0.783	0.783	0.783	0.783
	9	0.781	0.781	0.781	0.781	0.781
	10	0.781	0.781	0.781	0.781	0.781

Table 5.29. BIAS and RMSE: Case $a_{0|0} = 5$

	$\Sigma_{0|0}$ / t	0	1	2	3	5
	1	2.496	1.661	0.826	0.447	0.177
	2	1.007	0.632	0.301	0.163	0.069
B	3	0.403	0.257	0.130	0.077	0.042
I	4	0.164	0.108	0.060	0.040	0.026
A	5	0.062	0.040	0.022	0.014	0.009
S	6	0.029	0.021	0.014	0.011	0.009
	7	0.013	0.010	0.008	0.006	0.006
	8	0.004	0.003	0.002	0.002	0.001
	9	0.002	0.002	0.001	0.001	0.001
	10	0.005	0.005	0.005	0.005	0.005
	1	2.638	1.845	1.189	1.014	0.974
	2	1.287	1.014	0.856	0.825	0.819
R	3	0.887	0.829	0.799	0.793	0.792
M	4	0.807	0.798	0.793	0.792	0.792
S	5	0.791	0.789	0.789	0.788	0.788
E	6	0.786	0.785	0.785	0.785	0.785
	7	0.783	0.783	0.783	0.783	0.783
	8	0.783	0.783	0.783	0.783	0.783
	9	0.781	0.781	0.781	0.781	0.781
	10	0.781	0.781	0.781	0.781	0.781

Figure 5.3. $\mathrm{BIAS}_t(a_{0|0}, \Sigma_{0|0}) - \mathrm{BIAS}_t(0,1)$: Case $\Sigma_{0|0} = 1$

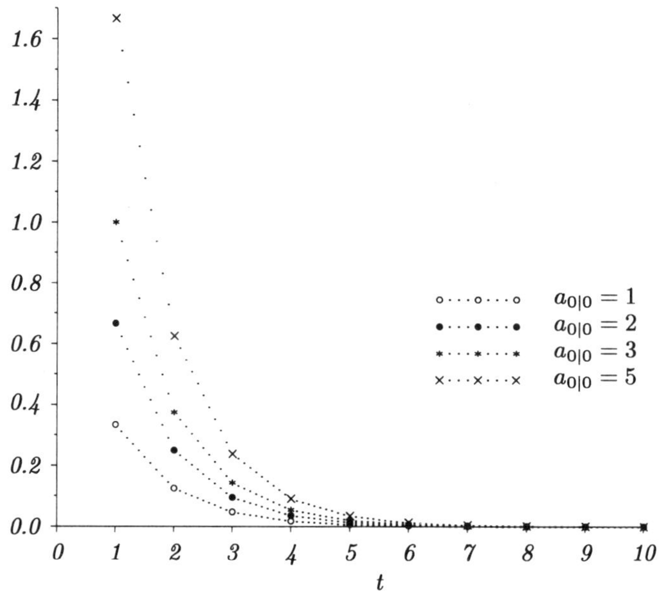

Figure 5.4. $\mathrm{RMSE}_t(a_{0|0}, \Sigma_{0|0}) - \mathrm{RMSE}_t(0,1)$: Case $\Sigma_{0|0} = 1$

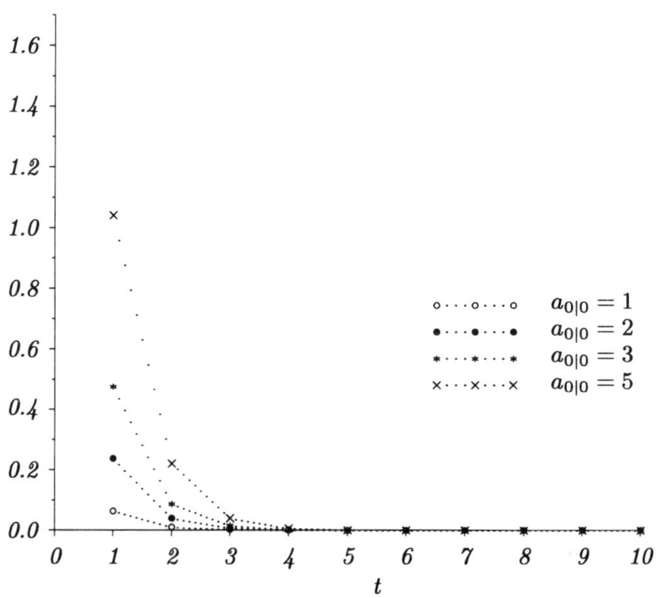

Figure 5.5. $\text{BIAS}_t(a_{0|0}, \Sigma_{0|0}) - \text{BIAS}_t(0, 1)$: Case $a_{0|0} = 5$

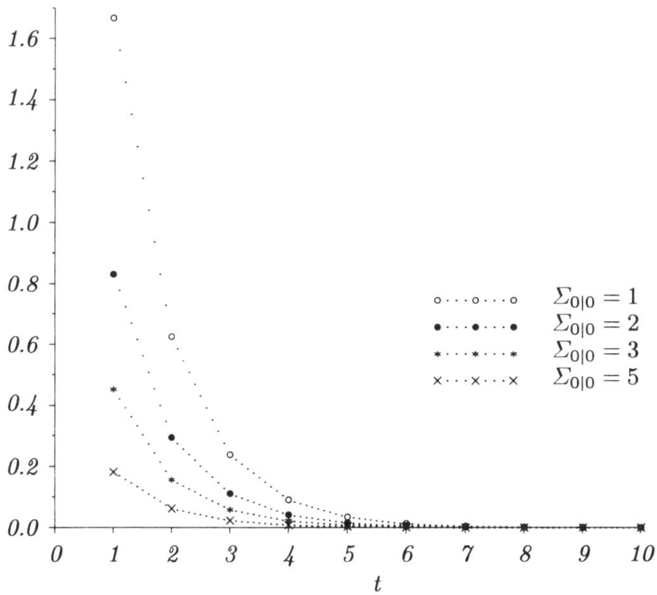

Figure 5.6. $\text{RMSE}_t(a_{0|0}, \Sigma_{0|0}) - \text{RMSE}_t(0, 1)$: Case $a_{0|0} = 5$

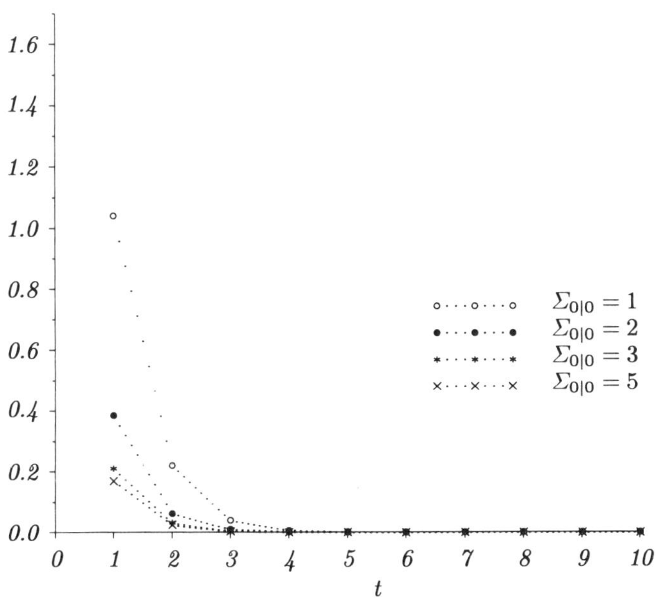

A5.2 Precision of Monte-Carlo Integration with Importance Sampling

Mariano and Tanizaki (1995) discussed about precision of Monte-Carlo integration with importance sampling, which is as follows. Let x be a random variable with distribution $P(x)$ and $z(x)$ be a function of x. The expectation of $z(x)$ is defined as follows.

$$\mathrm{E}\big(z(x)\big) = \int z(x)P(x)\mathrm{d}x.$$

When the above integration cannot be evaluated explicitly, Monte-Carlo integration with importance sampling is often used. The above expectation can be rewritten as:

$$\mathrm{E}\big(z(x)\big) = \int z(x)\omega(x)P_x(x)\mathrm{d}x,$$

where $\omega(x)$ is defined as:

$$\omega(x) = \frac{P(x)}{P_x(x)}.$$

Generating n random draws from the importance density $P_x(x)$, the expectation is approximated. Let $\overline{z}_n$ be the estimate of $\mathrm{E}\big(z(x)\big)$, which is computed as follows.

$$\overline{z}_n = \frac{1}{n}\sum_{i=1}^{n} z(x_i)\omega(x_i), \tag{5.12}$$

where x_i is a random number generated from $P_x(x)$.

Note as follows. Precision of the following approximation is much better than equation (5.12).

$$\overline{z}_n = \sum_{i=1}^{n} z(x_i)\omega(x_i) \Big/ \sum_{i=1}^{n} \omega(x_i),$$

where $\sum_{i=1}^{n} z(x_i)\omega(x_i)$ is divided by:

$$\sum_{i=1}^{n} \omega(x_i),$$

not n. In this book, all the results related to Monte-Carlo integration with importance sampling, including Tables 5.30 and 5.31, are computed using the above approximation.

Now, we take an example of $z(x) = (x - \mu)/\sigma$, where the underlying distribution of x is normal with mean μ and variance σ^2, i.e., $P(x) = N(\mu, \sigma^2)$. Note that $z(x)$ has a standard normal distribution, i.e., $z(x) \sim N(0,1)$.

We choose the bimodal distribution function for the importance density in this book, which is the importance density $P_\alpha(\alpha_t)$ such that $P_\alpha(\alpha_t)$ covers both $P(\alpha_{t-1}|Y_t)$ and $P(\alpha_t|Y_t)$, Therefore, in this appendix, we examine the importance density which is assumed to be:

$$P_x(x) = \frac{1}{2}\Phi(x - \mu_1, \sigma_1^2) + \frac{1}{2}\Phi(x - \mu_2, \sigma_2^2),$$

where $\mu_1 = -4$, $\mu_2 = 4$, $\sigma_1 = 1$ and $\sigma_2 = 4$. The number of random draws is $n = 50,\ 100,\ 200$. Moreover, we choose the parameters μ and σ as $\mu = \mu_i + c_\mu \sigma_i$ for $c_\mu = \pm 2.0,\ \pm 1.5,\ \pm 1.0,\ \pm 0.5,\ 0.0$ and $i = 1, 2$ and $\sigma = c_\sigma \sigma_i$ for $c_\sigma = 1/3,\ 1/2,\ 2/3,\ 1,\ 3/2,\ 2,\ 3$ and $i = 1, 2$.

Moreover, in order to examine precision of Monte-Carlo integration with importance sampling, we repeat the simulation run m times and obtain m estimates of $\overline{z}_n$, where $m = 10,000$ is taken. Thus, m simulation runs are performed. For each μ and σ, we compute the sample mean of $\overline{z}_n$ in Table 5.30 and the sample standard error of $\overline{z}_n$ in Table 5.31 from m estimates.

The asymptotic properties are as follows. Note that x_i is a random variable from $P_x(x)$ while x is from $P(x)$. Therefore, $\mathrm{E}\big(z(x_i)\omega(x_i)\big)$ implies taking the expectation with respect to x_i but $\mathrm{E}\big(z(x)\big)$ is with respect to x. $z(x_i)\omega(x_i)$, $i = 1, \cdots, n$, are random. Define:

$$\begin{aligned} m &= \mathrm{E}\big(z(x_i)\omega(x_i)\big) \\ &= \mathrm{E}\big(z(x)\big) \end{aligned}$$

and

$$\begin{aligned} s^2 &= \mathrm{Var}\big(z(x_i)\omega(x_i)\big) \\ &= \mathrm{E}\big(z(x_i)\omega(x_i)\big)^2 - m^2 \\ &= \mathrm{E}\Big(\big(z(x)\big)^2\omega(x)\Big) - m^2. \end{aligned}$$

By the central limit theorem, we can easily show that

$$\frac{\overline{z}_n - m}{s/\sqrt{n}} \longrightarrow N(0,1),$$

where $m = 0$ in this example. Thus, $\overline{z}_n$ is consistent. For any μ and σ^2, $\overline{z}_n$ gives us an asymptotically unbiased estimator as n goes to infinity. However, convergence is quite slow as $\sqrt{n}$.

In Table 5.30, the sample means of $\overline{z}_n$ are close to zero except for large σ or large μ in absolute value, which implies that $\overline{z}_n$ is asymptotically unbiased. In Table 5.31, the sample variances of $\overline{z}_n$ decrease as n is large. Moreover, around $\mu = -4,\ 4$ and $\sigma = 0.5 \sim 1,\ 2 \sim 4$, the sample variances of $\overline{z}_n$ are small.

When either of two peaks of $P_x(x)$ is close to center of $P(x)$, $\overline{z}_n$ is asymptotically unbiased. And when $P_x(x)$ are wider in range than $P(x)$, the sample variance of $\overline{z}_n$ is small. Therefore, it might be concluded from the simulation study that we should choose the importance density with the following conditions: (i) the importance density should have a wide range of distribution, compared with the original distribution and (ii) center of the importance density should be close to that of the original density but we do not have to pay too much attention to center of the importance density.

The importance density used in this book (i.e., $P_\alpha(\alpha_t)$) needs to cover $P(\alpha_t|Y_s)$, $s = t-1, t$. Peak and range of $P(\alpha_t|Y_{t-1})$ are different from those of $P(\alpha_t|Y_t)$. In general, range of $P(\alpha_t|Y_{t-1})$ is larger than that of $P(\alpha_t|Y_t)$. For the importance sampling filter (ISF), two densities $P(\alpha_t|Y_s)$, $s = t-1, t$, have to be approximated by one importance density $P_\alpha(\alpha_t)$. Thus, the simulation study in this appendix shows that the bimodal importance density is appropriate for the importance sampling filter.

Moreover, peak and range of $P(\alpha_t|Y_s)$ are not known, but mean and variance of α_t estimated by the extended Kalman filter (EKF) or the second-order nonlinear filter (SNF) are available. It might be plausible to consider that the true values are not too far from the extended Kalman filter (EKF) estimates or the second-order nonlinear filter (SNF) estimates. Utilizing the extended Kalman filter (EKF) estimates or the second-order nonlinear filter (SNF) estimates, the filtering estimates using importance sampling (i.e., the importance sampling filter (ISF)) would be improved. Therefore, in the empirical example of Chapter 6, we have chosen the bimodal density with $a^*_{t|t-1}$, $a^*_{t|t}$, $\Sigma^*_{t|t-1}$ and $\Sigma^*_{t|t}$, which are the extended Kalman filter (EKF) estimates (i.e., $a^*_{t|s} = \mathrm{E}(\alpha_t|Y_s)$ and $\Sigma^*_{t|s} = \mathrm{Var}(\alpha_t|Y_s)$ for $s = t-1, t$), defined as:

$$P_\alpha(\alpha_t) = \frac{1}{2}\Phi(\alpha - a^*_{t|t-1}, c\Sigma^*_{t|t-1}) + \frac{1}{2}\Phi(\alpha - a^*_{t|t}, c\Sigma^*_{t|t}),$$

where $\Phi(\alpha - a^*_{t|s}, \Sigma^*_{t|s})$ represents the normal distribution with mean $a^*_{t|s}$ and variance $\Sigma^*_{t|s}$, and c is a constant term. In practice, we take $c = 25$ in Chapters 5 and 6.

Table 5.30. Precision of Monte-Carlo Integration — Sample Mean of $\overline{z}_n$

μ	σ / n	1/2	2/3	1	4/3	3/2	2	8/3	3	4	6	8
	50	.4 10	.382	.397	.428	.444	.489	.534	.552	.590	.638	.662
-6.0	100	.173	.186	.232	.278	.300	.363	.431	.458	.517	.586	.623
	200	.076	.091	.129	.171	.192	.253	.332	.366	.443	.533	.582
	50	.112	.129	.184	.245	.273	.347	.422	.451	.514	.587	.622
-5.5	100	.046	.058	.096	.145	.170	.243	.329	.364	.444	.536	.584
	200	.022	.029	.051	.084	.102	.161	.244	.283	.375	.484	.545
	50	.035	.041	.071	.121	.149	.229	.321	.358	.442	.537	.582
-5.0	100	.017	.020	.035	.065	.084	.149	.240	.281	.376	.487	.546
	200	.008	.0 10	.019	.036	.049	.095	.170	.2 10	.312	.438	.508
	50	.012	.014	.025	.0 50	.068	.137	.233	.275	.374	.490	.544
-4.5	100	.006	.007	.012	.024	.035	.082	.165	.207	.313	.441	.509
	200	.004	.004	.007	.014	.020	.051	.113	.1 50	.255	.394	.473
	50	.001	.002	.005	.014	.023	.070	.158	.203	.313	.444	.506
-4.0	100	.000	.001	.002	.006	.0 10	.039	.106	.146	.256	.397	.473
	200	.001	.001	.001	.004	.006	.024	.070	.102	.204	.353	.438
	50	-.007	-.007	-.008	-.007	-.003	.025	.099	.142	.256	.400	.469
-3.5	100	-.004	-.005	-.005	-.005	-.003	.012	.062	.097	.204	.356	.438
	200	-.002	-.003	-.003	-.002	-.001	.009	.041	.065	.159	.313	.405
	50	-.019	-.020	-.023	-.022	-.020	-.002	.055	.093	.206	.359	.433
-3.0	100	-.011	-.012	-.013	-.012	-.011	-.002	.032	.059	.159	.316	.403
	200	-.008	-.007	-.007	-.006	-.005	.000	.021	.040	.121	.276	.372
	50	-.0 50	-.046	-.041	-.035	-.031	-.016	.024	.055	.161	.320	.397
-2.5	100	-.023	-.022	-.020	-.017	-.016	-.009	.013	.033	.121	.279	.370
	200	-.012	-.011	-.0 10	-.008	-.007	-.004	.009	.022	.090	.242	.341
	50	-.099	-.077	-.054	-.041	-.035	-.022	.006	.029	.123	.283	.362
-2.0	100	-.037	-.032	-.024	-.019	-.017	-.012	.002	.016	.089	.245	.337
	200	-.013	-.012	-.011	-.009	-.008	-.005	.003	.011	.064	.2 10	.310
	50	.056	.038	.019	.009	.006	.000	-.002	.000	.030	.157	.227
0.0	100	.019	.014	.008	.003	.002	-.001	-.002	-.001	.017	.131	.214
	200	.007	.005	.003	.001	.000	-.001	-.001	.000	.013	.109	.197
	50	.016	.013	.013	.013	.013	.011	.007	.005	.004	.061	.099
2.0	100	.005	.005	.005	.006	.006	.005	.004	.003	.002	.052	.099
	200	.003	.003	.002	.003	.003	.003	.002	.002	.003	.044	.095
	50	.000	.001	.002	.002	.002	.002	.001	.000	-.005	-.020	-.028
4.0	100	.001	.002	.002	.003	.003	.002	.002	.001	-.002	-.007	-.011
	200	.001	.002	.002	.002	.003	.003	.002	.002	.000	.000	-.001
	50	-.0 10	-.011	-.011	-.011	-.0 10	-.011	-.013	-.015	-.029	-.109	-.157
6.0	100	-.006	-.005	-.005	-.004	-.004	-.004	-.006	-.007	-.013	-.071	-.124
	200	-.001	-.001	-.001	-.001	-.001	-.001	-.002	-.002	-.005	-.046	-.098
	50	-.069	-.052	-.041	-.037	-.037	-.038	-.047	-.055	-.097	-.227	-.294
8.0	100	-.014	-.013	-.015	-.016	-.016	-.018	-.022	-.026	-.048	-.163	-.243
	200	-.008	-.008	-.008	-.008	-.008	-.009	-.011	-.012	-.024	-.117	-.203
	50	-.362	-.258	-.171	-.141	-.134	-.134	-.160	-.180	-.258	-.382	-.440
10.0	100	-.079	-.062	-.053	-.052	-.053	-.058	-.074	-.087	-.154	-.294	-.375
	200	-.026	-.023	-.023	-.025	-.025	-.028	-.036	-.042	-.089	-.227	-.321

Table 5.31. Precision of Monte-Carlo Integration — Sample Standard Error of $\bar{z}_n$

μ	σ / n	1/2	2/3	1	4/3	3/2	2	8/3	3	4	6	8
	50	.813	.714	.625	.579	.561	.506	.436	.408	.348	.280	.256
-6.0	100	.551	.531	.533	.535	.535	.518	.466	.438	.374	.300	.270
	200	.354	.368	.415	.452	.465	.492	.476	.454	.392	.313	.276
	50	.476	.465	.492	.504	.504	.488	.436	.4 10	.3 50	.282	.260
-5.5	100	.298	.312	.378	.431	.4 50	.477	.456	.434	.375	.302	.272
	200	.200	.213	.274	.339	.365	.425	.451	.440	.388	.313	.277
	50	.284	.283	.341	.406	.427	.455	.430	.408	.352	.284	.263
-5.0	100	.188	.190	.243	.321	.354	.421	.438	.425	.374	.303	.275
	200	.130	.133	.172	.238	.271	.352	.415	.418	.383	.313	.279
	50	.208	.199	.230	.304	.340	.408	.418	.403	.352	.286	.267
-4.5	100	.144	.137	.160	.225	.263	.355	.4 10	.4 10	.371	.303	.278
	200	.100	.096	.113	.162	.193	.281	.369	.388	.374	.312	.280
	50	.188	.175	.183	.231	.264	.352	.398	.393	.351	.288	.272
-4.0	100	.132	.123	.129	.165	.194	.288	.372	.387	.366	.304	.281
	200	.092	.086	.090	.117	.139	.218	.316	.3 50	.363	.3 10	.282
	50	.201	.186	.186	.203	.221	.296	.368	.376	.348	.289	.276
-3.5	100	.139	.131	.131	.144	.158	.228	.326	.355	.358	.303	.284
	200	.099	.093	.092	.101	.111	.167	.263	.306	.348	.308	.283
	50	.2 50	.231	.217	.209	.209	.248	.330	.351	.343	.290	.280
-3.0	100	.173	.161	.151	.145	.146	.182	.276	.316	.346	.303	.287
	200	.121	.112	.105	.101	.102	.130	.213	.259	.329	.304	.285
	50	.349	.304	.255	.223	.212	.213	.287	.319	.336	.291	.285
-2.5	100	.226	.202	.172	.152	.145	.151	.228	.273	.332	.301	.290
	200	.154	.138	.118	.105	.101	.106	.169	.213	.307	.300	.286
	50	.492	.391	.289	.233	.214	.191	.244	.282	.325	.292	.289
-2.0	100	.295	.245	.188	.156	.145	.132	.185	.229	.313	.300	.293
	200	.192	.163	.128	.108	.100	.092	.133	.172	.281	.296	.288
	50	.735	.536	.358	.274	.246	.187	.154	.161	.249	.295	.307
0.0	100	.378	.304	.226	.181	.164	.128	.106	.113	.208	.288	.305
	200	.230	.193	.1 50	.123	.112	.088	.074	.079	.161	.270	.294
	50	.502	.388	.292	.248	.234	.201	.169	.157	.171	.301	.324
2.0	100	.285	.238	.191	.167	.159	.139	.118	.1 10	.124	.275	.317
	200	.189	.161	.132	.117	.111	.097	.083	.077	.088	.241	.300
	50	.442	.349	.268	.229	.217	.192	.177	.173	.178	.318	.337
4.0	100	.261	.221	.179	.156	.149	.134	.125	.122	.126	.280	.326
	200	.177	.151	.124	.108	.103	.093	.087	.086	.088	.235	.306
	50	.488	.377	.285	.244	.232	.211	.205	.208	.248	.348	.346
6.0	100	.283	.238	.192	.167	.159	.146	.143	.145	.179	.3 10	.333
	200	.188	.161	.131	.116	.111	.102	.100	.101	.125	.266	.312
	50	.7 50	.559	.399	.337	.319	.292	.295	.308	.375	.379	.351
8.0	100	.377	.309	.247	.217	.208	.194	.201	.214	.295	.349	.337
	200	.234	.200	.164	.147	.142	.134	.139	.148	.218	.312	.317
	50	1.496	1.108	.749	.601	.561	.506	.500	.506	.489	.403	.353
10.0	100	.731	.553	.405	.348	.334	.322	.347	.373	.425	.379	.339
	200	.376	.308	.248	.223	.217	.213	.238	.264	.353	.349	.321

A5.3 On Random Draws by Importance Sampling

Tanizaki and Mariano (1994) pointed out as follows. It is known that the simulation techniques proposed by Brown and Mariano (1984, 1989), Mariano and Brown (1983, 1989), McFadden (1989), Tanizaki (1995a) and so on include *simulation errors* and accordingly give a less efficient estimator. Therefore, it is pertinent to note that evaluation of the expectation by Monte-Carlo integration with importance sampling clearly has not only *computation errors* but also *simulation errors*, and furthermore that convergence is quite slow as $\sqrt{n}$ for Monte-Carlo integration with importance sampling. These two problems give us the disadvantages of using Monte-Carlo integration. In order to reduce the *simulation errors*, we may choose $\alpha_{i,t}$, $i = 1, \cdots, n$, such that

$$\begin{aligned}\frac{1}{n} &= \int_{-\infty}^{\alpha_{i,t}} P_\alpha(\alpha_t)\mathrm{d}\alpha_t - \int_{-\infty}^{\alpha_{i-1,t}} P_\alpha(\alpha_t)\mathrm{d}\alpha_t \\ &= \int_{\alpha_{i-1,t}}^{\alpha_{i,t}} P_\alpha(\alpha_t)\mathrm{d}\alpha_t,\end{aligned}$$

given the terminal conditions:

$$\begin{aligned}&\int_{-\infty}^{\alpha_{1,t}} P_\alpha(\alpha_t)\mathrm{d}\alpha_t \\ &= \int_{\alpha_{n,t}}^{\infty} P_\alpha(\alpha_t)\mathrm{d}\alpha_t \\ &= \frac{1}{2n},\end{aligned}$$

which implies that $\alpha_{i,t}$ is fixed and nonstochastic. There is no randomization for α_t.

Note that the *computation errors* used here include the round-off error, the error generated from the random number generator, the truncation errors and so on, while the *simulation errors* come from taking the random samples out of the population. The former depends on a computer and a program but the latter implies the simulation errors conventionally used when applying Monte-Carlo experiments. Numerical integration has the *computation errors* only, while Monte-Carlo integration has both the *computation errors* and the *simulation errors*.

Traditionally, it is well known that numerical integration is associated with the fixed nodes and Monte-Carlo integration is with random draws from the importance density. In such a case, Monte-Carlo integration is clearly less efficient than numerical integration, because numerical integration depends on the errors independent of simulation (i.e., *computation errors*) while Monte-Carlo integration depends on both the errors due to the simulation (i.e., *simulation errors*) and those independent of the simulation (i.e., *computation errors*). In order for Monte-Carlo integration to reduce the *simulation errors*, we may utilize the Monte-Carlo integration procedure with the fixed nodes discussed above. The appropriate random numbers imply that each event occurs with equal probability, i.e., $1/n$. For the sorted appropriate random draws $\alpha_{i,t}$, $i = 1, \cdots, n$, therefore, the probability that α_t lies on the interval between $\alpha_{i,t}$ and $\alpha_{i-1,t}$ must be equal to $1/n$ for all i. For the Monte-Carlo integration procedure, furthermore, the order of random draws does not matter because we take the weighted average of the random numbers, i.e.,

$$a_{t|s} = \sum_{i=1}^{n} \alpha_{i,t} \omega_{i,t|s}.$$

See equations (4.31) – (4.34) for the filtering estimates by the importance sampling filter (ISF). Therefore, we can regard $\alpha_{i,t}$, $i = 1, \cdots, n$, as the sorted random draws which are generated from the importance density function $P_\alpha(\alpha_t)$.

Take an example of integration of the standard normal density, i.e.,

$$P(x) = \Phi(x, 1).$$

The importance density is given by:

$$P_x(x) = \Phi(x - \mu, \sigma^2),$$

which is also normal with mean μ and variance σ^2. We consider the following integration:

$$\int P(x) \mathrm{d}x,$$

which is theoretically equal to one, because $P(x)$ is a density function and is approximated by Monte-Carlo integration (5.13) or numerical integration (5.14). In Table 5.32, we evaluate the above integration by Monte-Carlo integration with importance sampling, i.e.,

$$\frac{1}{n} \sum_{i=1}^{n} \omega(x_i), \tag{5.13}$$

where $\omega(x_i)$ is defined as the weight function, i.e., $\omega(x_i) \equiv \dfrac{P(x_i)}{P_x(x_i)}$, and x_i denotes the fixed nodes obtained from the importance density $P_x(x)$, i.e.,

$$\int_{x_{i-1}}^{x_i} P_x(x) \mathrm{d}x = \frac{1}{n},$$

for all i, given the terminal conditions. In the case where $P_x(x)$ is normal, x_i is easily obtained from the statistical table found in a usual textbook (see, for example, Kokoska and Nevison (1989)).

Table 5.32. Monte-Carlo Integration with Fixed Nodes

σ	μ / n	0.0	0.5	1.0	1.5	2.0	3.0	4.0
	5	0.1560	0.1379	0.0951	0.0512	0.0216	0.0018	0.0001
	10	0.1771	0.1566	0.1082	0.0584	0.0247	0.0021	0.0001
0.1	50	0.2200	0.1947	0.1351	0.0734	0.0312	0.0027	0.0001
	100	0.2364	0.2094	0.1454	0.0792	0.0339	0.0030	0.0001
	500	0.2709	0.2402	0.1675	0.0918	0.0395	0.0036	0.0001
	5	0.4940	0.4420	0.3166	0.1811	0.0826	0.0086	0.0004
	10	0.5493	0.4942	0.3595	0.2109	0.0994	0.0112	0.0005
1/3	50	0.6517	0.5931	0.4461	0.2757	0.1389	0.0185	0.0010
	100	0.6870	0.6281	0.4788	0.3021	0.1562	0.0221	0.0013
	500	0.7540	0.6963	0.5462	0.3600	0.1967	0.0318	0.0021
	5	0.6920	0.6293	0.4721	0.2902	0.1450	0.0187	0.0010
	10	0.7503	0.6895	0.5328	0.3429	0.1814	0.0269	0.0016
0.5	50	0.8454	0.7926	0.6488	0.4555	0.2685	0.0514	0.0041
	100	0.8738	0.8252	0.6895	0.4995	0.3063	0.0644	0.0057
	500	0.9206	0.8818	0.7673	0.5914	0.3922	0.0996	0.0109
	5	1.0000	0.9697	0.8687	0.6905	0.4671	0.1173	0.0119
	10	1.0000	0.9836	0.9211	0.7893	0.5920	0.1942	0.0270
1.0	50	1.0000	0.9962	0.9768	0.9196	0.8002	0.4059	0.0982
	100	1.0000	0.9980	0.9865	0.9481	0.8576	0.4975	0.1452
	500	1.0000	0.9995	0.9962	0.9820	0.9388	0.6831	0.2846
	5	0.9977	1.0027	1.0269	1.0792	1.1265	0.8995	0.3299
	10	0.9997	1.0000	1.0032	1.0169	1.0477	1.0525	0.6634
2.0	50	1.0000	1.0000	1.0000	1.0002	1.0015	1.0190	1.0166
	100	1.0000	1.0000	1.0000	1.0000	1.0003	1.0064	1.0234
	500	1.0000	1.0000	1.0000	1.0000	1.0000	1.0003	1.0049
	5	1.0011	1.0020	0.9971	0.9748	0.9597	1.2102	1.3840
	10	1.0000	1.0000	1.0001	0.9999	0.9972	0.9925	1.1546
3.0	50	1.0000	1.0000	1.0000	1.0000	1.0000	1.0000	0.9991
	100	1.0000	1.0000	1.0000	1.0000	1.0000	1.0000	0.9999
	500	1.0000	1.0000	1.0000	1.0000	1.0000	1.0000	1.0000
	5	2.0000	1.7650	1.2133	0.6514	0.2826	0.2073	1.0592
	10	0.9166	0.9765	1.0727	1.0686	0.9630	0.9702	1.1179
10.0	50	1.0000	1.0000	1.0000	1.0000	1.0000	1.0000	1.0000
	100	1.0000	1.0000	1.0000	1.0000	1.0000	1.0000	1.0000
	500	1.0000	1.0000	1.0000	1.0000	1.0000	1.0000	1.0000

Table 5.33. Numerical Integration

σ	μ / n	0.0	0.5	1.0	1.5	2.0	3.0	4.0
	5	0.1019	0.0900	0.0620	0.0333	0.0140	0.0012	0.0000
	10	0.1306	0.1154	0.0796	0.0429	0.0180	0.0015	0.0000
0.1	50	0.1839	0.1627	0.1126	0.0609	0.0258	0.0022	0.0001
	100	0.2033	0.1799	0.1246	0.0676	0.0287	0.0025	0.0001
	500	0.2427	0.2150	0.1495	0.0816	0.0350	0.0031	0.0001
	5	0.3295	0.2932	0.2066	0.1152	0.0508	0.0049	0.0002
	10	0.4159	0.3716	0.2649	0.1506	0.0682	0.0070	0.0003
1/3	50	0.5618	0.5073	0.3732	0.2229	0.1077	0.0130	0.0006
	100	0.6094	0.5529	0.4122	0.2515	0.1248	0.0161	0.0008
	500	0.6970	0.6390	0.4908	0.3135	0.1648	0.0244	0.0015
	5	0.4746	0.4263	0.3089	0.1803	0.0846	0.0096	0.0004
	10	0.5877	0.5322	0.3949	0.2393	0.1179	0.0150	0.0008
0.5	50	0.7552	0.6981	0.5493	0.3641	0.2006	0.0333	0.0023
	100	0.8023	0.7477	0.6020	0.4132	0.2377	0.0436	0.0034
	500	0.8778	0.8317	0.7016	0.5167	0.3243	0.0729	0.0071
	5	0.7877	0.7336	0.5915	0.4103	0.2423	0.0493	0.0044
	10	0.9014	0.8569	0.7314	0.5513	0.3596	0.0919	0.0107
1.0	50	0.9827	0.9658	0.9075	0.7935	0.6246	0.2502	0.0486
	100	0.9917	0.9815	0.9431	0.8582	0.7154	0.3347	0.0784
	500	0.9984	0.9956	0.9824	0.9446	0.8619	0.5344	0.1817
	5	1.0310	1.0019	0.9167	0.7892	0.6445	0.3555	0.1143
	10	1.0221	1.0199	1.0058	0.9613	0.8709	0.5752	0.2582
2.0	50	1.0012	1.0016	1.0029	1.0048	1.0044	0.9510	0.7203
	100	1.0003	1.0004	1.0008	1.0018	1.0032	0.9910	0.8659
	500	1.0000	1.0000	1.0000	1.0001	1.0003	1.0014	0.9901
	5	1.0732	1.0762	1.0673	1.0044	0.8690	0.6013	0.4882
	10	1.0162	1.0183	1.0248	1.0351	1.0407	0.9460	0.6873
3.0	50	1.0006	1.0007	1.0009	1.0014	1.0024	1.0080	1.0200
	100	1.0002	1.0002	1.0002	1.0003	1.0006	1.0020	1.0078
	500	1.0000	1.0000	1.0000	1.0000	1.0000	1.0001	1.0003
	5	2.0921	1.8463	1.2692	0.6815	0.2964	0.2294	1.1799
	10	0.9269	0.9875	1.0848	1.0810	0.9749	0.9850	1.1367
10.0	50	1.0004	1.0004	1.0004	1.0005	1.0005	1.0006	1.0007
	100	1.0001	1.0001	1.0001	1.0001	1.0001	1.0001	1.0002
	500	1.0000	1.0000	1.0000	1.0000	1.0000	1.0000	1.0000

In Table 5.33, the integration is evaluated by numerical integration, which utilizes the following trapezoid rule:

$$\sum_{i=2}^{n} \frac{1}{2}\big(P(x_i) + P(x_{i-1})\big)(x_i - x_{i-1}), \tag{5.14}$$

where for comparison the exactly same values x_i, $i = 1, \cdots, n$, are used for both numerical integration and Monte-Carlo integration.

The results are in Tables 5.32 and 5.33. For almost the cases in Tables 5.32 and 5.33, Monte-Carlo integration (5.13) performs better than numerical integration (5.14), because Monte-Carlo integration with fixed nodes is closer to one than numerical integration. Thus, the fixed nodes can be applied to improve precision of Monte-Carlo integration. In this book, however, the above procedure is not utilized for the importance sampling filter (ISF), where the random draws of the state-variable are artificially generated by a computer.

A5.4 Rejection Sampling

In this appendix, we consider the random number generation by rejection sampling, especially, computational time and precision of random draws.

Consider the scalar case. Define the following two densities:

$$P(\alpha) = \Phi(\alpha - \mu_\alpha, \sigma_\alpha^2),$$
$$P(y|\alpha) = \Phi(y - \alpha, \sigma_y^2).$$

$P(\alpha)$ corresponds to the prior distribution of α. Based on the above two densities, we have the following posterior density of α:

$$\begin{aligned} P(\alpha|y) &= \frac{P(y|\alpha)P(\alpha)}{\displaystyle\int P(y|\alpha)P(\alpha)\mathrm{d}\alpha} \\ &= \frac{\Phi(y-\alpha, \sigma_y^2)\Phi(\alpha - \mu_\alpha, \sigma_\alpha^2)}{\displaystyle\int \Phi(y-\alpha, \sigma_y^2)\Phi(\alpha - \mu_\alpha, \sigma_\alpha^2)\mathrm{d}\alpha}. \end{aligned}$$

The followings, i.e., (a) and (b), are computed.

(a) Theoretically, the conditional distribution of α is:

$$P(\alpha|y) = \Phi\left(\alpha - \frac{y + \mu_\alpha}{\sigma_y^2}\frac{1}{1/\sigma_y^2 + 1/\sigma_\alpha^2}, \frac{1}{1/\sigma_y^2 + 1/\sigma_\alpha^2}\right)$$

Accordingly, conditional mean and variance are obtained as:

$$\mathrm{E}(\alpha|y) = \frac{y + \mu_\alpha}{\sigma_y^2}\frac{1}{1/\sigma_y^2 + 1/\sigma_\alpha^2},$$

$$\mathrm{Var}(\alpha|y) = \frac{1}{1/\sigma_y^2 + 1/\sigma_\alpha^2}.$$

Given μ_α, σ_α^2, σ_y^2 and y, the conditional mean and variance of α are explicitly computed. No Monte-Carlo simulation is used here.

(b) For rejection sampling, the acceptance probability is defined as:

$$\omega(\alpha; y) = \exp\left(-\frac{1}{2\sigma_y^2}(y-\alpha)^2\right),$$

which corresponds to the exponential part of the normal density $P(y|\alpha) = \Phi(y-\alpha, \sigma_y^2)$. Generate a random draw from $P(\alpha)$ and accept it with probability $\omega(\alpha; y)$.
10,000 random draws are generated in this case and the arithmetic average and variance are computed.

Table 5.34. Random Number Generation by Rejection Sampling

y	σ_y	0.1		0.2		0.5		1.0		2.0		4.0	
		Ave	Var	Ave	Var	Ave	Var	Ave	Var	Ave	Var	Ave	Var
	(a)	.000	.100	.000	.196	.000	.447	.000	.707	.000	.894	.000	.970
0.0	(b)	-.001	.101	.000	.196	-.002	.445	-.008	.707	.007	.893	.004	.966
	(c)	10.042		5.082		2.213		1.414		1.117		1.030	
	(a)	.099	.100	.096	.196	.080	.447	.0 50	.707	.020	.894	.006	.970
0.1	(b)	.100	.100	.092	.196	.076	.446	.040	.707	.025	.892	.0 10	.965
	(c)	10.137		5.168		2.220		1.418		1.118		1.031	
	(a)	.198	.100	.192	.196	.160	.447	.100	.707	.040	.894	.012	.970
0.2	(b)	.197	.100	.190	.196	.155	.445	.089	.708	.043	.892	.016	.965
	(c)	10.354		5.245		2.254		1.425		1.121		1.032	
	(a)	.495	.100	.481	.196	.400	.447	.2 50	.707	.100	.894	.029	.970
0.5	(b)	.495	.100	.481	.195	.395	.447	.246	.709	.105	.889	.032	.966
	(c)	11.360		5.734		2.477		1.498		1.145		1.039	
	(a)	.990	.100	.962	.196	.800	.447	.500	.707	.200	.894	.059	.970
1.0	(b)	.990	.100	.963	.198	.799	.4 50	.497	.7 10	.194	.891	.062	.965
	(c)	16.690		8.385		3.344		1.824		1.229		1.059	
	(a)	1.980	.100	1.923	.196	1.600	.447	1.000	.707	.400	.894	.118	.970
2.0	(b)	1.981	.099	1.922	.196	1.601	.449	1.003	.716	.393	.901	.117	.964
	(c)	72.364		34.818		11.092		3.875		1.681		1.161	
	(a)	3.960	.100	3.846	.196	3.200	.447	2.000	.707	.800	.894	.235	.970
4.0	(b)	3.960	.099	3.847	.198	3.197	.444	2.002	.707	.798	.904	.234	.968
	(c)	27177.6		11166.7		1347.98		76.728		5.512		1.665	

Assume that $\mu_\alpha = 0$ and $\sigma_\alpha = 1$, which implies that the prior density of α is a standard normal distribution. $y = 0.0, 0.1, 0.2, 0.5, 1, 2, 4$ and $\sigma_y^2 =$

$0.1, 0.2, 0.5, 1, 2, 4$ are taken. Given y and σ_y, we compute mean (Ave) and variance (Var) of α.

In Table 5.34, (a) denotes theoretical mean and variance given y and σ_y, (b) indicates mean and variance obtained by rejection sampling, and (c) denotes number of rejection required to generate one random draw by rejection sampling, which is proportional to computational time.

The results are obtained as follows. For all y and σ_y, the theoretical value (a) is very close to the simulated value (b). Accordingly, rejection sampling works well. See Appendix A4.6 for the theoretical reason on equality between (a) and (b). For computational time (i.e., number of rejection required to obtain one random draw), we can have the following results (see (c) in Table 5.34):

(i) Given y, computational time increases as σ_y is small.
(ii) Given σ_y, computational time increases as y is large.

The above two results imply that rejection sampling takes a lot of time computationally when the acceptance probability, i.e., $\omega(\alpha; y)$, is small.

6. Application of Nonlinear Filters

6.1 Introduction

As an application, in this chapter, we take an example of estimating per capita permanent consumption. In the last decade, a large amount of macroeconomic research has been devoted to various aspects of the life cycle permanent income hypothesis under rational expectations. See Hall (1978), Hall and Mishkin (1982), Mankiw and Shapiro (1985), Campbell (1987), Campbell and Mankiw (1987), and West (1988). For the most concise and useful survey, see Diebold and Nerlove (1989).

Numerous papers deal with testing the permanent income hypothesis. For example, Flavin (1981), Hall and Mishkin (1982) and Campbell and Mankiw (1987) examined the permanent income hypothesis taking into account transitory consumption. Hayashi (1985a, 1985b) considered testing it based on liquidity constraints and durable goods. Mankiw (1981) and Muellbauer (1983) introduced the variable interest rate into the model. Thus, the permanent income hypothesis has been tested from various aspects. For another research on asset pricing models, we have Hamori (1992a, 1992b, 1993), Hamori and Kitasaka (1996), Hamori, Kitasaka and Tanizaki (1996) and so on.

Moreover, Tanizaki (1993), Tanizaki and Mariano (1994) and Mariano and Tanizaki (1995, 1996) applied the nonlinear filtering technique to consumer behavior of the representative agent. In this chapter, based on their papers, we also test whether the permanent income hypothesis proposed by Hall (1978) holds in reality, including the variable interest rate, nonlinearity of the Euler equation and transitory consumption. We formulate the model to test the permanent income hypothesis, taking into account all of the three issues simultaneously, where the nonlinear filtering technique is applied. Annual data are used for estimation. To test the permanent income hypothesis, we now consider the state-space model

Hall (1978) proposed a new approach to the permanent income hypothesis. The essence is the following. If consumption is based on the permanent income hypothesis, by which kind of model can a sequence of consumption be written? According to the approach proposed by Hall (1978), the permanent income hypothesis can be tested without estimating the consumption function as conventionally done. We sometimes have the criticism that in the long run the random walk hypothesis of consumption is unrealistic and only in the short run the hypothesis holds in reality. In Chapter 6, we test whether the permanent income hypothesis, taking into account transitory consumption, the variable interest rate and nonlinearity of the Euler equation simultaneously. There, we consider estimating permanent consumption, transitory consumption and the other factor which is independent of the permanent income hypothesis and is a function of income. If consumption depends on the other factor independent of the permanent income hypothesis, we can conclude that the life cycle permanent income hypothesis does not hold in reality. We test the permanent income hypothesis and estimate a ratio of per capita permanent consumption relative to per capita total consumption for Japan, United States, United Kingdom, France, Spain , Italy, Canada and Germany. We see for almost all the countries that consumption significantly depends on income according to the likelihood ratio test, which implies that the permanent income hypothesis is unrealistic.

6.2 State-Space Form

In this section, we formulate the state-space model to test the permanent income hypothesis, taking into account the variable interest rate, nonlinearity of the Euler equation and transitory consumption simultaneously.

The state-space model to the problem is represented in Sections 6.2.1 (i.e., measurement equation) and 6.2.2 (i.e., transition equation), which is as follows:

$$\text{(Measurement equation)} \qquad c_t = c_t^p + c_t^* + \epsilon_t, \tag{6.1}$$

$$\text{(Transition equation)} \qquad \frac{\beta R_{t-1} u'(c_t^p)}{u'(c_{t-1}^p)} = 1 + \eta_t. \tag{6.2}$$

where (ϵ_t, η_t) are bivariate normal, independent over time, with mean zero and covariance matrix

$$\begin{pmatrix} (\sigma_\epsilon Y_t)^2 & 0 \\ 0 & \sigma_\eta^2 \end{pmatrix}.$$

In the above model, behavior of the representative agent is given and accordingly c_t, c_t^p, c_t^T and c_t^* denote per capita real data. c_t^p denotes permanent consumption. $u'(\cdot)$ represents the first derivative of the underlying utility function. We introduce c_t^* into the model, which indicates a factor independent of the permanent income hypothesis. $c_t^T \equiv \epsilon_t$ denotes transitory consumption, which is assumed to be normal with zero and variance $(\sigma_\epsilon Y_t)^2$, where Y_t and L_t denote income and population, respectively. Note that c_t^p and c_t^T are unobservable while c_t is observed. In this section, the Kalman filtering technique is utilized to estimate the unobservable components, i.e., c_t^p.

The notations used in the above equations are summarized as follows:

c_t: per capita total consumption,
c_t^p: per capita permanent consumption,
c_t^T: per capita transitory consumption,
c_t^*: per capita other consumption independent of the permanent income hypothesis,
Y_t: per capita income,
L_t: population,
R_t: gross rate of return on savings between time t and $t+1$.

And therefore, the notations in the previous chapters (i.e., Chapters 1 – 5) are changed as follows:

Chapters 1 – 5		Chapter 6
α_t	$\Longrightarrow$	c_t^p
y_t	$\Longrightarrow$	c_t
H_t	$\Longrightarrow$	$(\sigma_\epsilon Y_t)^2$
Q_t	$\Longrightarrow$	σ_η^2
$Y_t = \{y_1, y_2, \cdots, y_t\}$	$\Longrightarrow$	$C_t = \{c_1, c_2, \cdots, c_t\}$

Note that Y_t in Chapters 1 – 5 is different from Y_t in Chapter 6, i.e., the former denotes the information set available at time t while the latter indicates income at time t.

The measurement equation (6.1) and the transition equation (6.2) are derived in Sections 6.2.1 and 6.2.2, respectively.

6.2.1 Measurement Equation

The measurement equation is simply the identity that total consumption is the sum of permanent consumption, transitory consumption and the other part of consumption which is independent of the permanent income hypothesis. Each component of consumption is represented as follows.

The assumption on transitory consumption is based on earlier author's treatment of the subject. In a cross-sectional framework,

$$\sum_i c_{it}^T = 0$$

is assumed by Friedman (1957), where c_{it}^T denotes transitory consumption of the i-th household at time t. Furthermore, Friedman (1957) assumed that c_{it}^T is independent of the permanent income, transitory income and permanent consumption (see Branson (1979)). Average of aggregate transitory consumption, i.e., per capita transitory consumption, is represented as:

$$c_t^T = \frac{1}{L_t} \sum_i c_{it}^T.$$

Assuming that c_{it}^T are independently distributed with mean zero for all i and t, the transitory consumption of the representative agent (i.e., c_t^T) is given by a random shock with mean zero. It might be plausible to assume that the transitory consumption increases (decreases) as the income level increases (decreases). Therefore, simply, we assume that standard error of per capita transitory consumption c_t^T is proportional to per capita income Y_t, i.e., $\mathrm{Var}(c_t^T) = (\sigma_\epsilon Y_t)^2$ where $c_t^T \equiv \epsilon_t$.

c_t^* represents an exogenous factor of consumption, which does not depend on the permanent income hypothesis proposed by Hall (1978). It is well known that variables other than lagged consumption appear to play a significant role in the determination of current consumption (see Diebold and Nerlove (1989)). In other words, it is known that current consumption depends on not only lagged consumption but also income or time trend. Accordingly, c_t^* is a part of consumption which depends on other variables such as income or trend. Therefore, it is assumed in this chapter that c_t^* is a function of per capita income, i.e.,

$$c_t^* = \delta_1 \widehat{Y}_t + \delta_2 Y_{t-1}, \tag{6.3}$$

where δ_1 and δ_2 are unknown parameters to be estimated. Under the permanent income hypothesis, we have the equality: $c_t^* = 0$, which is equivalent to $\delta_1 = \delta_2 = 0$ in equation (6.3). $\widehat{Y}_t$ denotes an instrument variable of income Y_t, because income Y_t is correlated with consumption c_t. Thus, because of correlation between income Y_t and consumption c_t, we use a proxy variable of income Y_t as $\widehat{Y}_t$ to exclude correlation between Y_t and c_t. In this chapter, the instrument variable $\widehat{Y}_t$ is taken as the predicted value of Y_t, where Y_t is regressed on constant term, time trend, Y_{t-1} and c_{t-1} using the ordinary least squares estimation.

In Section 6.3, to test whether the permanent income hypothesis holds, we use the likelihood ratio test for the null hypothesis:

$$H_0 : \delta_1 = \delta_2 = 0. \tag{6.4}$$

Thus, we take the measurement equation as the identity equation, where consumption is given by a sum of permanent consumption, transitory consumption and the other part of consumption, i.e.,

$$c_t = c_t^p + c_t^* + c_t^T.$$

6.2.2 Transition Equation

In this application, the transition equation corresponds to the Euler equation, which is derived as follows. Consider the problem of choosing a consumption sequence $\{c_t^p\}$ by the representative agent which maximizes the following expected utility:

$$\mathrm{E}_t\Big(\sum_t \beta^t u(c_t^p)\Big),$$

subject to:

$$\begin{aligned} A_{t+1} &= R_t(A_t + y_t - c_t), \\ c_t &= c_t^p + c_t^* + c_t^T, \end{aligned} \tag{6.5}$$

where A_0 is given. The representative utility function $u(\cdot)$ is twice continuously differentiable, bounded, increasing and concave. A_{t+1} is the stock of assets at the beginning of time $t+1$, y_t is noncapital or labor income at t, and R_t is the gross rate of return on savings between t and $t+1$. $\mathrm{E}_t(\cdot)$ denotes the mathematical expectation, given information known at t. β is the discount rate. It might be plausible in empirical studies that the discount rate β is less than one. However, Kocherlakota (1990) showed that well-behaved competitive equilibria with positive interest rates may exist in infinite horizon growth economies even though individuals have discount factors larger than one, which implies that when per capita consumption is growing over time, it is possible for equilibria to exist in representative consumer endowment economies even though $\beta > 1$. Therefore, in the empirical studies, we do not have to pay much attention to the possibility of the discount rate being greater than one.

Thus, under the above setup, maximizing the expected utility with respect to the permanent consumption sequence $\{c_t^p\}$, we obtain the Euler equation:

$$\mathrm{E}_{t-1}\left(\frac{\beta R_{t-1} u'(c_t^p)}{u'(c_{t-1}^p)}\right) = 1,$$

which is rewritten as:

$$\frac{\beta R_{t-1} u'(c_t^p)}{u'(c_{t-1}^p)} = 1 + \eta_t, \tag{6.6}$$

where the error term η_t is assumed to be normal. $u'(\cdot)$ represents the first derivative of the underlying utility function.

Usually, the utility function is maximized with respect to total consumption c_t, not permanent consumption c_t^p. We assume in this chapter that the utility function depends on c_t as well as μ_t, where μ_t denotes transitory consumption and the other part of consumption independent of the permanent income hypothesis, i.e., $\mu_t = c_t^* + c_t^T$. When we take the utility function given by $u(c_t - \mu_t)$ and maximize it subject to the constraints given by (6.5), the Euler equation (6.6) can be derived.

Note that we have the restriction: $1 + \eta_t > 0$ from equation (6.6). Therefore, the exact distribution of η_t is represented as the following truncated normal distribution:

$$P_\eta(\eta_t) = \frac{(2\pi\sigma_\eta^2)^{-1/2} \exp\left(-\dfrac{1}{2\sigma_\eta}\eta_t^2\right)}{\displaystyle\int_{-1}^{\infty} (2\pi\sigma_\eta^2)^{-1/2} \exp\left(-\dfrac{1}{2\sigma_\eta^2}\eta_t^2\right) \mathrm{d}\eta_t}.$$

If $-1/\sigma_\eta$ is small enough, we can approximate the density function of η_t as the normal density.

Moreover, the utility function is taken as:

$$u(c_t^p) = \begin{cases} \dfrac{(c_t^p)^{1-\gamma} - 1}{1-\gamma}, & \text{if } \gamma \neq 1, \\ \log(c_t^p), & \text{if } \gamma = 1, \end{cases} \tag{6.7}$$

where γ denotes a measure of relative risk aversion. The Euler equation corresponding to this particular form of the utility function reduces to the transition equation in (6.2), which represents the transition equation in the state-space model. ϵ_t and η_t are assumed to be mutually independently distributed.

Accordingly, from equation (6.7), the state-space model to the problem is represented as follows:

$$\begin{aligned} &\text{(Measurement equation)} \quad && c_t = c_t^p + c_t^* + \epsilon_t, \\ &\text{(Transition equation)} \quad && (c_t^p)^{-\gamma} = (\beta R_{t-1})^{-1}(c_{t-1}^p)^{-\gamma}(1+\eta_t), \end{aligned} \tag{6.8}$$

where

$$\begin{pmatrix} \epsilon_t \\ \eta_t \end{pmatrix} \sim N\left(\begin{pmatrix} 0 \\ 0 \end{pmatrix}, \begin{pmatrix} (\sigma_\epsilon Y_t)^2 & 0 \\ 0 & \sigma_\eta^2 \end{pmatrix} \right).$$

The likelihood function of the innovation form is maximized by a simple grid search with respect to the unknown parameters.

6.3 Estimation

For each nonlinear filter, the appropriate likelihood function of the innovation form (i.e., equations (2.30), (4.17), (4.26), (4.36), (4.42) and (4.44)) is maximized by a simple grid search with respect to the unknown parameters, which are discussed in this section.

We test the permanent income hypothesis (i.e., the null hypothesis H_0 : $\delta_1 = \delta_2 = 0$) and estimate a ratio of per capita permanent consumption relative to per capita total consumption for Japan, United States, United Kingdom, France, Spain , Italy, Canada and Germany. The results are in Tables 6.1– 6.10 and Figures 6.4 – 6.12.

6.3.1 Data

Annual data from 1955 to 1993 are used for Japan, United States (U.S.), United Kingdom (U.K.), France, Spain , Italy, Canada and Germany. The estimation period is from 1957 to 1993 for each country. The gross rate of return on savings between time t and $t+1$ (i.e., R_t) is defined as $R_t = (1+r_t/100)P_t/P_{t+1}$, where r_t is the interest rate and P_t denotes the price index. All the data for each country are taken from *International Financial Statistics Yearbook* (International Monetary Fund), which are summarized as follows.

- Japan

c_t: Private Consumption (Yen, 1990 prices, Per capita),
National Accounts data are divided by Consumer Prices times Population, i.e., P_tL_t.

Y_t: Gross Domestic Product (Yen, 1990 prices, Per Capita),
National Accounts data are divided by Consumer Prices times Population, i.e., P_tL_t.

P_t: Consumer Prices (1990=1.00),

L_t: Population (Midyear Estimates),

r_t: Deposit Rate (Percent Per Annum),

- United States

c_t: Private Consumption (US Dollars, 1990 prices, Per Capita),
National Accounts data are divided by Consumer Prices times Population, i.e., P_tL_t.

Y_t: Gross Domestic Product (US Dollars, 1990 prices, Per Capita),
National Accounts data are divided by Consumer Prices times Population, i.e., P_tL_t.

P_t: Consumer Prices (1990=1.00),

L_t: Population (Midyear Estimates),

r_t: Government Bond Yield, Long Term (Percent Per Annum),

- United Kingdom

c_t: Private Consumption (Yen, 1990 prices, Per Capita),
National Accounts data are divided by Consumer Prices times Population, i.e., P_tL_t.

Y_t: Gross Domestic Product (Yen, 1990 prices, Per Capita),
National Accounts data are divided by Consumer Prices times Population, i.e., P_tL_t.

P_t: Consumer Prices (1990=1.00),

L_t: Population (Midyear Estimates),

r_t: Deposit Rate (Percent Per Annum),

- France

c_t: Private Consumption (Pounds, 1990 prices, Per Capita), National Accounts data are divided by Consumer Prices times Population, i.e., P_tL_t.

Y_t: Gross Domestic Product (Pounds, 1990 prices, Per Capita), National Accounts data are divided by Consumer Prices times Population, i.e., P_tL_t.

P_t: Consumer Prices (1990=1.00),

L_t: Population (Midyear Estimates),

r_t: Government Bond Yield, Long Term (Percent Per Annum),

- Spain

c_t: Private Consumption (Yen, 1990 prices, Per Capita), National Accounts data are divided by Consumer Prices times Population, i.e., P_tL_t.

Y_t: Gross Domestic Product (Yen, 1990 prices, Per Capita), National Accounts data are divided by Consumer Prices times Population, i.e., P_tL_t.

P_t: Consumer Prices (1990=1.00),

L_t: Population (Midyear Estimates),

r_t: Deposit Rate (Percent Per Annum),

- Italy

c_t: Private Consumption (Pesetas, 1990 prices, Per Capita), National Accounts data are divided by Consumer Prices times Population, i.e., P_tL_t.

Y_t: Gross Domestic Product (Pesetas, 1990 prices, Per Capita), National Accounts data are divided by Consumer Prices times Population, i.e., P_tL_t.

P_t: Consumer Prices (1990=1.00),

L_t: Population (Midyear Estimates),

r_t: Bank of Spain Rate (End of Period, Percent Per Annum),

- Canada

c_t: Private Consumption (Canadian Dollars, 1990 prices, Per Capita), National Accounts data are divided by Consumer Prices times Population, i.e., P_tL_t.

Y_t: Gross Domestic Product (Canadian Dollars, 1990 prices, Per Capita), National Accounts data are divided by Consumer Prices times Population, i.e., P_tL_t.

P_t: Consumer Prices (1990=1.00),

L_t: Population (Midyear Estimates),

r_t: Government Bond Yield, Long Term (Percent Per Annum),

- Germany

c_t: Private Consumption (Deutsche Mark, 1990 prices, Per Capita), National Accounts data are divided by Consumer Prices times Population, i.e., $P_t L_t$.

Y_t: Gross Domestic Product (Deutsche Mark, 1990 prices, Per Capita), National Accounts data are divided by Consumer Prices times Population, i.e., $P_t L_t$.

P_t: Consumer Prices (1990=1.00),

L_t: Population (Midyear Estimates),

r_t: Government Bond Yield (Percent Per Annum),

Thus, for all the countries, c_t, Y_t, P_t and L_t denote per capita real private consumption, per capita real GDP, consumer price index and population, respectively. Note that Gross Domestic Product (GDP) is deflated by Consumer Price Index, not GDP Deflator.

Using the above data, the nonlinear filtering techniques introduced in Chapters 3 and 4 are implemented to illustrate the empirical application discussed in Section 6.2.

In Figures 6.1 – 6.3, inflationary rate of consumer price index, growth rate of per capita real GDP and interest rate are displayed for each country to describe each economy.

In Figure 6.1 (inflationary rate), except for Spain, there are three peaks; around 1974 (the first oil shock) , about 1981 (the second oil shock) and around 1990 (crash of stock market). There is only one peak around 1977 in Spain, which is different from the other countries. Japan, United Kingdom, France and Germany have less damages from the second oil shock than the first oil shock. Germany shows relatively small inflationary rate over time, while the inflationary rates in Spain and Italy are quite large during the period from 1965 to 1994.

In Figure 6.2 (growth rate of per capita real GDP), there are three bottoms, which are 1974 – 1975, 1979 – 1982 and 1990 – 1991. The first two bottoms correspond to the two oil shocks. The growth rate of per capita GDP in Germany is -16.76% in 1991, which is due to unification of West Germany and East Germany.

Figure 6.1. Inflationary Rate ($i.e.$, $100 \times \dfrac{P_t - P_{t-1}}{P_{t-1}}$)

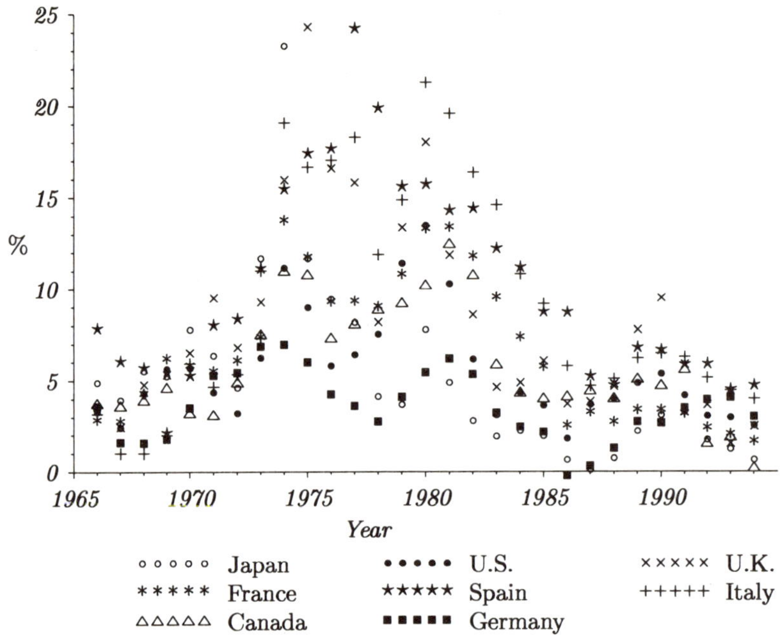

Figure 6.2. Growth Rate of Per Capita Real GDP ($i.e.$, $100 \times \dfrac{Y_t - Y_{t-1}}{Y_{t-1}}$)

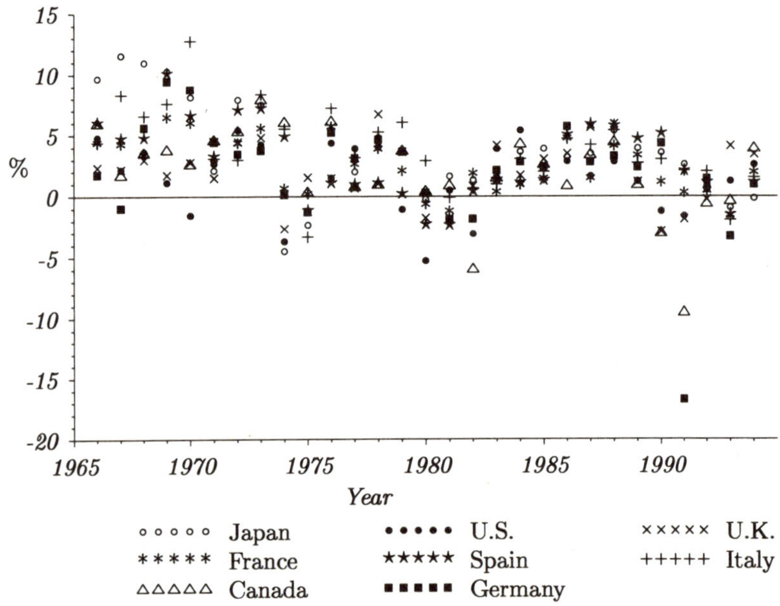

Figure 6.3. Interest Rate (i.e., r_t)

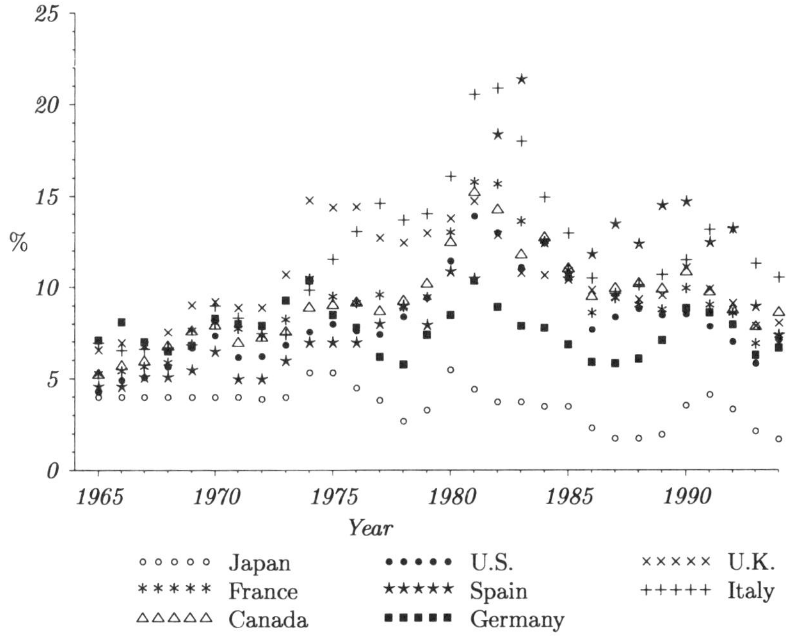

In Figure 6.3 (interest rate), we can find that interest rate in Japan is extremely low compared with the other countries for all the periods. Also we can observe the high interest rates all over the world after 1974 (the first oil shock), i.e., more than 10% during the periods:

- 1980 – 1985 for U.S.,
- 1973 – 1985 and 1990 for U.K.,
- 1974 and 1980 – 1985 for France,
- 1980 – 1992 for Spain,
- 1975 – 1986 and 1988 – 1994 for Italy,
- 1979 – 1985, 1988 and 1990 for Canada, and
- 1974 and 1981 for Germany.

The interest rates in all the countries except for Japan and Germany are more than 10% around 1980 – 1985.

6.3.2 Nonlinear Filtering Techniques

In Tables 6.1 – 6.9, each parameter is estimated by a simple grid search method, i.e., the log-likelihood function is maximized by changing the parameter value by 0.00001 for σ_ϵ, 0.001 for β, σ_η, δ_1 and δ_2, and 0.1 for γ, respectively. $\log L$ in Tables 6.1 – 6.9 denotes the maximized log-likelihood function.

We estimate the state-space model (6.8) by the following nonlinear filtering methods:

- Extended Kalman Filter (EKF) in Table 6.1 and Figure 6.4,
- Second-Order Nonlinear Filter (SNF) in Table 6.2 and Figure 6.5,
- Monte-Carlo Simulation Filter (MSF) in Table 6.3 and Figure 6.6,
- Single-Stage Iteration Filter with Monte-Carlo approximation (SIFc) in Table 6.4 and Figure 6.7,
- Gaussian Sum Filter (GSF) in Table 6.5 and Figure 6.8,
- Numerical Integration Filter (NIF) in Table 6.6 and Figure 6.9,
- Importance Sampling Filter (ISF) in Table 6.7 and Figure 6.10,
- Density-Based Monte-Carlo Filter (DMF) in Table 6.8 and Figure 6.11,
- Rejection Sampling Filter (RSF) in Table 6.9 and Figure 6.12.

The parameter estimates are similar for Tables 6.1 – 6.9.

The setup of each nonlinear filter is as follows.

(1) For the extended Kalman filter (EKF), the second-order nonlinear filter (SNF), the Monte-Carlo simulation filter (MSF), the single-stage iteration filter (SIFc), the density-based Monte-Carlo filter (DMF) and the rejection sampling filter (RSF), we take the following initial values:

$$a_{0|0} = c_0 - (\delta_1 Y_0 + \delta_2 Y_{-1}),$$
$$\Sigma_{0|0} = (\sigma_\eta Y_0)^2,$$

where $a_{0|0}$ denotes the initial value of the filtering estimates and $\Sigma_{0|0}$ represents variance of the initial value c_0^p ($a_{0|0}$ comes from the measurement equation and $\Sigma_{0|0}$ is from variance of η_t). c_0 is per capita total consumption in 1956, and Y_0 and Y_{-1} denote per capita gross domestic product in 1956 and 1955, respectively. The state-variable is given by c_t^p in this application. Accordingly, note that the initial value of c_t^p is denoted by c_0^p and the filtering value of c_0^p is given by $a_{0|0}$, i.e., $c_0^p \sim N(a_{0|0}, \Sigma_{0|0})$. δ_1 and δ_2 are the unknown parameters to be estimated.

(2) For the Monte-Carlo simulation filter (MSF), $n = 1,000$ are chosen, i.e., n normal random numbers are generated for $\alpha_{i,t|t-1}$, $\alpha_{i,t-1|t-1}$, $\epsilon_{i,t}$ and $\eta_{i,t}$, i.e.,

$$y_{i,t|t-1} = \alpha_{i,t|t-1} + c_t^* + \epsilon_{i,t},$$
$$a_{i,t|t-1} = \left(\beta R_{t-1}\right)^{1/\gamma} \alpha_{i,t-1|t-1}(1 + \eta_{i,t})^{-1/\gamma},$$

for $i = 1, \cdots, n$, where

$$\begin{pmatrix} \alpha_{i,t|t-1} \\ \epsilon_{i,t} \end{pmatrix} \sim N\left(\begin{pmatrix} a_{t|t-1} \\ 0 \end{pmatrix}, \begin{pmatrix} \Sigma_{t|t-1} & 0 \\ 0 & \sigma_\epsilon^2 \end{pmatrix} \right),$$

$$\begin{pmatrix} \alpha_{i,t-1|t-1} \\ \eta_{i,t} \end{pmatrix} \sim N\left(\begin{pmatrix} a_{t-1|t-1} \\ 0 \end{pmatrix}, \begin{pmatrix} \Sigma_{t-1|t-1} & 0 \\ 0 & (\sigma_\eta Y_t)^2 \end{pmatrix} \right).$$

$a_{t|t-1}$ and $\Sigma_{t|t-1}$ are obtained by equations (3.31) and (3.32) while $a_{t|t}$ and $\Sigma_{t|t}$ are given by equations (3.16) and (3.15).

As mentioned in (1), the initial value of the state-variable (i.e., c_0^p) is assumed to be normal, i.e.,

$$c_0^p \sim N(a_{0|0}, \Sigma_{0|0}),$$

where

$$a_{0|0} = c_0 - (\delta_1 Y_0 + \delta_2 Y_{-1}),$$
$$\Sigma_{0|0} = (\sigma_\eta Y_0)^2.$$

c_0 is per capita total consumption in 1956, and Y_0 and Y_{-1} denote per capita gross domestic product in 1956 and 1955, respectively.

(3) For the single-stage iteration filter (SIFc), the transition equation is approximated by the Monte-Carlo stochastic simulations (SIFc), where we take $n = 1,000$.

(4) For the Gaussian sum filter (GSF), $a_{i,0|0}$ is generated from a normal random number with mean:

$$a_{0|0} = c_0 - (\delta_1 Y_0 + \delta_2 Y_{-1}),$$

and variance:

$$\Sigma_{0|0} = (\sigma_\eta Y_0)^2,$$

for $i = 1, \cdots, n$, where $n = 200$ is taken. That is, $a_{i,0|0} \sim N(a_{0|0}, \Sigma_{0|0})$ is taken for i. For variance, we choose $\Sigma_{i,0|0} = \Sigma_{0|0} \equiv (\sigma_\eta Y_0)^2$, which is fixed for all i.

(5) In the numerical integration filter (NIF), we take $n = 200$. The first half of n nodes (i.e., m nodes) are obtained from

$$[a^*_{t|t-1} - \sqrt{c\Sigma^*_{t|t-1}}, a^*_{t|t-1} + \sqrt{c\Sigma^*_{t|t-1}}],$$

and the second half of n nodes (i.e., $n - m$ nodes) are from

$$[a^*_{t|t} - \sqrt{c\Sigma^*_{t|t}}, a^*_{t|t} + \sqrt{c\Sigma^*_{t|t}}],$$

where $c = 25$ is taken. For both intervals, the distance between the nodes are equal. That is, the first m nodes are from

$$a^*_{t|t} + \frac{2i - 1 - m}{m} \sqrt{c\Sigma^*_{t|t}},$$

and the last $n - m$ nodes are from

$$a^*_{t|t-1} + \frac{2i-1-m}{m}\sqrt{c\Sigma^*_{t|t-1}},$$

where $i = 1, \cdots, m$ and $m = 100$ is taken. $a^*_{t|t-1}$, $\Sigma^*_{t|t-1}$, $a^*_{t|t}$ and $\Sigma^*_{t|t}$ are obtained from the extended Kalman filter (EKF) algorithm, where $a^*_{t|t}$ denotes the filtering estimate of the state-variable c^p_t and $\Sigma^*_{t|t}$ represents its variance.

The initial density of the state-variable (i.e., c^p_0) is assumed to be normal with mean $a_{0|0}$ and variance $\Sigma_{0|0}$, i.e.,

$$P(c^p_0) = \Phi(c^p_0 - a_{0|0}, \Sigma_{0|0}).$$

(6) In the importance sampling filter (ISF), $n = 200$ random draws are generated from the following importance density:

$$P_\alpha(\alpha_t) = \frac{1}{2}\Phi(\alpha_t - a^*_{t|t-1}, c\Sigma^*_{t|t-1}) + \frac{1}{2}\Phi(\alpha_t - a^*_{t|t}, c\Sigma^*_{t|t}),$$

where $c = 25$ is taken and $a^*_{t|t-1}$, $\Sigma^*_{t|t-1}$, $a^*_{t|t}$ and $\Sigma^*_{t|t}$ are obtained from the extended Kalman filter (EKF) algorithm, where $a^*_{t|t}$ denotes the filtering estimate of the state-variable c^p_t and $\Sigma^*_{t|t}$ represents its variance. For density approximation, the importance density should have broader tails than the prediction density and the filtering density (see Tanizaki (1993), Tanizaki and Mariano (1994) and Mariano and Tanizaki (1995)). Therefore, for the importance density, we choose a larger variance than the variance obtained from the extended Kalman filter (EKF).

The initial density of the state-variable (i.e., c^p_0) is assumed to be normal with mean $a_{0|0}$ and variance $\Sigma_{0|0}$, i.e.,

$$P(c^p_0) = \Phi(c^p_0 - a_{0|0}, \Sigma_{0|0}).$$

(7) In the density-based Monte-Carlo filter (DMF), n=10,000 random draws are generated from the transition equation (6.2).

The initial value of the state-variable is assumed to be normal, i.e.,

$$c^p_0 \sim N(a_{0|0}, \Sigma_{0|0}),$$

where

$$a_{0|0} = c_0 - (\delta_1 Y_0 + \delta_2 Y_{-1}),$$
$$\Sigma_{0|0} = (\sigma_\eta Y_0)^2.$$

c_0 is per capita total consumption in 1956, and Y_0 and Y_{-1} denote per capita gross domestic product in 1956 and 1955, respectively.

(8) In the rejection sampling filter (RSF), the random draws are generated from the transition equation (6.2), where the number of random draws is taken as $n = 1,000$. The acceptance probability is given by the exponential part of the normal density obtained from the measurement equation (6.1), i.e.,

$$\omega_1(c_t^p;c_t) = \exp\left(-\frac{1}{2(\sigma_\epsilon Y_t)^2}(c_t - c_t^p - c_t^*)^2\right).$$

A random draw of c_t^p, say $\alpha_{i,t}$, is generated from the transition equation in the system (6.8). We take $\alpha_{i,t}$ as a random draw from the filtering density (denoted by $\alpha_{i,t|t}$ in Chapter 4) if $\alpha_{i,t}$ is accepted with probability $\omega_1(\alpha_{i,t};c_t)$, and generate another random draw of c_t^p otherwise.
The initial random draws of the state-variable are generated from the following normal density:

$$\alpha_{i,0|0} \sim N(a_{0|0}, \Sigma_{0|0}),$$

where

$$a_{0|0} = c_0 - (\delta_1 Y_0 + \delta_2 Y_{-1}),$$
$$\Sigma_{0|0} = (\sigma_\eta Y_0)^2.$$

c_0 is per capita total consumption in 1956, and Y_0 and Y_{-1} denote per capita gross domestic product in 1956 and 1955, respectively.

Under the above setup of each nonlinear filter, the unknown parameters are estimated in Tables 6.1 – 6.9. The unknown parameters β, γ, σ_η, δ_1, δ_2 and σ_ϵ are estimated by the maximum likelihood estimation method.

The maximization procedure of the parameters is as follows.

- For the extended Kalman filter (EKF), in the first iteration, the values of the unknown parameters β, γ, σ_η, δ_1, δ_2 and σ_ϵ are taken as: 1.0, 1.0, 0.1, 0.0, 0.0 and 0.01, respectively. In Table 6.1, each parameter is estimated by a simple grid search method, given the other parameter values, i.e., the log-likelihood function is maximized by changing the parameter value by 0.00001 for σ_ϵ, 0.001 for β, σ_η, δ_1 and δ_2, and 0.1 for γ, respectively. When the log-likelihood function is close to that in the last iteration (i.e., 0.001% convergence criterion is taken), we judge that the log-likelihood function reaches the maximum value.
- The other nonlinear filters (SNF, MSF, SIFc, GSF, NIF, ISF, DMF and RSF) are based on the extended Kalman filter (EKF). The parameter values obtained from the extended Kalman filter (EKF) are used in the first iteration. In Tables 6.2 – 6.9, each parameter is estimated by a simple grid search method, given the other parameter values, i.e., the log-likelihood function is maximized by changing the parameter value by 0.00001 for σ_ϵ, 0.001 for β, σ_η, δ_1 and δ_2, and 0.1 for γ, respectively. When the log-likelihood function is close to that in the last iteration (i.e., 0.001% convergence criterion is taken), we judge that the log-likelihood function reaches the maximum value.

Thus, the maximization procedure used in the extended Kalman filter (EKF) is different from that in the other nonlinear filters (SNF, MSF, SIFc, GSF,

NIF, ISF, DMF and RSF), which is because we want to avoid too much computational time. In other words, with respect to the parameter values in the first iteration, the extended Kalman filter (EKF) is different from the other nonlinear filters.

6.3.3 Estimation Results

For each country in Tables 6.1 – 6.9, we have three models (1) – (3), which are as follows:

(1) $\delta_1 = 0$ and $\delta_2 = 0$,
(2) $\delta_1 \neq 0$ and $\delta_2 = 0$,
(3) $\delta_1 \neq 0$ and $\delta_2 \neq 0$.

Each log-likelihood function is reported for each country in each table. In Tables 6.1 – 6.8 (i.e., EKF, SNF, MSF, SIFc, GSF, NIF, ISF and DMF), the estimation results in all the countries of Japan, U.S., U.K., France, Spain, Italy, Canada and Germany are represented. In Table 6.9 (i.e., RSF), the estimation results in Japan, U.S., U.K., France and Spain are reported, i.e., those in Italy, Canada and Germany are not partly available.

Table 6.10 is based on Tables 6.1 – 6.9, the likelihood ratio tests are performed for the following hypotheses.

(a) Null hypothesis: $H_0 : \delta_1 = 0$,
Alternative hypothesis: $H_1 : \delta_1 \neq 0$.

(b) Null hypothesis: $H_0 : \delta_2 = 0|\delta_1 \neq 0$,
Alternative hypothesis: $H_1 : \delta_2 \neq 0|\delta_1 \neq 0$.

(c) Null hypothesis: $H_0 : \delta_1 = 0$ and $\delta_2 = 0$,
Alternative hypothesis: $H_1 : \delta_1 \neq 0$ or $\delta_2 \neq 0$.

(a), (b) and (c) denote the single test, the conditional test and the joint test, respectively. Theoretically, we have the equation: (a)+(b)=(c). Note that the likelihood ratio test statistic of the hypothesis (a) or (b) is asymptotically distributed as a Chi-square random variable with one degree of freedom and that of (c) is asymptotically distributed as a Chi-square random variable with two degrees of freedom. The critical values of a Chi-square distribution with one degree of freedom are given by 3.84 for 0.05 % and 6.63 for 0.01 %, and those of a Chi-square distribution with two degrees of freedom are given by 5.99 for 0.05 % and 9.21 for 0.01 %, respectively.

Technical Comments. The simulation-based procedures cannot follow up the sudden shift such as Germany from 1990 to 1991 (i.e., integration of West Germany and East Germany in 1990) and Canada from 1981 to 1982 and from 1990 to 1991 (I got rid of the estimation results in Canada and Germany from Tables 6.6 – 6.9 since the obtained estimates of the state-variable were too unreasonable). The nonlinear filters with simulation techniques sometimes have difficulty in evaluating the likelihood function because of simulation errors (in Tables 6.10, — denotes the negative likelihood ratio test statistics while N.A. represents Not Available). Thus, for the simulation-based nonlinear filters (i.e., MSF in Table 6.3 and Figure 6.6, NIF in Table 6.6 and Figure 6.9, ISF in Table 6.7 and Figure 6.10, DMF in Table 6.8 and Figure 6.11, and RSF in Table 6.9 and Figure 6.12), there is a possibility that the two problems of — and N.A. in Table 6.10 come out.

Empirical Results. The parameter estimates are in Tables 6.1 – 6.9. γ is small for U.S., U.K. and Germany but large for France and Spain. MSF is unreliable from the results of Table 6.3. The standard error of ϵ_t (i.e., σ_ϵ) is 0.1% – 0.5% of income.

The likelihood ratio statistics are given in Table 6.10. For all the estimation methods and almost all the the countries, the hypothesis $H_0 : \delta_1 = 0$ or $H_0 : \delta_2 = 0$ is rejected according to the likelihood ratio test. For almost all the countries, except for U.S., current consumption depends on lagged consumption as well as income. Current consumption depends on current income for Japan, France, Italy, Canada and Germany and lagged income for U.K. and Spain. This indicates that the permanent income hypothesis does not hold for almost all the countries. This test takes into account transitory consumption, nonlinearity of the Euler equation and the variable interest rate. Numerous earlier papers testing the permanent income hypothesis consider some of these three issues, but none of them deals with the three issues simultaneously. The obtained results show that we still find significant evidence against the permanent income hypothesis even when all the three issues are incorporated in the model.

Next, in Figures 6.4 – 6.12, we estimate the ratio of per capita permanent consumption relative to per capita total consumption, i.e., $100a_{t|t}/c_t$, where $a_{t|t}$ denotes the filtering estimate of per capita permanent consumption at time t, i.e., $a_{t|t} = \mathrm{E}(c_t^p|C_t)$ and $C_t = \{c_1, c_2, \cdots, c_t\}$. The estimated ratio of permanent consumption is calculated for Japan, U.S., U.K., France, Spain, Italy, Canada and Germany. A high ratio of per capita permanent consumption implies that a large amount of people rationally behave along the life cycle permanent income hypothesis under rational expectations. All the Figures 6.4 – 6.12 are similar. From the figures, it is easily found that a ratio of per capita permanent consumption relative to per capita total consumption is very different, depending on each country.

Table 6.1. Extended Kalman Filter (EKF)

	Model	β	γ	σ_η	δ_1	δ_2	σ_ϵ	$\log L$
	(1)	1.255	6.6	0.158	—	—	0.00201	-321.748
Japan	(2)	1.686	12.8	0.729	0.424	—	0.00188	-309.884
	(3)	1.495	10.5	0.520	0.346	0.091	0.00442	-309.837
	(1)	1.025	3.0	0.056	—	—	0.00131	-186.551
U.S.	(2)	1.023	2.2	0.087	0.358	—	0.00138	-185.112
	(3)	1.009	1.5	0.063	0.289	0.107	0.00242	-184.024
	(1)	1.035	2.7	0.064	—	—	0.00168	-166.930
U.K.	(2)	1.034	2.6	0.072	0.091	—	0.00169	-166.852
	(3)	1.012	1.5	0.052	0.014	0.256	0.00355	-163.801
	(1)	1.881	28.7	0.460	—	—	0.00113	-221.963
France	(2)	1.583	19.1	0.659	0.396	—	0.00310	-216.724
	(3)	1.553	19.2	0.586	0.319	0.087	0.00413	-216.582
	(1)	1.397	15.1	0.379	—	—	0.00267	-300.804
Spain	(2)	1.385	16.0	0.573	0.205	—	0.00263	-300.151
	(3)	1.221	14.0	0.499	-0.056	0.418	0.00582	-294.421
	(1)	1.665	15.5	0.375	—	—	0.00335	-374.931
Italy	(2)	1.146	4.3	0.548	0.543	—	0.00426	-366.891
	(3)	1.484	13.2	0.088	0.309	0.211	0.01072	-366.331
	(1)	1.768	33.1	0.886	—	—	0.00152	-195.084
Canada	(2)	1.118	4.9	0.496	0.444	—	0.00144	-191.592
	(3)	1.051	3.3	0.259	0.304	0.113	0.00329	-191.269
	(1)	1.009	2.5	0.101	—	—	0.00658	-215.016
Germany	(2)	1.038	1.6	0.282	0.467	—	0.00263	-205.709
	(3)	1.067	1.7	0.364	0.422	0.071	0.00577	-205.554

Figure 6.4. Ratio of Permanent Consumption: Case $\delta_1 \neq 0$ and $\delta_2 \neq 0$
Extended Kalman Filter (EKF)

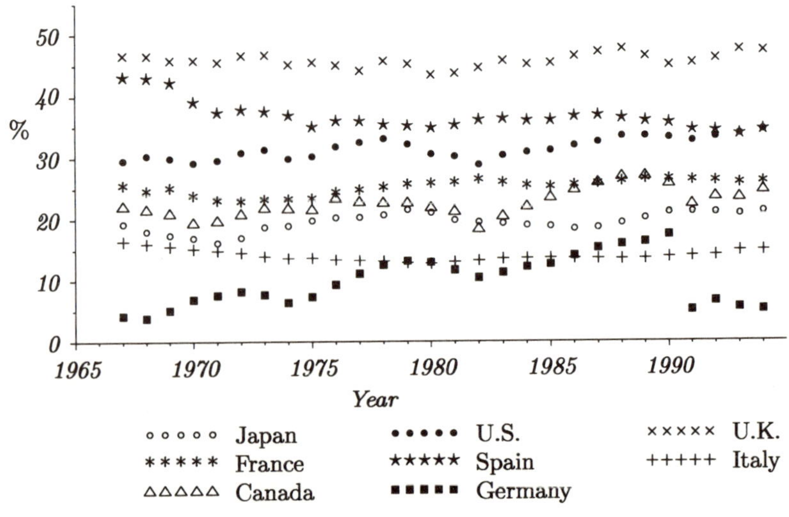

Table 6.2. Second-Order Nonlinear Filter (SNF)

	Model	β	γ	σ_η	δ_1	δ_2	σ_ϵ	$\log L$
	(1)	1.255	6.6	0.158	—	—	0.00199	-321.741
Japan	(2)	1.687	12.8	0.734	0.424	—	0.00183	-309.867
	(3)	1.502	10.6	0.529	0.348	0.089	0.00433	-309.807
	(1)	1.025	3.0	0.056	—	—	0.00130	-186.550
U.S.	(2)	1.023	2.2	0.087	0.358	—	0.00136	-185.108
	(3)	1.009	1.5	0.063	0.289	0.107	0.00240	-184.020
	(1)	1.035	2.7	0.064	—	—	0.00167	-166.928
U.K.	(2)	1.034	2.6	0.072	0.091	—	0.00167	-166.848
	(3)	1.012	1.5	0.052	0.014	0.256	0.00353	-163.797
	(1)	1.884	28.8	0.462	—	—	0.00110	-221.955
France	(2)	1.586	19.2	0.663	0.396	—	0.00306	-216.715
	(3)	1.554	19.2	0.588	0.319	0.087	0.00411	-216.577
	(1)	1.397	15.1	0.379	—	—	0.00266	-300.801
Spain	(2)	1.385	16.0	0.573	0.205	—	0.00262	-300.148
	(3)	1.220	14.0	0.500	-0.056	0.419	0.00581	-294.419
	(1)	1.675	15.7	0.380	—	—	0.00325	-374.891
Italy	(2)	1.146	4.3	0.553	0.543	—	0.00421	-366.872
	(3)	1.484	13.2	0.087	0.309	0.211	0.01072	-366.327
	(1)	1.769	33.1	0.886	—	—	0.00151	-195.082
Canada	(2)	1.118	4.9	0.496	0.444	—	0.00142	-191.588
	(3)	1.051	3.3	0.259	0.304	0.113	0.00328	-191.267
	(1)	1.009	2.5	0.101	—	—	0.00656	-215.012
Germany	(2)	1.038	1.6	0.282	0.467	—	0.00261	-205.703
	(3)	1.067	1.7	0.364	0.424	0.069	0.00570	-205.535

Figure 6.5. Ratio of Permanent Consumption: Case $\delta_1 \neq 0$ and $\delta_2 \neq 0$
Second-Order Nonlinear Filter (SNF)

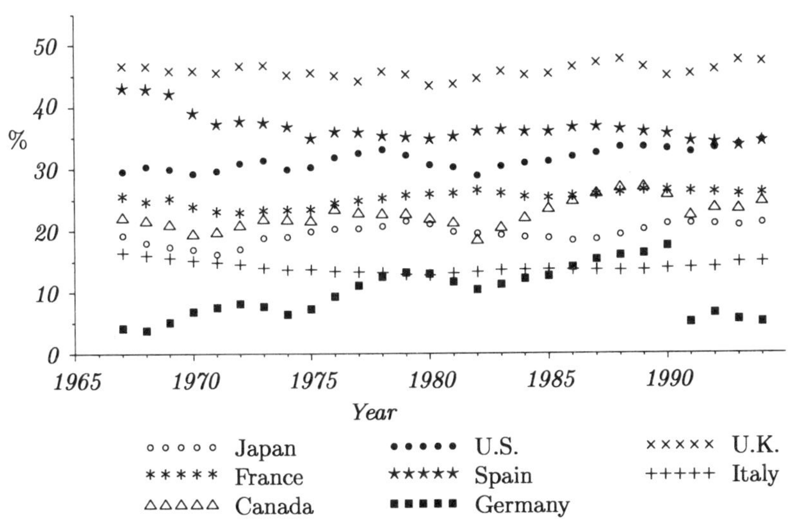

Table 6.3. Monte-Carlo Simulation Filter (MSF)

	Model	β	γ	σ_η	δ_1	δ_2	σ_ϵ	$\log L$
	(1)	1.260	5.3	0.188	—	—	0.00231	-369.695
Japan	(2)	2.299	26.5	0.728	0.516	—	0.00910	-325.994
	(3)	1.932	26.4	0.521	0.332	0.170	0.00808	-320.254
	(1)	1.025	2.5	0.042	—	—	0.00193	-218.201
U.S.	(2)	1.075	3.4	0.121	0.516	—	0.00256	-198.174
	(3)	1.045	2.4	0.003	0.245	0.290	0.00760	-187.483
	(1)	1.062	3.7	0.031	—	—	0.00146	-174.254
U.K.	(2)	1.061	3.7	0.035	0.071	—	0.00201	-174.143
	(3)	1.034	2.5	0.023	-0.119	0.406	0.00499	-165.286
	(1)	2.358	34.7	0.457	—	—	0.00327	-259.742
France	(2)	2.538	47.1	0.658	0.430	—	0.00739	-232.797
	(3)	2.488	52.2	0.585	0.144	0.308	0.00520	-224.603
	(1)	1.679	24.1	0.379	—	—	0.00000	-330.645
Spain	(2)	1.300	29.3	0.564	0.335	—	0.00174	-320.160
	(3)	1.478	59.0	0.496	-0.090	0.489	0.00873	-300.135
	(1)	1.931	17.4	0.375	—	—	0.00601	-412.258
Italy	(2)	1.408	20.2	0.547	0.540	—	0.00322	-368.932
	(3)	1.491	13.4	0.014	0.294	0.229	0.00862	-365.051
	(1)	2.510	46.1	0.887	—	—	0.00000	-222.849
Canada	(2)	1.971	36.0	0.492	0.379	—	0.01145	-206.898
	(3)	1.067	5.7	0.113	0.036	0.392	0.00471	-191.086
	(1)	1.035	2.5	0.071	—	—	0.00625	-240.100
Germany	(2)	1.129	6.1	0.263	0.439	—	0.01170	-222.944
	(3)	1.120	8.2	0.367	0.184	0.314	0.00500	-211.448

Figure 6.6. Ratio of Permanent Consumption: Case $\delta_1 \neq 0$ and $\delta_2 \neq 0$
Monte-Carlo Simulation Filter (MSF)

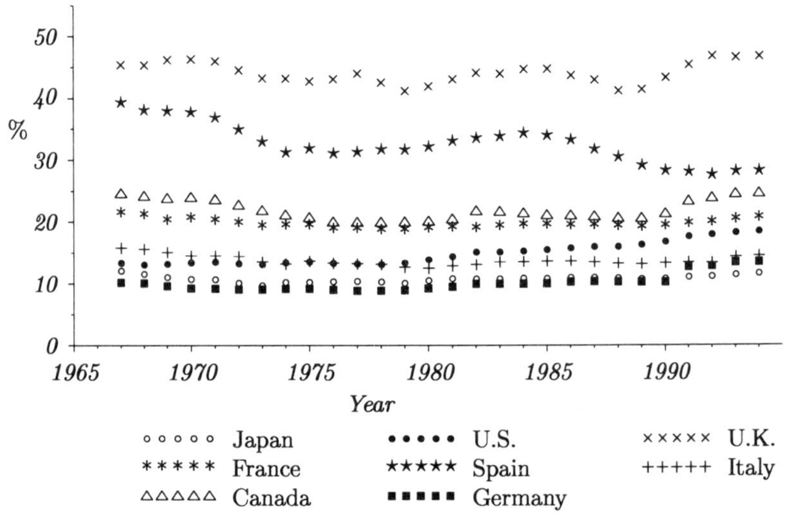

Table 6.4. Single-Stage Iteration Filter with Monte-Carlo Approximation (SIFc)

	Model	β	γ	σ_η	δ_1	δ_2	σ_ϵ	$\log L$
	(1)	1.254	6.7	0.157	—	—	0.00200	-321.751
Japan	(2)	1.630	18.3	0.729	0.417	—	0.00132	-309.642
	(3)	1.458	15.3	0.519	0.363	0.077	0.00393	-309.317
	(1)	1.002	1.7	0.038	—	—	0.00162	-190.069
U.S.	(2)	1.018	2.2	0.085	0.359	—	0.00132	-184.546
	(3)	1.006	1.5	0.062	0.295	0.100	0.00226	-183.496
	(1)	1.032	2.7	0.064	—	—	0.00164	-166.868
U.K.	(2)	1.030	2.6	0.071	0.088	—	0.00164	-166.777
	(3)	1.009	1.5	0.052	0.013	0.256	0.00323	-163.488
	(1)	1.926	37.8	0.459	—	—	0.00000	-223.889
France	(2)	1.506	26.8	0.663	0.388	—	0.00233	-216.772
	(3)	1.497	25.4	0.586	0.348	0.063	0.00361	-216.935
	(1)	1.448	20.7	0.376	—	—	0.00161	-300.815
Spain	(2)	1.296	23.6	0.570	0.221	—	0.00174	-299.433
	(3)	1.152	20.9	0.502	-0.053	0.427	0.00513	-294.093
	(1)	1.767	20.8	0.371	—	—	0.00186	-374.319
Italy	(2)	1.052	6.3	0.549	0.553	—	0.00332	-366.044
	(3)	1.497	13.4	0.072	0.308	0.209	0.01072	-366.333
	(1)	1.695	40.1	0.886	—	—	0.00079	-195.025
Canada	(2)	1.048	7.6	0.495	0.461	—	0.00074	-190.790
	(3)	1.054	4.0	0.253	0.327	0.091	0.00143	-189.831
	(1)	1.004	2.6	0.103	—	—	0.00653	-215.102
Germany	(2)	0.987	2.1	0.279	0.464	—	0.00212	-206.782
	(3)	1.005	2.9	0.363	0.425	0.070	0.00523	-205.850

Figure 6.7. Ratio of Permanent Consumption: Case $\delta_1 \neq 0$ and $\delta_2 \neq 0$
Single-Stage Iteration Filter with Monte-Carlo Approximation (SIFc)

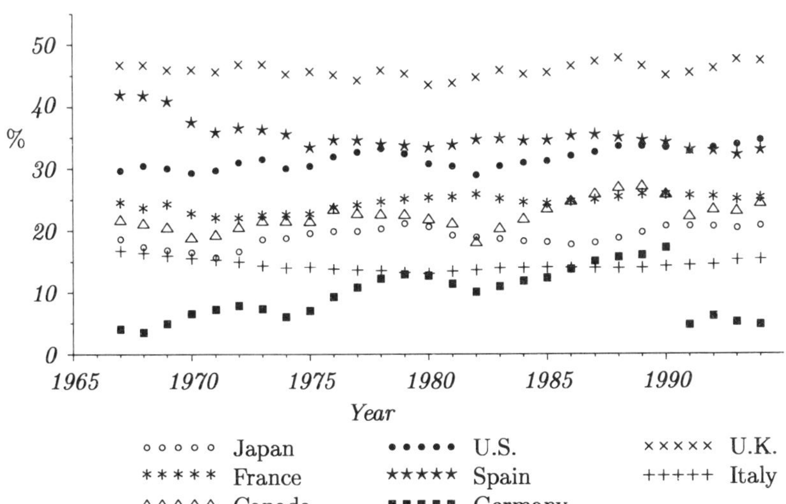

Table 6.5. Gaussian Sum Filter (GSF)

	Model	β	γ	σ_η	δ_1	δ_2	σ_ϵ	$\log L$
	(1)	1.255	6.6	0.158	—	—	0.00199	-321.717
Japan	(2)	1.689	12.8	0.732	0.424	—	0.00185	-309.864
	(3)	1.502	10.4	0.527	0.351	0.090	0.00434	-309.755
	(1)	1.025	3.0	0.056	—	—	0.00130	-186.550
U.S.	(2)	1.023	2.2	0.087	0.358	—	0.00137	-185.112
	(3)	1.009	1.5	0.063	0.289	0.106	0.00225	-184.010
	(1)	1.035	2.7	0.064	—	—	0.00167	-166.929
U.K.	(2)	1.034	2.6	0.072	0.091	—	0.00167	-166.848
	(3)	1.012	1.5	0.052	0.014	0.256	0.00353	-163.798
	(1)	1.884	28.8	0.462	—	—	0.00110	-221.953
France	(2)	1.588	19.0	0.664	0.400	—	0.00305	-216.635
	(3)	1.554	18.7	0.602	0.334	0.079	0.00397	-216.505
	(1)	1.397	15.1	0.379	—	—	0.00265	-300.796
Spain	(2)	1.385	16.0	0.573	0.205	—	0.00262	-300.149
	(3)	1.220	14.0	0.500	-0.057	0.419	0.00581	-294.418
	(1)	1.668	15.6	0.377	—	—	0.00332	-374.899
Italy	(2)	1.149	4.3	0.551	0.543	—	0.00423	-366.852
	(3)	1.490	13.3	0.080	0.313	0.205	0.01075	-366.700
	(1)	1.769	33.1	0.886	—	—	0.00151	-195.083
Canada	(2)	1.118	4.9	0.496	0.444	—	0.00142	-191.586
	(3)	1.051	3.3	0.259	0.304	0.113	0.00327	-191.258
	(1)	1.010	2.5	0.101	—	—	0.00644	-214.999
Germany	(2)	1.038	1.6	0.282	0.467	—	0.00261	-205.702
	(3)	1.066	1.7	0.352	0.425	0.062	0.00447	-205.356

Figure 6.8. Ratio of Permanent Consumption: Case $\delta_1 \neq 0$ and $\delta_2 \neq 0$
Gaussian Sum Filter (GSF)

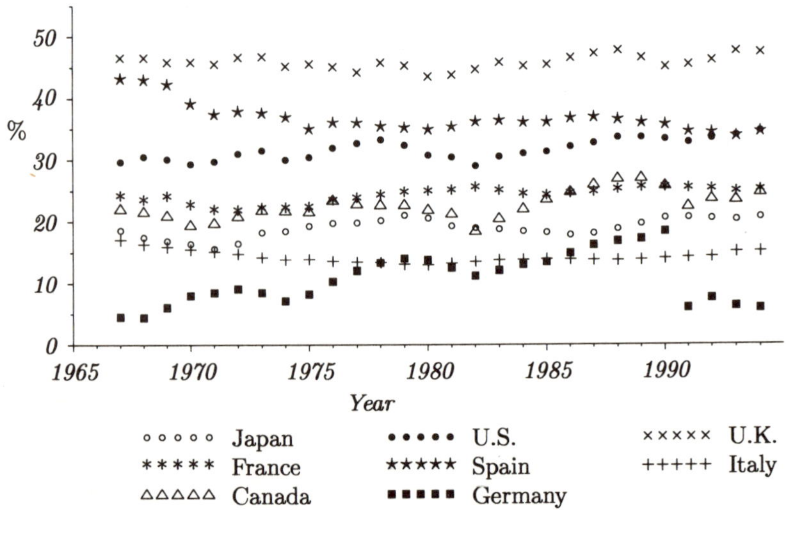

Table 6.6. Numerical Integration Filter (NIF)

	Model	β	γ	σ_η	δ_1	δ_2	σ_ϵ	$\log L$
	(1)	1.248	6.9	0.157	—	—	0.00194	-320.794
Japan	(2)	1.625	11.1	0.790	0.416	—	0.00153	-308.459
	(3)	1.285	7.9	0.423	0.365	0.063	0.00232	-307.490
	(1)	1.000	1.6	0.036	—	—	0.00168	-190.953
U.S.	(2)	1.012	1.9	0.075	0.352	—	0.00059	-184.503
	(3)	1.005	1.5	0.063	0.282	0.103	0.00048	-183.347
	(1)	1.032	2.7	0.064	—	—	0.00152	-166.942
U.K.	(2)	1.028	2.5	0.068	0.083	—	0.00017	-161.539
	(3)	1.007	1.4	0.053	0.034	0.252	0.00044	-162.430
	(1)	1.875	32.1	0.507	—	—	0.00057	-220.719
France	(2)	1.539	15.8	0.703	0.379	—	0.00354	-217.624
	(3)	1.526	22.7	0.619	0.304	0.094	0.00448	-216.833
	(1)	1.321	15.2	0.373	—	—	0.00191	-300.287
Spain	(2)	1.420	18.6	0.811	0.193	—	0.00025	-297.828
	(3)	1.234	11.2	0.484	-0.069	0.399	0.00569	-293.978
	(1)	1.574	14.7	0.377	—	—	0.00426	-377.674
Italy	(2)	1.168	3.4	0.514	0.519	—	0.00478	-369.113
	(3)	1.483	13.3	0.087	0.309	0.211	0.01072	-366.329
	(1)	1.447	15.8	0.815	—	—	0.00557	-202.932
Canada	(2)	0.994	1.5	0.174	0.437	—	0.00037	-193.818
	(3)	1.067	1.9	0.200	0.285	0.094	0.00161	-192.529
	(1)	0.990	1.7	0.078	—	—	0.00687	-219.305
Germany	(2)	1.053	1.8	0.185	0.454	—	0.00024	-231.903
	(3)	1.113	0.7	0.276	0.492	-0.032	0.00547	-213.098

Figure 6.9. Ratio of Permanent Consumption: Case $\delta_1 \neq 0$ and $\delta_2 \neq 0$
Numerical Integration Filter (NIF)

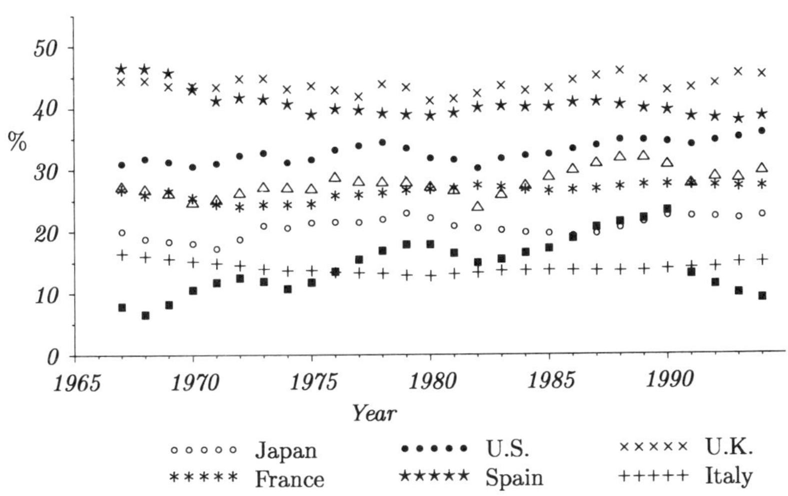

Table 6.7. Importance Sampling Filter (ISF)

	Model	β	γ	σ_η	δ_1	δ_2	σ_ϵ	$\log L$
	(1)	1.247	6.9	0.159	—	—	0.00221	-321.607
Japan	(2)	1.583	11.8	0.832	0.433	—	0.00182	-307.687
	(3)	1.452	12.9	0.564	0.351	0.096	0.00411	-307.829
	(1)	1.018	2.7	0.051	—	—	0.00227	-187.263
U.S.	(2)	1.009	1.8	0.074	0.361	—	0.00099	-184.385
	(3)	1.003	1.4	0.063	0.292	0.109	0.00094	-182.646
	(1)	1.031	2.6	0.063	—	—	0.00239	-167.677
U.K.	(2)	1.029	2.5	0.071	0.094	—	0.00105	-165.842
	(3)	1.007	1.4	0.052	0.019	0.258	0.00197	-162.619
	(1)	1.850	32.2	0.494	—	—	0.00148	-221.533
France	(2)	1.597	24.2	0.736	0.393	—	0.00446	-218.235
	(3)	1.537	23.6	0.543	0.314	0.086	0.00461	-216.792
	(1)	1.362	16.6	0.411	—	—	0.00302	-301.575
Spain	(2)	1.282	17.5	0.676	0.212	—	0.00160	-298.147
	(3)	1.130	13.3	0.345	-0.059	0.416	0.00746	-294.751
	(1)	1.509	13.7	0.332	—	—	0.00491	-378.044
Italy	(2)	1.034	3.3	0.300	0.536	—	0.00688	-369.351
	(3)	1.496	13.2	0.049	0.310	0.210	0.01068	-365.398
	(1)	1.343	17.8	0.742	—	—	0.00665	-203.585
Canada	(2)	0.991	1.4	0.156	0.435	—	0.00065	-192.400
	(3)	0.990	1.5	0.139	0.293	0.131	0.00262	-193.563
	(1)	0.990	1.7	0.076	—	—	0.00808	-220.303
Germany	(2)	1.053	1.7	0.213	0.471	—	0.00146	-237.163
	(3)	0.967	0.7	0.157	0.474	-0.015	0.00647	-215.495

Figure 6.10. Ratio of Permanent Consumption: Case $\delta_1 \neq 0$ and $\delta_2 \neq 0$
Importance Sampling Filter (ISF)

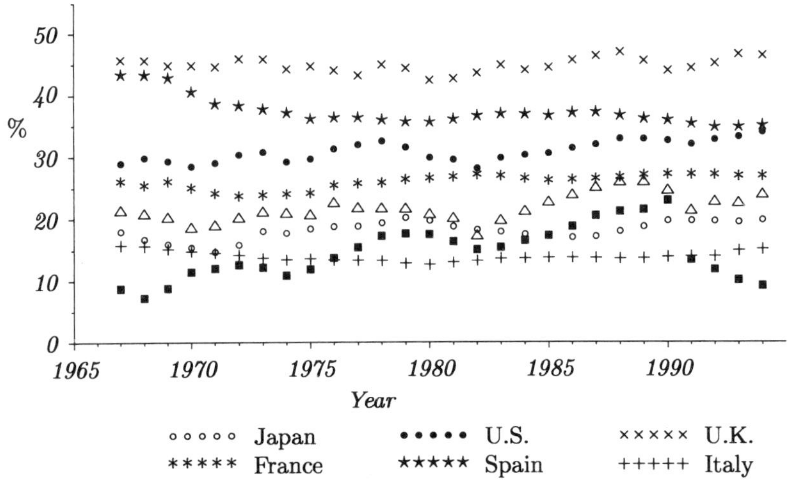

Table 6.8. Density-Based Monte-Carlo Filter (DMF)

	Model	β	γ	σ_η	δ_1	δ_2	σ_ϵ	$\log L$
	(1)	1.256	6.6	0.158	—	—	0.00201	-332.586
Japan	(2)	1.687	12.9	0.729	0.424	—	0.00188	-320.932
	(3)	1.495	10.6	0.521	0.346	0.091	0.00442	-320.723
	(1)	1.024	2.9	0.055	—	—	0.00131	-192.792
U.S.	(2)	1.022	2.2	0.087	0.358	—	0.00138	-192.658
	(3)	1.009	1.5	0.063	0.289	0.107	0.00242	-190.971
	(1)	1.035	2.7	0.064	—	—	0.00167	-172.665
U.K.	(2)	1.035	2.6	0.072	0.091	—	0.00169	-173.217
	(3)	1.011	1.5	0.052	0.014	0.256	0.00355	-169.663
	(1)	1.880	28.8	0.460	—	—	0.00113	-229.582
France	(2)	1.582	19.0	0.658	0.394	—	0.00310	-229.190
	(3)	1.554	19.2	0.586	0.319	0.087	0.00414	-227.027
	(1)	1.397	15.2	0.379	—	—	0.00267	-311.430
Spain	(2)	1.384	16.0	0.573	0.205	—	0.00263	-312.450
	(3)	1.222	14.0	0.498	-0.055	0.418	0.00581	-307.450
	(1)	1.664	15.4	0.375	—	—	0.00336	-392.818
Italy	(2)	1.146	4.3	0.549	0.543	—	0.00426	-387.955
	(3)	1.483	13.3	0.087	0.309	0.211	0.01073	-380.677
	(1)	1.767	33.1	0.886	—	—	0.00152	-216.242
Canada	(2)	1.117	4.8	0.495	0.444	—	0.00145	-228.765
	(3)	1.049	3.4	0.259	0.303	0.114	0.00328	-203.854
	(1)	1.008	2.5	0.101	—	—	0.00658	-233.296
Germany	(2)	1.039	1.5	0.282	0.467	—	0.00263	-210.720
	(3)	1.068	1.5	0.363	0.421	0.072	0.00577	-214.229

Figure 6.11. Ratio of Permanent Consumption: Case $\delta_1 \neq 0$ and $\delta_2 \neq 0$
Density-Based Monte-Carlo Filter (DMF)

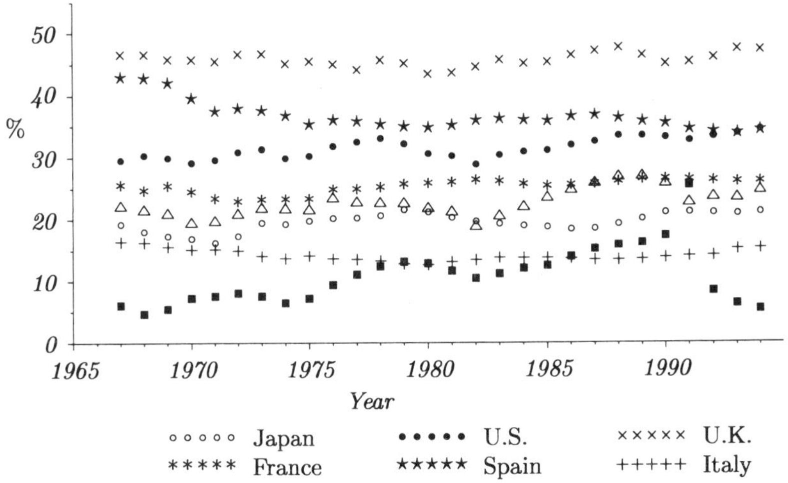

Table 6.9. Rejection Sampling Filter (RSF)

	Model	β	γ	σ_η	δ_1	δ_2	σ_ϵ	$\log L$
	(1)	1.254	6.6	0.158	—	—	0.00201	-332.280
Japan	(2)	1.685	12.9	0.728	0.425	—	0.00188	-321.086
	(3)	1.494	10.5	0.520	0.346	0.092	0.00441	-322.043
	(1)	1.025	3.0	0.055	—	—	0.00131	-192.861
U.S.	(2)	1.022	2.1	0.087	0.358	—	0.00138	-192.076
	(3)	1.008	1.5	0.063	0.288	0.106	0.00242	-190.329
	(1)	1.034	2.7	0.064	—	—	0.00168	-171.819
U.K.	(2)	1.034	2.6	0.072	0.091	—	0.00169	-171.921
	(3)	1.011	1.5	0.052	0.013	0.256	0.00355	-169.202
	(1)	1.881	28.7	0.461	—	—	0.00113	-229.096
France	(2)	1.582	19.0	0.659	0.395	—	0.00310	-228.707
	(3)	1.553	19.3	0.586	0.317	0.086	0.00413	-227.979
	(1)	1.398	15.1	0.379	—	—	0.00266	-311.756
Spain	(2)	1.386	15.9	0.572	0.205	—	0.00264	-311.775
	(3)	1.220	14.0	0.499	-0.056	0.419	0.00582	-307.187
	(1)	1.665	15.7	0.375	—	—	0.00335	-391.273
Italy	(2)	1.145	4.4	0.548	0.543	—	0.00426	-385.754
	(3)	Not Available						

Figure 6.12. Ratio of Permanent Consumption: Case $\delta_1 \neq 0$ and $\delta_2 \neq 0$
Rejection Sampling Filter (RSF)

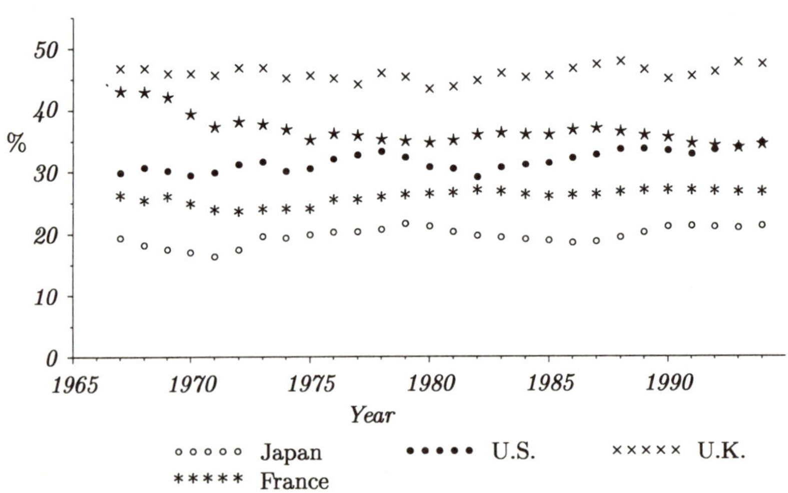

Table 6.10. Likelihood Ratio Tests

	Hypoth-esis	Japan	U.S.	U.K.	France	Spain	Italy	Canada	Germany
	(a)	23.728	2.878	0.156	10.478	1.306	16.080	6.984	18.614
EKF	(b)	0.094	2.176	6.102	0.284	11.460	1.120	0.646	0.310
	(c)	23.822	5.054	6.258	10.762	12.766	17.200	7.630	18.924
	(a)	23.748	2.884	0.160	10.480	1.306	16.038	6.988	18.618
SNF	(b)	0.120	2.176	6.102	0.276	11.458	1.090	0.642	0.336
	(c)	23.868	5.060	6.262	10.756	12.764	17.128	7.630	18.954
	(a)	87.402	40.054	0.222	53.890	20.970	86.652	31.902	34.312
MSF	(b)	11.480	21.382	17.714	16.388	40.050	7.762	31.624	22.992
	(c)	98.882	61.436	17.936	70.278	61.020	94.414	63.526	57.304
	(a)	24.218	11.046	0.182	14.234	2.764	16.550	8.470	16.640
SIFc	(b)	0.650	2.100	6.578	—	10.680	—	1.918	1.864
	(c)	24.868	13.146	6.760	13.908	13.444	15.972	10.388	18.504
	(a)	23.706	2.876	0.162	10.636	1.294	16.094	6.994	18.594
GSF	(b)	0.218	2.204	6.100	0.260	11.462	0.304	0.656	0.692
	(c)	23.924	5.080	6.262	10.896	12.756	16.398	7.650	19.286
	(a)	24.670	12.900	10.806	6.190	4.918	17.122	18.228	—
NIF	(b)	1.938	2.312	—	1.582	7.700	5.568	2.578	37.610
	(c)	26.608	15.212	9.024	7.772	12.618	22.690	20.806	12.414
	(a)	27.840	5.756	3.670	6.596	6.856	17.386	22.370	—
ISF	(b)	—	3.478	6.446	2.886	6.792	7.906	—	43.336
	(c)	27.556	9.234	10.116	9.482	13.648	25.292	20.044	9.616
	(a)	23.308	0.268	—	0.784	—	9.726	—	45.152
DMF	(b)	0.418	3.374	7.108	4.326	10.000	14.556	49.822	—
	(c)	23.726	3.642	6.004	5.110	7.960	24.282	24.776	38.134
	(a)	22.388	1.570	—	0.778	—	11.308	N.A.	N.A.
RSF	(b)	—	3.494	5.438	1.456	9.176	N.A.	N.A.	N.A.
	(c)	20.474	5.064	5.234	2.234	9.138	N.A.	N.A.	N.A.

Taken an example of Figure 6.4 (EKF). The ratio in U.K. is around 45% – 50% during the estimation period, which is the largest of all the countries. In Germany, the ratio continues to increase until 1990 and there is a sudden jump from 1990 (i.e., 20%) to 1991 (i.e., 5% – 10%), which corresponds to integration of West Germany and East Germany. In Spain, there is a tendency to decrease from 45% to 35% over time. U.S. shows 30% – 35% while Canada and France are about 25%. The ratio in Japan indicates around 20%. Italy is low and about 15%. We can see that there is a big recession around 1974 (the first oil crisis) for Italy, Germany, U.S. and Spain, around 1982 (the second oil crisis) for Germany, Canada and U.S., and in 1991 for Germany and Canada. In 1974, 1981 and 1991, we can observe a downward tendency for almost all the countries.

6.4 Summary

In this section, for Japan, U.S., U.K., France, Spain, Italy, Canada and Germany, we have tested the permanent income hypothesis taking into account transitory consumption, nonlinearity of the Euler equation and the variable interest rate simultaneously, and estimated the ratio of per capita permanent consumption relative to per capita total consumption. The state-space model is constructed from the identity equation and the Euler equation derived from a utility maximization problem of the representative agent. Under the model setup, nonlinear filtering techniques discussed in Chapters 3 and 4 are used for estimation. Annual consumption data from 1965 to 1994 are used for each country.

In this example, we have to estimate the unknown parameters β, δ_1, δ_2, γ, σ_ϵ and σ_η, and at the same time we perform the nonlinear filtering techniques. The approximated likelihood function, given by (2.30), (4.17), (4.26), (4.36), (4.42) and (4.44), is maximized with respect to the unknown parameters for estimation. By using the approximated likelihood function such as equation (2.30), the bias arises for the estimated parameters. In the estimation methods based on the likelihood functions (4.26), (4.36), (4.42) and (4.44), the estimates of unknown parameters might be asymptotically unbiased, because the nonlinear functions are not linearized in the likelihood functions (4.26), (4.36), (4.42) and (4.44).

Taking into account transitory consumption, nonlinearity of the Euler equation and the variable interest rate, the permanent income hypothesis was tested in this section. Numerous papers deal with testing the permanent income hypothesis, taking one of the above three issues. However, none of the papers deals with the model including the three issues. All the nonlinear filtering techniques used here produced similar results. Taking into account transitory consumption, nonlinearity of the Euler equation and the variable interest rate simultaneously, our results reject the permanent income hypothesis (except for U.S.) and point out differences in degree of rationality in consumer behavior in each country. Also, we see that permanent consumption decreases for the periods including the two oil crises and crash of stock market.

A6 Appendix

A6.1 Alternative Interpretation

The state-space model represented by equations (6.1) and (6.2) can be interpreted in a different way. Simply, we write equation (6.1) as equation (6.9), where c_t^* is erased from equation (6.1).

$$\begin{aligned} &\text{(Measurement equation)} \quad && c_t = c_t^p + \epsilon_t, \\ &\text{(Transition equation)} \quad && \frac{\beta R_{t-1} u'(c_t^p)}{u'(c_{t-1}^p)} = 1 + \eta_t. \end{aligned} \tag{6.9}$$

where (ϵ_t, η_t) are bivariate normal, independent over time, with mean zero and covariance matrix

$$\begin{pmatrix} (\sigma_\epsilon Y_t)^2 & 0 \\ 0 & \sigma_\eta^2 \end{pmatrix}.$$

The state-space model (6.9) is taken as the errors-in-variable model. Usually, the published data include measurement errors. If we estimate an econometric model with such data, an appropriate result might not be obtained. Therefore, we need to incorporate the errors-in-variables into the model directly. There, c_t^p denotes the true value of per capita consumption while c_t represents the observed data of per capita consumption. It might be appropriate to consider that an economic theory is related to true data (i.e., c_t^p), not actually obtained data (i.e., c_t). Note that it is assumed in the model that R_{t-1} does not have the measurement errors.

Thus, the state-space model represented by equations (6.1) and (6.2) can be interpreted as the errors-in-variables model. See, for example, Mariano and Tanizaki (1995).

7. Prediction and Smoothing

7.1 Introduction

In the state-space model shown in equations (3.1) and (3.2), we estimate the unobserved variable α_t. Filtering is one of the estimation methods, which has been treated in this book. In this chapter, we consider the other two estimation methods, i.e., prediction and smoothing. Filtering is used to estimate the present state-variable at the present time, prediction is helpful to estimate the future state-variable at the present time, and smoothing is an estimation of the past state-variable at the present time.

To show the three estimation methods (i.e., prediction, filtering and smoothing) mathematically, consider evaluating the following conditional mean and variance of the state-variable:

$$\mathrm{E}(\alpha_r|Y_s) \equiv a_{r|s},$$
$$\mathrm{Var}(\alpha_r|Y_s) \equiv \Sigma_{r|s},$$

where α_r is the state-variable at time r and Y_s denotes the information set available at time s.

Depending on relationship between r and s, we have prediction, filtering and smoothing as shown below.

Prediction: $r > s$,
Filtering: $r = s$,
Smoothing: $r < s$.

The relationship among prediction, filtering and smoothing is as follows.

- The initial value of the prediction estimate is given by the filtering estimate.
- The one-step ahead prediction estimate is utilized to obtain the filtering estimate.
- We need the filtering estimate to derive the smoothing estimate.

In Section 7.2, the prediction algorithms in the case of $(r,s) = (t+L,t)$ are derived based on the density functions, which is L-step ahead prediction. Also, we have the smoothing algorithms in the case of $(r,s) = (t,T)$ in Section 7.3, which corresponds to fixed-interval smoothing.

The procedures introduced in Sections 4.3 – 4.6 are directly applied to prediction and smoothing algorithms.

7.2 Prediction

In this section, we consider prediction problem, i.e., the following conditional expectations:

$$\mathrm{E}(\alpha_{t+L}|Y_t) = a_{t+L|t},$$
$$\mathrm{Var}(\alpha_{t+L}|Y_t) = \Sigma_{t+L|t},$$

for $L = 1, 2, \cdots$, which represent the L-step ahead prediction estimates.

The L-step ahead prediction algorithm based on the density function is known as the following recursion (see, for example, Harvey (1989)):

$$\begin{aligned} P(\alpha_{t+L}|Y_t) &= \int P(\alpha_{t+L}, \alpha_{t+L-1}|Y_t)\mathrm{d}\alpha_{t+L-1} \\ &= \int P(\alpha_{t+L}|\alpha_{t+L-1}, Y_t)P(\alpha_{t+L-1}|Y_t)\mathrm{d}\alpha_{t+L-1} \\ &= \int P(\alpha_{t+L}|\alpha_{t+L-1})P(\alpha_{t+L-1}|Y_t)\mathrm{d}\alpha_{t+L-1}, \end{aligned} \tag{7.1}$$

for $L = 1, 2, \cdots$, which is a generalization of equation (2.16) with respect to L, known as the density-based prediction algorithm. Equation (2.16) corresponds to the case $L = 1$ in equation (7.1). The reason why the third equality, i.e.,

$$P(\alpha_{t+L}|\alpha_{t+L-1}, Y_t) = P(\alpha_{t+L}|\alpha_{t+L-1}),$$

holds is because the righthand side of the transition equation does not depend on the information set available at time t, i.e., Y_t. Note that the transition equation is given by $\alpha_{t+1} = g_{t+1}(\alpha_t, \eta_{t+1})$ in this case. See Kitagawa (1987) and Harvey (1989) for the density-based prediction algorithm. Also, see Appendix A2.1 for the derivation, where the case of $L = 1$ is described. In the above prediction algorithm (7.1), the filtering density $P(\alpha_t|Y_t)$ is assumed to be known, which is the initial density of the density-based L-step ahead prediction algorithm. In equation (7.1), the density function $P(\alpha_{t+L}|\alpha_{t+L-1})$ is obtained from the transition equation (3.2) for $L = 1, 2, \cdots$.

The L-step ahead prediction density is obtained as follows. From the two density functions $P(\alpha_t|Y_t)$ and $P(\alpha_{t+1}|\alpha_t)$, we can obtain $P(\alpha_{t+1}|Y_t)$. Similarly, $P(\alpha_{t+1}|Y_t)$ and $P(\alpha_{t+2}|\alpha_{t+1})$ yield $P(\alpha_{t+2}|Y_t)$. Thus, $P(\alpha_{t+L}|Y_t)$, $L = 1, 2, \cdots$, is recursively computed. Once the L-step ahead prediction density $P(\alpha_{t+L}|Y_t)$ is obtained, the conditional expectations of prediction are given by:

$$a_{t+L|t} = \int \alpha_{t+L}P(\alpha_{t+L}|Y_t)\mathrm{d}\alpha_{t+L},$$
$$\Sigma_{t+L|t} = \int (\alpha_{t+L} - a_{t+L|t})(\alpha_{t+L} - a_{t+L|t})'P(\alpha_{t+L}|Y_t)\mathrm{d}\alpha_{t+L},$$

which is a definition of mean and variance.

Moreover, when the system is linear and normal (i.e., equations (2.1) and (2.2)), the density-based prediction algorithm given by equation (7.1) yields the following linear recursive algorithm:

$$a_{t+L|t} = T_{t+L}a_{t+L-1|t} + c_{t+L} \tag{7.2}$$

$$\Sigma_{t+L|t} = T_{t+L}\Sigma_{t+L-1|t}T'_{t+L} + R_{t+L}Q_{t+L}R'_{t+L}, \tag{7.3}$$

for $L = 1, 2, \cdots$. Both equations are derived from the first- and the second-moments of the normal distribution. When Proof I in Appendix A2.2 is applied to equation (7.1), the predictions equations (7.2) and (7.3) can be obtained.

Given $a_{t|t}$ and $\Sigma_{t|t}$, from equations (7.2) and (7.3), one-step ahead prediction estimate $a_{t+1|t}$ and its variance $\Sigma_{t+1|t}$ are obtained. Similarly, from $a_{t+1|t}$ and $\Sigma_{t+1|t}$, we have $a_{t+2|t}$ and $\Sigma_{t+2|t}$. Thus, the recursive calculation represented by equations (7.2) and (7.3) gives us $a_{t+L|t}$ and $\Sigma_{t+L|t}$ for $L = 1, 2, \cdots$. $a_{t|t}$ and $\Sigma_{t|t}$ are the filtering estimates.

Equation (7.1) is evaluated to obtain the density-based algorithms such as numerical integration, importance sampling and so on.

7.2.1 Numerical Integration Prediction

Kitagawa (1987) and Kramer and Sorenson (1988) proposed a nonlinear and nonnormal filter using numerical integration. Consider a scalar case of the state-vector (it is easily extended to the higher-dimensional cases of the state-vector). Numerical integration requires the nodes which are denoted by $\alpha_{i,t}$, $i = 1, \cdots, n$. For all t, $\alpha_{i,t}$, $i = 1, \cdots, n$, are sorted by size with respect to i, i.e., $\alpha_{1,t}$ is the smallest value and $\alpha_{n,t}$ the largest one for all t. There are some methods to evaluate integration numerically, i.e., a rectangle rule, a sum of trapezoids, Simpson's formula and so on. Here, for simplicity of discussion, we take the numerical integration method by a sum of rectangles.

By numerical integration, equation (7.1) is evaluated as:

$$\begin{aligned} &P(\alpha_{t+L}|Y_t) \\ &= \int P(\alpha_{t+L}|\alpha_{t+L-1})P(\alpha_{t+L-1}|Y_t)\mathrm{d}\alpha_{t+L-1} \\ &\approx \sum_{j=1}^{n} P(\alpha_{t+L}|\alpha_{j,t+L-1})P(\alpha_{j,t+L-1}|Y_t)(\alpha_{j,t+L-1} - \alpha_{j-1,t+L-1}). \end{aligned}$$

Moreover, evaluating the state-variable at $\alpha_{t+L} = \alpha_{i,t+L}$, we have the following L-step ahead prediction algorithm:

$$\begin{aligned} &P(\alpha_{i,t+L}|Y_t) \\ &= \sum_{j=1}^{n} P(\alpha_{i,t+L}|\alpha_{j,t+L-1})P(\alpha_{j,t+L-1}|Y_t)(\alpha_{j,t+L-1} - \alpha_{j-1,t+L-1}), \end{aligned} \tag{7.4}$$

for $L = 1, 2, \cdots$.

Suppose that $P(\alpha_{i,t}|Y_t)$, $i = 1, \cdots, n$, are available, where $P(\alpha_{i,t}|Y_t)$ denotes the filtering density at time t evaluated at $\alpha_t = \alpha_{i,t}$ and $\alpha_{i,t}$ represents the node of α_t. $P(\alpha_{i,t+1}|\alpha_{j,t})$, $i, j = 1, \cdots, n$, are obtained from the transition equation (3.2). Based on $P(\alpha_{i,t}|Y_t)$ and $P(\alpha_{i,t+1}|\alpha_{j,t})$, one-step ahead prediction density $P(\alpha_{i,t+1}|Y_t)$, $i = 1, \cdots, n$, can be computed from equation (7.4). Similarly, $P(\alpha_{i,t+2}|Y_t)$, $i = 1, \cdots, n$, are obtained from $P(\alpha_{i,t+1}|Y_t)$ and $P(\alpha_{i,t+2}|\alpha_{j,t+1})$, $i, j = 1, \cdots, n$. Thus, given the two densities $P(\alpha_{i,t}|Y_t)$ and $P(\alpha_{i,t+L}|\alpha_{j,t+L-1})$, the prediction densities $P(\alpha_{i,t+L}|Y_t)$ for $L = 1, 2, \cdots$ are computed recursively.

Given $P(\alpha_{i,t+L}|Y_t)$ for $i = 1, \cdots, n$ and $L = 1, 2, \cdots$, the prediction estimates are obtained as:

$$
\begin{aligned}
a_{t+L|t} &= \int \alpha_{t+L} P(\alpha_{t+L}|Y_t) \mathrm{d}\alpha_{t+L} \\
&\approx \sum_{i=1}^{n} \alpha_{i,t+L} P(\alpha_{i,t+L}|Y_t)(\alpha_{i,t+L} - \alpha_{i-1,t+L}),
\end{aligned}
$$

$$
\begin{aligned}
\Sigma_{t+L|t} &= \int (\alpha_{t+L} - a_{t+L|t})(\alpha_{t+L} - a_{t+L|t})' P(\alpha_{t+L}|Y_t) \mathrm{d}\alpha_{t+L} \\
&\approx \sum_{i=1}^{n} (\alpha_{i,t+L} - a_{t+L|t})(\alpha_{i,t+L} - a_{t+L|t})' \\
&\qquad\qquad \times P(\alpha_{i,t+L}|Y_t)(\alpha_{i,t+L} - \alpha_{i-1,t+L}),
\end{aligned}
$$

for $L = 1, 2, \cdots$.

In the filtering algorithm with numerical integration, equation (4.20) corresponds to equation (7.4) when $L = 1$.

7.2.2 Importance Sampling Prediction

Tanizaki and Mariano (1994) and Mariano and Tanizaki (1995) derived the prediction algorithm using Monte-Carlo integration with importance sampling.

Define the weight function as:

$$
\omega_{t+L|t} \equiv \frac{P(\alpha_{t+L}|Y_t)}{P_\alpha(\alpha_{t+L})}.
$$

Using the weight functions $\omega_{t+L|t}$ and $\omega_{t+L-1|t}$, the prediction equation (7.1) is transformed into:

$$
\begin{aligned}
\omega_{t+L|t} &\equiv \frac{P(\alpha_{t+L}|Y_t)}{P_\alpha(\alpha_{t+L})} \\
&= \int \frac{P(\alpha_{t+L}|\alpha_{t+L-1})}{P_\alpha(\alpha_{t+L})} \frac{P(\alpha_{t+L-1}|Y_t)}{P_\alpha(\alpha_{t+L-1})} P_\alpha(\alpha_{t+L-1}) \mathrm{d}\alpha_{t+L-1}
\end{aligned}
$$

$$
\begin{aligned}
&= \int \frac{P(\alpha_{t+L}|\alpha_{t+L-1})}{P_\alpha(\alpha_{t+L})} \omega_{t+L-1|t} P_\alpha(\alpha_{t+L-1}) \mathrm{d}\alpha_{t+L-1} \\
&\approx \frac{1}{n} \sum_{j=1}^{n} \frac{P(\alpha_{t+L}|\alpha_{j,t+L-1})}{P_\alpha(\alpha_{t+L})} \omega_{j,t+L-1|t},
\end{aligned}
\tag{7.5}
$$

for $L = 1, 2, \cdots$, where we define $\omega_{i,t+L|t}$ as:

$$
\omega_{i,t+L|t} \equiv \frac{P(\alpha_{i,t+L}|Y_t)}{P_\alpha(\alpha_{i,t+L})},
$$

which is $\omega_{t+L|t}$ evaluated at $\alpha_{t+L} = \alpha_{i,t+L}$. $\alpha_{i,t+L}$ is a random number generated from the importance density $P_\alpha(\alpha_{t+L})$, which is appropriately chosen by a researcher.

Therefore, generating n random numbers of α_{t+L} (i.e., $\alpha_{i,t+L}$ for $i = 1, \cdots, n$) from the importance density $P_\alpha(\alpha_{t+L})$, equation (7.1) is approximated from equation (7.5) to the following:

$$
\omega_{i,t+L|t} = \frac{1}{n} \sum_{j=1}^{n} \frac{P(\alpha_{i,t+L}|\alpha_{j,t+L-1})}{P_\alpha(\alpha_{i,t+L})} \omega_{j,t+L-1|t}.
\tag{7.6}
$$

$P(\alpha_{i,t+L}|\alpha_{j,t+L-1})$ can be derived from the transition equation (3.2) and $P_\alpha(\alpha_{i,t+L})$ is appropriately assumed by a researcher. Accordingly, given $\omega_{j,t+L-1|t}$, we can compute $\omega_{j,t+L|t}$. Thus, the weight functions $\omega_{j,t+L|t}$, $L = 1, 2, \cdots$, are recursively obtained.

The prediction estimates, i.e., mean and variance, are computed as:

$$
\begin{aligned}
a_{t+L|t} &= \int \alpha_{t+L} P(\alpha_{t+L}|Y_t) \mathrm{d}\alpha_{t+L} \\
&\approx \frac{1}{n} \sum_{i=1}^{n} \alpha_{i,t+L} \omega_{i,t+L|t},
\end{aligned}
$$

$$
\begin{aligned}
\Sigma_{t+L|t} &= \int (\alpha_{t+L} - a_{t+L|t})(\alpha_{t+L} - a_{t+L|t})' P(\alpha_{t+L}|Y_t) \mathrm{d}\alpha_{t+L} \\
&\approx \frac{1}{n} \sum_{i=1}^{n} (\alpha_{i,t+L} - a_{t+L|t})(\alpha_{i,t+L} - a_{t+L|t})' \omega_{i,t+L|t},
\end{aligned}
$$

for $L = 1, 2, \cdots$.

In the filtering algorithm with importance sampling, equation (4.29) corresponds to equation (7.6) when $L = 1$, which represents one-step ahead prediction.

7.2.3 Density-Based Monte-Carlo Prediction

Tanizaki and Mariano (1995a, 1995b) and Mariano and Tanizaki (1996) proposed the prediction algorithm generating random draws from the transition equation.

Consider the conditional mean and variance of α_{t+L}. Let us define A_{t+L} and Y_t as:

$$A_{t+L} = \{\alpha_0, \alpha_1, \cdots, \alpha_{t+L}\},$$
$$Y_t = \{y_1, y_2, \cdots, y_t\}.$$

The L-step ahead prediction of α_{t+L}, i.e., $\mathrm{E}(\alpha_{t+L}|Y_t) \equiv a_{t+L|t}$, is represented as:

$$a_{t+L|t} = \frac{\displaystyle\int \alpha_{t+L} P(Y_t|A_t) P(A_{t+L}) \mathrm{d}A_{t+L}}{\displaystyle\int P(Y_t|A_t) P(A_{t+L}) \mathrm{d}A_{t+L}}, \tag{7.7}$$

which is a definition of the conditional mean of α_{t+L}. The conditional variance $\mathrm{Var}(\alpha_{t+L}|Y_t) \equiv \Sigma_{t+L|t}$, is given by:

$$\Sigma_{t+L|t} = \frac{\displaystyle\int (\alpha_{t+L} - a_{t+L|t})(\alpha_{t+L} - a_{t+L|t})' P(Y_t|A_t) P(A_{t+L}) \mathrm{d}A_{t+L}}{\displaystyle\int P(Y_t|A_t) P(A_{t+L}) \mathrm{d}A_{t+L}}, \tag{7.8}$$

which is also a definition of the conditional variance of α_{t+L}.

In equations (7.7) and (7.8), the two densities $P(A_{t+L})$ and $P(Y_t|A_t)$ are defined as:

$$P(A_{t+L}) \equiv \prod_{s=1}^{t+L} P(\alpha_s|\alpha_{s-1}),$$
$$P(Y_t|A_t) \equiv \prod_{s=1}^{t} P(y_s|\alpha_s),$$

where $P(\alpha_s|\alpha_{s-1})$ and $P(y_s|\alpha_s)$ are derived from the transition equation (3.2) and the measurement equation (3.1).

Generate the random draws of α_{t+L} (i.e., $\alpha_{i,t+L}$) from $P(A_{t+L})$, or equivalently, from the transition equation:

$$\alpha_{i,s} = g_t(\alpha_{i,s-1}, \eta_{i,s}),$$

where $s = 1, 2, \cdots, t+L$. Moreover, $A_{i,t+L}$ denotes:

$$A_{i,t+L} = \{\alpha_{i,0}, \alpha_{i,1}, \cdots, \alpha_{i,t+L}\}.$$

Evaluating equation (7.7) by the unconditional random draws $A_{i,t+L}$, the prediction mean obtained by the Monte-Carlo simulation is given by:

$$\begin{aligned} a_{t+L|t} &= \frac{\displaystyle\int \alpha_{t+L} P(Y_t|A_t) P(A_{t+L}) \mathrm{d}A_{t+L}}{\displaystyle\int P(Y_t|A_t) P(A_{t+L}) \mathrm{d}A_{t+L}} \\ &\approx \frac{\displaystyle\frac{1}{n}\sum_{i=1}^{n} \alpha_{i,t+L} P(Y_t|A_{i,t})}{\displaystyle\frac{1}{n}\sum_{i=1}^{n} P(Y_t|A_{i,t})} \\ &= \sum_{i=1}^{n} \alpha_{i,t+L}\omega_{i,t}. \end{aligned}$$

Similarly, equation (7.8) is approximately obtained as:

$$\Sigma_{t+L|t} = \sum_{i=1}^{n} (\alpha_{i,t+L} - a_{t+L|t})(\alpha_{i,t+L} - a_{t+L|t})'\omega_{i,t},$$

where $L = 1, 2, \cdots$. $\omega_{i,t}$ is defined as the weight function (4.40) with the initial condition (4.41), which is represented as:

$$\begin{aligned} \omega_{i,t} &= \frac{P(Y_t|A_{i,t})}{\displaystyle\sum_{i=1}^{n} P(Y_t|A_{i,t})} \\ &= \frac{\displaystyle\prod_{s=1}^{t} P(y_t|\alpha_{i,t})}{\displaystyle\sum_{i=1}^{n}\prod_{s=1}^{t} P(y_t|\alpha_{i,t})} \\ &= \frac{P(y_t|\alpha_{i,t})\omega_{i,t-1}}{\displaystyle\sum_{i=1}^{n} P(y_t|\alpha_{i,t})\omega_{i,t-1}}, \end{aligned}$$

where the initial weight function $\omega_{i,0}$ is given by:

$$\omega_{i,0} = \frac{1}{n}.$$

Thus, the Monte-Carlo simulation technique is utilized based on the random draws of the state-variable, which are generated from the transition equation.

Note that $\omega_{i,s}$ represents the weight function such that:

$$\sum_{i=1}^{n} \omega_{i,s} = 1,$$

for all s.

7.2.4 Rejection Sampling Prediction

The filtering algorithm with the rejection sampling procedure has been discussed in Section 4.6. In the same framework, we consider the prediction algorithm. The prediction estimate with rejection sampling is very simple and easy (see Tanizaki and Mariano (1995b)), which is obtained as follows. Suppose that $\alpha_{i,t+L-1|t}$, $i = 1, \cdots, n$, are available, which are the random draws from $(L-1)$-step ahead prediction density. Consider generating the random draws from L-step ahead prediction density, given those from $(L-1)$-step ahead prediction density. The L-step ahead prediction algorithm (7.1) is approximately represented as:

$$\begin{aligned} P(\alpha_{t+L}|Y_t) &= \int P(\alpha_{t+L}|\alpha_{t+L-1}) P(\alpha_{t+L-1}|Y_t) \mathrm{d}\alpha_{t+L-1} \\ &\approx \frac{1}{n} \sum_{i=1}^{n} P(\alpha_{t+L}|\alpha_{i,t+L-1|t}). \end{aligned} \tag{7.9}$$

Based on equation (7.9), we can generate random draws of α_{t+L} given Y_t. Pick up $\alpha_{i,t+L-1|t}$ randomly (i.e., pick up i with equal probability) and generate random numbers of η_{t+L} (i.e., $\alpha_{j,t+L}$), and we have random draws of α_{t+L} (i.e., $\alpha_{j,t+L|t}$) from the transition equation:

$$\alpha_{j,t+L|t} = g_{t+L}(\alpha_{i,t+L-1|t}, \eta_{j,t+L}),$$

where $\eta_{j,t+L}$ denotes the j-th random draw of η_{t+L} and $\alpha_{j,t+L|t}$ represents the j-th random draw of α_{t+L} given Y_t. That is, we generate $\alpha_{j,t+L|t}$ for $j = 1, \cdots, n$.

Thus, given $\alpha_{i,t|t}$, $\alpha_{i,t+L|t}$ is recursively obtained for $L = 1, 2, \cdots$. Note that $\alpha_{i,t|t}$ is generated in the filtering algorithm shown in Section 4.6. As we can see, the rejection sampling approach is not utilized in the prediction algorithm in spite of nonlinearity and nonnormality.

Once the random draws from the prediction density (i.e., $\alpha_{i,t+L|t}$) are available, the conditional mean is computed as:

$$\begin{aligned} a_{t+L|t} &= \int \alpha_{t+L} P(\alpha_{t+L}|Y_t) \mathrm{d}\alpha_{t+L} \\ &\approx \frac{1}{n} \sum_{i=1}^{n} \alpha_{i,t+L|t}, \end{aligned}$$

and the conditional variance is:

$$\begin{aligned}\Sigma_{t+L|t} &= \int (\alpha_{t+L} - a_{t+L|t})(\alpha_{t+L} - a_{t+L|t})' P(\alpha_{t+L}|Y_t) \mathrm{d}\alpha_{t+L} \\ &\approx \frac{1}{n}\sum_{i=1}^{n}(\alpha_{i,t+L|t} - a_{t+L|t})(\alpha_{i,t+L|t} - a_{t+L|t})',\end{aligned}$$

for $L = 1, 2, \cdots$.

7.3 Smoothing

Moreover, next, we consider smoothing estimation. There are three kinds of smoothing algorithms, i.e., fixed-point smoothing, fixed-lag smoothing and fixed-interval smoothing (see, for example, Anderson and Moore (1979)). In the case of economics, fixed-interval smoothing might be useful for estimation of unobservable variables, rather than fixed-lag smoothing or fixed-point smoothing. Kitagawa (1987), Tanizaki and Mariano (1993) and Tanizaki and Mariano (1995b) originally proposed the fixed-interval smoothing algorithm based on density approximation by numerical integration, Monte-Carlo integration with importance sampling. and rejection sampling, respectively.

There are three kinds of smoothing problems (see Anderson and Moore (1979) and Harvey (1989)):

- **Fixed-Point Smoothing**, i.e., $\mathrm{E}(\alpha_L|Y_t) \equiv a_{L|t}$, for $t = L{+}1, L{+}2, \cdots, T$, where L is a fixed integer and T represents the sample size.
- **Fixed-Lag Smoothing**, i.e., $\mathrm{E}(\alpha_t|Y_{t+L}) \equiv a_{t|t+L}$, for $t = 1, 2, \cdots, T - L$, where L is a fixed integer and T denotes the sample size.
- **Fixed-Interval Smoothing**, i.e., $\mathrm{E}(\alpha_t|Y_T) \equiv a_{t|T}$, for $t = 1, 2, \cdots, T$, where T is the sample size.

The fixed-point smoothing is used to estimate the initial condition from the available data. The fixed-lag smoothing computes smoothed estimates with a fixed delay. In the case where we analyze the past condition from all the data available at the present time, the fixed-interval smoothing is useful. In a field of economics, economic situations in the past are often analyzed using the data available at the present time. Accordingly, in this book, we consider the fixed-interval smoothing, which is the most useful of the three in a field of economics. That is, we consider the following conditional mean and variance:

$$\begin{aligned}\mathrm{E}(\alpha_t|Y_T) &= a_{t|T}, \\ \mathrm{Var}(\alpha_t|Y_T) &= \Sigma_{t|T},\end{aligned}$$

for $t = 1, 2, \cdots, T$. For the three kinds of smoothing algorithms, see, for example, Anderson and Moore (1979) and Harvey (1990).

The density-based fixed-interval smoothing algorithm is represented as (see, for example, Kitagawa (1987) and Harvey (1989)):

$$
\begin{aligned}
P(\alpha_t|Y_T) &= \int P(\alpha_t, \alpha_{t+1}|Y_T)\mathrm{d}\alpha_{t+1} \\
&= \int P(\alpha_{t+1}|Y_T)P(\alpha_t|\alpha_{t+1}, Y_T)\mathrm{d}\alpha_{t+1} \\
&= \int P(\alpha_{t+1}|Y_T)P(\alpha_t|\alpha_{t+1}, Y_t)\mathrm{d}\alpha_{t+1} \\
&= \int P(\alpha_{t+1}|Y_T)\frac{P(\alpha_t, \alpha_{t+1}|Y_t)}{P(\alpha_{t+1}|Y_t)}\mathrm{d}\alpha_{t+1} \\
&= P(\alpha_t|Y_t)\int \frac{P(\alpha_{t+1}|Y_T)P(\alpha_{t+1}|\alpha_t)}{P(\alpha_{t+1}|Y_t)}\mathrm{d}\alpha_{t+1}, \qquad (7.10)
\end{aligned}
$$

for $t = T-1, T-2, \cdots, 1$. The density function $P(\alpha_{t+1}|\alpha_t)$ obtained from the transition equation (3.2) and the density functions $P(\alpha_t|Y_t)$ and $P(\alpha_{t+1}|Y_t)$ which are obtained from equations (2.16) and (2.17) in the filtering algorithm yield the above density-based fixed-interval smoothing algorithm (7.10), which is a backward recursion from $P(\alpha_{t+1}|Y_T)$ to $P(\alpha_t|Y_T)$. Note that the smoothing density at time $t = T$ (i.e., the endpoint case in the smoothing algorithm) is equivalent to the filtering density at time $t = T$. Thus, the fixed-interval smoothing is derived together with the filtering algorithm given by equations (2.16) and (2.17).

Once the smoothing densities $P(\alpha_t|Y_T)$ for $t = 1, \cdots, T$ are available from equation (7.10), the fixed-interval smoothing mean of α_t is defined as:

$$
a_{t|T} = \int \alpha_t P(\alpha_t|Y_T)\mathrm{d}\alpha_t,
$$

and the conditional variance of α_t is:

$$
\Sigma_{t|T} = \int (\alpha_t - a_{t|T})(\alpha_t - a_{t|T})' P(\alpha_t|Y_T)\mathrm{d}\alpha_t,
$$

for $t = 1, \cdots, T$.

As a special case, under linearity and normality assumptions which are represented by equations (2.1) and (2.2), from the first- and the second-moments in equation (7.10), the fixed-interval smoothing algorithm can be derived as:

$$
C_t = \Sigma_{t|t} T'_{t+1} \Sigma^{-1}_{t+1|t}, \qquad (7.11)
$$

$$
a_{t|T} = a_{t|t} + C_t(a_{t+1|T} - a_{t+1|t}), \qquad (7.12)
$$

$$
\Sigma_{t|T} = \Sigma_{t|t} + C_t(\Sigma_{t+1|T} - \Sigma_{t+1|t})C'_t, \qquad (7.13)
$$

for $t = T-1, T-2, \cdots, 1$, which is also a backward recursion. Given $\Sigma_{t+1|t}$ and $\Sigma_{t|t}$, C_t is obtained from equation (7.11). From $\Sigma_{t+1|t}$, $\Sigma_{t|t}$, C_t, $a_{t+1|t}$, $a_{t|t}$, $\Sigma_{t+1|T}$ and $a_{t+1|T}$ we can have $\Sigma_{t|T}$ and $a_{t|T}$ through equations (7.12) and (7.13). Thus, the smoothing algorithm is a backward recursion based on one-step ahead prediction and filtering algorithms.

In the case where we take the case: $T_t = I_k$ (a $k \times k$ identity matrix, where k is a dimension of α_t), $c_t = 0$ and $R_t = 0$, the Kalman filter algorithm represented as equations (2.9) – (2.15) is equivalent to the recursive ordinary least squares estimation method. In the case of the Kalman smoothed algorithm given by equations (7.11) – (7.13), $a_{t|T}$ and $\Sigma_{t|T}$ are constant over time t and are equivalent to the ordinary least squares estimates with the sample size T.

7.3.1 Numerical Integration Smoothing

Kitagawa (1987) proposed the fixed-interval smoothing algorithm using the numerical integration procedure. To derive the smoothing algorithm based on numerical integration, equation (7.10) is rewritten as:

$$\begin{aligned}
&P(\alpha_t|Y_T) \\
&= P(\alpha_t|Y_t) \int \frac{P(\alpha_{t+1}|Y_T)P(\alpha_{t+1}|\alpha_t)}{P(\alpha_{t+1}|Y_t)} \mathrm{d}\alpha_{t+1} \\
&\approx P(\alpha_t|Y_t) \sum_{j=1}^{n} \frac{P(\alpha_{j,t+1}|Y_T)P(\alpha_{j,t+1}|\alpha_t)}{P(\alpha_{j,t+1}|Y_t)} (\alpha_{j,t+1} - \alpha_{j-1,t+1}),
\end{aligned}$$

and accordingly, replacing α_t by the node $\alpha_{i,t}$, we have the following smoothing algorithm based on numerical integration:

$$\begin{aligned}
&P(\alpha_{i,t}|Y_T) \\
&= P(\alpha_{i,t}|Y_t) \sum_{j=1}^{n} \frac{P(\alpha_{j,t+1}|Y_T)P(\alpha_{j,t+1}|\alpha_{i,t})}{P(\alpha_{j,t+1}|Y_t)} (\alpha_{j,t+1} - \alpha_{j-1,t+1}), \qquad (7.14)
\end{aligned}$$

for $t = T-1, T-2, \cdots, 1$.

$P(\alpha_{i,t}|Y_t)$, $i = 1, \cdots, n$, are also computed in the filtering algorithm represented by equations (4.20) and (4.21). $P(\alpha_{j,t+1}|Y_t)$, $j = 1, \cdots, n$, denote the one-step ahead prediction density, which is also derived from the filtering algorithm. $P(\alpha_{j,t+1}|\alpha_{i,t})$, $i, j = 1, \cdots, n$, can be obtained from the transition equation (3.2). Thus, we have all the density functions of $P(\alpha_{i,t}|Y_t)$, $P(\alpha_{j,t+1}|\alpha_{i,t})$ and $P(\alpha_{j,t+1}|Y_t)$. Here, suppose that $P(\alpha_{j,t+1}|Y_T)$, $j = 1, \cdots, n$, are available. Then, $P(\alpha_{i,t}|Y_T)$, $i = 1, \cdots, n$, are computed from equation (7.14). Note that we have the density $P(\alpha_{j,t+1}|Y_T)$ for $t = T-1$, because $P(\alpha_{j,t+1}|Y_T)$ for $t = T-1$ is equivalent to the filtering density function. Accordingly, $P(\alpha_{i,t}|Y_T)$, $i = 1, \cdots, n$, are obtained from $P(\alpha_{j,t+1}|Y_T)$, $j = 1, \cdots, n$, which is a backward recursion.

Once we have the smoothing densities $P(\alpha_{i,t}|Y_T)$ for $t = 1, \cdots, T$, the conditional mean and variance of the state-variable is easily computed as follows:

$$a_{t|T} = \sum_{i=1}^{n} \alpha_{i,t} P(\alpha_{i,t}|Y_T)(\alpha_{i,t} - \alpha_{i-1,t}),$$

$$\Sigma_{t|T} = \sum_{i=1}^{n} (\alpha_{i,t} - a_{t|T})(\alpha_{i,t} - a_{t|T})' P(\alpha_{i,t}|Y_T)(\alpha_{i,t} - \alpha_{i-1,t}),$$

for $t = 1, \cdots, T$.

7.3.2 Importance Sampling Smoothing

Tanizaki and Mariano (1994) and Mariano and Tanizaki (1995) proposed the smoothing algorithm with importance sampling. Using Monte-Carlo integration with importance sampling, equation (7.10) is rewritten as:

$$\begin{aligned}
\omega_{t|T} &\equiv \frac{P(\alpha_t|Y_T)}{P_\alpha(\alpha_t)} \\
&= \frac{P(\alpha_t|Y_t)}{P_\alpha(\alpha_t)} \int \frac{\dfrac{P(\alpha_{t+1}|Y_T)}{P_\alpha(\alpha_{t+1})} P(\alpha_{t+1}|\alpha_t)}{\dfrac{P(\alpha_{t+1}|Y_t)}{P_\alpha(\alpha_{t+1})} P_\alpha(\alpha_{t+1})} P_\alpha(\alpha_{t+1}) \mathrm{d}\alpha_{t+1} \\
&= \omega_{t|t} \int \frac{\omega_{t+1|T} P(\alpha_{t+1}|\alpha_t)}{\omega_{t+1|t} P_\alpha(\alpha_{t+1})} P_\alpha(\alpha_{t+1}) \mathrm{d}\alpha_{t+1} \\
&\approx \omega_{t|t} \frac{1}{n} \sum_{j=1}^{n} \frac{\omega_{j,t+1|T} P(\alpha_{j,t+1}|\alpha_t)}{\omega_{j,t+1|t} P_\alpha(\alpha_{j,t+1})},
\end{aligned}$$

where the weight functions $\omega_{r|s}$ and $\omega_{i,r|s}$ are defined as:

$$\omega_{r|s} \equiv \frac{P(\alpha_r|Y_s)}{P_\alpha(\alpha_r)}, \qquad \omega_{i,r|s} \equiv \frac{P(\alpha_{i,r}|Y_s)}{P_\alpha(\alpha_{i,r})},$$

for $(r, s) = (t, T), (t, t), (t + 1, T), (t + 1, t)$. $\alpha_{i,r}$, $i = 1, \cdots, n$, denote the random draws generated from the importance density $P_\alpha(\alpha_r)$.

Evaluating $\omega_{t|t}$ at $\alpha_t = \alpha_{i,t}$ in the above equation, the following smoothing algorithm based on importance sampling is obtained:

$$\omega_{i,t|T} = \omega_{i,t|t} \frac{1}{n} \sum_{j=1}^{n} \frac{\omega_{j,t+1|T} P(\alpha_{j,t+1}|\alpha_{i,t})}{\omega_{j,t+1|t} P_\alpha(\alpha_{j,t+1})}, \tag{7.15}$$

for $t = T - 1, T - 2, \cdots, \cdots, 1$, which implies a backward recursive algorithm of the weight functions.

$\omega_{i,t|t}$ and $\omega_{j,t+1|t}$, $i = 1, \cdots, n$, are recursively computed from the filtering algorithm shown in equations (4.29) and (4.30), respectively. $P(\alpha_{j,t+1}|\alpha_{i,t})$, $i, j = 1, \cdots, n$, are derived from the transition equation (3.2). $P_\alpha(\alpha_{j,t+1})$, $j = 1, \cdots, n$, denote the appropriately chosen importance density function. Thus, $\omega_{i,t|t}$, $\omega_{j,t+1|t}$, $P(\alpha_{j,t+1}|\alpha_{i,t})$ and $P_\alpha(\alpha_{j,t+1})$ are available. Accordingly, $\omega_{i,t|T}$ is computed from $\omega_{j,t+1|T}$, which indicates the backward recursive algorithm of the weight functions. Thus, using the importance sampling procedure, the recursive algorithm of the weight functions, rather than the density functions, is obtained.

Once we have the smoothing weight functions $\omega_{i,t|T}$ for $t = 1, \cdots, T$, the conditional mean and variance of the state-variable is easily computed as follows:

$$a_{t|T} = \frac{1}{n}\sum_{i=1}^{n} \alpha_{i,t}\omega_{i,t|T},$$

$$\Sigma_{t|T} = \frac{1}{n}\sum_{i=1}^{n} (\alpha_{i,t} - a_{t|T})(\alpha_{i,t} - a_{t|T})'\omega_{i,t|T},$$

for $t = 1, \cdots, T$.

We sometimes need the shape of the smoothing density function. In such a case, we can obtain it as follows:

$$P(\alpha_{i,t}|Y_T) = \omega_{i,t|T}P_\alpha(\alpha_{i,t}),$$

for $i = 1, \cdots, n$ and $t = 1, \cdots, T$.

7.3.3 Density-Based Monte-Carlo Smoothing

The density-based Monte-Carlo smoothing algorithm which also utilizes the method of Monte-Carlo stochastic simulations was developed by Tanizaki and Mariano (1995a). First, note that the joint density function of (Y_T, A_T) is written as:

$$P(Y_T, A_T) = P(Y_T|A_T)P(A_T),$$

where A_T, Y_T, $P(Y_T|A_T)$ and $P(A_T)$ are denoted by:

$$A_T = \{\alpha_0, \alpha_1, \cdots, \alpha_T\},$$
$$Y_T = \{y_1, y_2, \cdots, y_T\},$$
$$P(Y_T|A_T) = \prod_{t=1}^{T} P(y_t|\alpha_t),$$
$$P(A_T) = P_\alpha(\alpha_0)\prod_{t=1}^{T} P(\alpha_t|\alpha_{t-1}).$$

$P(y_t|\alpha_t)$ and $P(\alpha_t|\alpha_{t-1})$ are obtained from the measurement and transition equations, i.e., equations (3.1) and (3.2). Therefore, the conditional density of α_t given information Y_T is given by:

$$P(\alpha_t|Y_T) = \frac{\displaystyle\int P(Y_T|A_T)P(A_T)\mathrm{d}A_{T,-t}}{\displaystyle\int P(Y_T|A_T)P(A_T)\mathrm{d}A_T}, \tag{7.16}$$

where $A_{T,-t}$ denotes the following information set:

$$A_{T,-t} = \{\alpha_0, \cdots, \alpha_{t-1}, \alpha_{t+1}, \cdots, \alpha_T\},$$

which is the set that α_t is excluded from A_T. That is, we have $A_{t,-t} \equiv A_{t-1}$. Note that equation (7.16) is equivalent to equation (7.10) although the latter indicates the recursive algorithm. The conditional mean and variance of α_t given Y_T, denoted by $a_{t|T}$ and $\Sigma_{t|T}$, are obtained as follows:

$$\begin{aligned} a_{t|T} &= \int \alpha_t P(\alpha_t|Y_T) \mathrm{d}\alpha_t \\ &= \frac{\displaystyle\int \alpha_t P(Y_T|A_T) P(A_T) \mathrm{d}A_T}{\displaystyle\int P(Y_T|A_T) P(A_T) \mathrm{d}A_T}, \end{aligned} \tag{7.17}$$

$$\begin{aligned} \Sigma_{t|T} &= \int (\alpha_t - a_{t|T})(\alpha_t - a_{t|T})' P(\alpha_t|Y_T) \mathrm{d}\alpha_t \\ &= \frac{\displaystyle\int (\alpha_t - a_{t|T})(\alpha_t - a_{t|T})' P(Y_T|A_T) P(A_T) \mathrm{d}A_T}{\displaystyle\int P(Y_T|A_T) P(A_T) \mathrm{d}A_T}, \end{aligned} \tag{7.18}$$

which are equivalent to the smoothing estimates.

Let us define $A_{i,T}$ as:

$$A_{i,T} - \{\alpha_{i,0}, \alpha_{i,1}, \cdots, \alpha_{i,T}\},$$

for $i = 1, \cdots, n$, which are the random draws of A_T generated from $P(A_T)$. That is, $A_{i,T}$ is a set of random draws from the transition equation:

$$\alpha_{i,t} = g_t(\alpha_{i,t-1}, \eta_{i,t}),$$

where $\eta_{i,t}$ is a random number generated from the underlying density for $t = 1, \cdots, T$ and $i = 1, \cdots, n$.

Given n sets of random draws of A_T, (i.e., $A_{i,T}$ for $i = 1, \cdots, n$), from equations (7.17) and (7.18), $a_{t|T}$ and $\Sigma_{t|T}$ are evaluated as:

$$\begin{aligned} a_{t|T} &= \frac{\displaystyle\int \alpha_t P(Y_T|A_T) P(A_T) \mathrm{d}A_T}{\displaystyle\int P(Y_T|A_T) P(A_T) \mathrm{d}A_T} \\ &\approx \frac{\displaystyle\frac{1}{n}\sum_{i=1}^{n} \alpha_{i,t} P(Y_T|A_{i,T})}{\displaystyle\frac{1}{n}\sum_{j=1}^{n} P(Y_T|A_{j,T})}, \end{aligned} \tag{7.19}$$

$$\Sigma_{t|T} = \int (\alpha_t - a_{t|T})(\alpha_t - a_{t|T})' P(\alpha_t|Y_T) d\alpha_t$$

$$\approx \frac{\frac{1}{n}\sum_{i=1}^{n}(\alpha_{i,t} - a_{t|T})(\alpha_{i,t} - a_{t|T})' P(Y_T|A_{i,T})}{\frac{1}{n}\sum_{j=1}^{n} P(Y_T|A_{j,T})}. \quad (7.20)$$

The following formula is easy and convenient to compute the smoothing estimate of α_t, i.e., $a_{t|T}$. Since we have the following equation:

$$P(Y_T|A_{i,T}) = \prod_{t=1}^{T} P(y_t|\alpha_{i,t}),$$

using the weight function $\omega_{i,T}$, equations (7.19) and (7.20) are rewritten as:

$$a_{t|T} = \sum_{i=1}^{n} \alpha_{i,t}\omega_{i,T},$$

$$\Sigma_{t|T} = \sum_{i=1}^{n} (\alpha_{i,t} - a_{t|T})(\alpha_{i,t} - a_{t|T})'\omega_{i,T},$$

where, as mentioned above, the weight function $\omega_{i,s}$ is recursively obtained as follows:

$$\omega_{i,t} = \frac{P(y_t|\alpha_{i,t})\omega_{i,t-1}}{\sum_{j=1}^{n} P(y_t|\alpha_{j,t})\omega_{j,t-1}}.$$

Note that $\omega_{i,T}$ is the endpoint case of $\omega_{i,t}$ for $t = T$. The initial values of the weight function are given by:

$$\omega_{i,0} = \frac{1}{n},$$

for $i = 1, \cdots, n$.

In the density-based Monte-Carlo smoothing algorithm, once we compute the weight function $\omega_{i,T}$, $i = 1, \cdots, n$, all the smoothing estimates $a_{t|T}$ and $\Sigma_{t|T}$ for $t = 1, \cdots, T$ are obtained. In other words, in spite of time t, the smoothing estimates $a_{t|T}$ and $\Sigma_{t|T}$ utilize $\omega_{i,T}$.

Note that $\omega_{i,T}$ represents the weight function such that

$$\sum_{i=1}^{n} \omega_{i,T} = 1,$$

which implies that a sum of the weight functions is equal to one.

7.3.4 Rejection Sampling Smoothing

Tanizaki and Mariano (1995b) proposed the fixed-interval smoothing algorithm using rejection sampling, where a backward recursive algorithm of the random draws from the smoothing densities is derived. The fixed-interval smoothing algorithm with rejection sampling is represented as follows. Let $\alpha_{i,t-1|t-1}$ and $\alpha_{j,t+1|T}$ be the random draws of α_{t-1} given Y_{t-1} and α_{t+1} given Y_T, respectively. Suppose that $\alpha_{i,t|t}$ for $t = 1,\cdots,T$ and $i = 1,\cdots,n$ are available, which are obtained from the rejection sampling filter shown in Section 4.6.

Again, write the smoothing algorithm (7.10) as follows:

$$P(\alpha_t|Y_T) = P(\alpha_t|Y_t)\int \frac{P(\alpha_{t+1}|Y_T)P(\alpha_{t+1}|\alpha_t)}{P(\alpha_{t+1}|Y_t)}\mathrm{d}\alpha_{t+1},$$

for $t = T-1, T-2, \cdots, 1$, which is a backward recursion.

In order to obtain the rejection sampling smoothing, first, note that each component in the above equation (7.10) is transformed as follows:

$$\begin{aligned} P(\alpha_t|Y_t) &= \frac{1}{P(y_t|Y_{t-1})}\int P(y_t|\alpha_t)P(\alpha_t|\alpha_{t-1})P(\alpha_{t-1}|Y_{t-1})\mathrm{d}\alpha_{t-1} \\ &\approx \frac{1}{P(y_t|Y_{t-1})}\left(\frac{1}{n}\sum_{i=1}^{n} P(y_t|\alpha_t)P(\alpha_t|\alpha_{i,t-1|t-1})\right), \end{aligned}$$

$$\int \frac{P(\alpha_{t+1}|Y_T)P(\alpha_{t+1}|\alpha_t)}{P(\alpha_{t+1}|Y_t)}\mathrm{d}\alpha_{t+1} \approx \frac{1}{n}\sum_{j=1}^{n}\frac{P(\alpha_{j,t+1|T}|\alpha_t)}{P(\alpha_{j,t+1|T}|Y_t)}.$$

Moreover, $P(\alpha_{t+1}|Y_t)$, which is in the denominator of equation (7.10), is rewritten as:

$$\begin{aligned} P(\alpha_{t+1}|Y_t) &= \int P(\alpha_{t+1}|\alpha_t)P(\alpha_t|Y_t)\mathrm{d}\alpha_t \\ &\approx \frac{1}{n}\sum_{m=1}^{n} P(\alpha_{t+1}|\alpha_{m,t|t}). \end{aligned}$$

Accordingly, the smoothing density (7.10) is approximated as:

$$\begin{aligned} P(\alpha_t|Y_T) &= P(\alpha_t|Y_t)\int \frac{P(\alpha_{t+1}|Y_T)P(\alpha_{t+1}|\alpha_t)}{P(\alpha_{t+1}|Y_t)}\mathrm{d}\alpha_{t+1} \\ &\approx \frac{1}{P(y_t|Y_{t-1})}\left(\frac{1}{n}\sum_{i=1}^{n} P(y_t|\alpha_t)P(\alpha_t|\alpha_{i,t-1|t-1})\right) \\ &\qquad \times \left(\frac{1}{n}\sum_{j=1}^{n}\frac{P(\alpha_{j,t+1|T}|\alpha_t)}{P(\alpha_{j,t+1|T}|Y_t)}\right) \end{aligned}$$

$$
\begin{aligned}
&\propto \left(\frac{1}{n}\sum_{i=1}^{n} P(y_t|\alpha_t)P(\alpha_t|\alpha_{i,t-1|t-1})\right)\left(\frac{1}{n}\sum_{j=1}^{n}\frac{P(\alpha_{j,t+1|T}|\alpha_t)}{P(\alpha_{j,t+1|T}|Y_t)}\right) \\
&\propto \frac{1}{n}\sum_{i=1}^{n}\sum_{j=1}^{n}\Bigg(\frac{1}{P(\alpha_{j,t+1|T}|Y_t)} \\
&\qquad \times P(y_t|\alpha_t)P(\alpha_{j,t+1|T}|\alpha_t)P(\alpha_t|\alpha_{i,t-1|t-1})\Bigg) \\
&\approx \frac{1}{n}\sum_{i=1}^{n}\sum_{j=1}^{n}\Bigg(\frac{1}{\frac{1}{n}\sum_{m=1}^{n} P(\alpha_{j,t+1|T}|\alpha_{m,t|t})} \\
&\qquad \times P(y_t|\alpha_t)P(\alpha_{j,t+1|T}|\alpha_t)P(\alpha_t|\alpha_{i,t-1|t-1})\Bigg) \\
&\propto \frac{1}{n}\sum_{i=1}^{n}\sum_{j=1}^{n}\frac{1}{q_{j,t}}P(y_t|\alpha_t)P(\alpha_{j,t+1|T}|\alpha_t)P(\alpha_t|\alpha_{i,t-1|t-1}) \\
&\propto \frac{1}{n}\sum_{i=1}^{n}\sum_{j=1}^{n}\frac{1}{q_{j,t}}\omega_1(\alpha_t;y_t)\omega_2(\alpha_t;\alpha_{j,t+1|T})P(\alpha_t|\alpha_{i,t-1|t-1}),
\end{aligned}
\tag{7.21}
$$

for $t = T-1, T-2, \cdots, 1$, where $q_{j,t}$ satisfies the following two conditions:

$$
\begin{aligned}
\text{(i)}\quad q_{j,t} &\propto P(\alpha_{j,t+1|T}|Y_t) \\
&= \int P(\alpha_{j,t+1|T}|\alpha_t)P(\alpha_t|Y_t)\mathrm{d}\alpha_t \\
&= \frac{1}{n}\sum_{m=1}^{n} P(\alpha_{j,t+1|T}|\alpha_{m,t|t}),
\end{aligned}
$$

$$
\text{(ii)}\quad \sum_{j=1}^{n}\frac{1}{q_{j,t}} = 1.
$$

Also, $\omega_1(\alpha_t;y_t)$ and $\omega_2(\alpha_t;\alpha_{j,t+1|T})$ satisfy:

$$
\begin{aligned}
&\omega_1(\alpha_t;y_t) \propto P(y_t|\alpha_t), \\
&\omega_2(\alpha_t;\alpha_{j,t+1|T}) \propto P(\alpha_{j,t+1|T}|\alpha_t), \\
&0 \le \omega_1(\alpha_t;y_t)\omega_2(\alpha_t;\alpha_{j,t+1|T}) \le 1.
\end{aligned}
$$

In equation (7.21), $\omega_1(\alpha_t;y_t)\omega_2(\alpha_t;\alpha_{j,t+1|T})$ is represented as the acceptance probability.

Thus, given $\alpha_{i,t-1|t-1}$ and $\alpha_{j,t+1|T}$, the smoothing density $P(\alpha_t|Y_T)$ is proportional to:

$$
\frac{1}{n}\sum_{i=1}^{n}\sum_{j=1}^{n}\frac{1}{q_{j,t}}\omega_1(\alpha_t;y_t)\omega_2(\alpha_t;\alpha_{j,t+1|T})P(\alpha_t|\alpha_{i,t-1|t-1}).
$$

When we cannot explicitly obtain the functional form of the posterior density (i.e., smoothing density $P(\alpha_t|Y_T)$), we may utilize rejection sampling to obtain a random observation from the smoothing density $P(\alpha_t|Y_T)$ for $t = T-1, T-2, \cdots, 1$, which is a backward recursive algorithm. Therefore, based on $\omega_1(\alpha_t; y_t)\omega_2(\alpha_t; \alpha_{j,t+1|T})P(\alpha_t|\alpha_{i,t-1|t-1})$, we consider generating the random numbers of α_t from $P(\alpha_t|Y_T)$. That is, for $t = T-1, T-2, \cdots, 1$, we choose $\alpha_{i,t-1|t-1}$ with probability $1/n$ and $\alpha_{j,t+1|T}$ with probability $1/q_{j,t}$ (i.e, we choose i and j randomly), sample the $P(\alpha_t|\alpha_{i,t-1|t-1})$ random variable and accept it with probability $\omega_1(\alpha_t; y_t)\omega_2(\alpha_t; \alpha_{j,t+1|T})$, which indicates the acceptance probability in the case of the smoothing algorithm (recall that the acceptance probability in the filtering algorithm is given by $\omega_1(\alpha_t; y_t)$). Thus, the random draw of α_t given information Y_T can be generated. Note that $\alpha_{i,T|T}$ is the random draw from the filtering density.

Summarizing the procedure, we have the following computational procedure.

(i) Choose one out of n random draws generated from $P(\alpha_{t-1}|Y_{t-1})$ randomly and one out of n random draws generated from $P(\alpha_{t+1}|Y_T)$ randomly, respectively i.e., choose one of $\alpha_{i,t-1|t-1}$, $i = 1, \cdots, n$, with probability $1/n$ and one of $\alpha_{j,t+1|T}$, $j = 1, \cdots, n$, with probability $1/q_{j,t}$.

(ii) Generating a random draw of η_t (i.e., $\eta_{m,t}$), compute $\alpha_{m,t}$ from the transition equation:

$$\alpha_{m,t} = g_t(\alpha_{i,t-1|t-1}, \eta_{m,t}),$$

for m and fixed i (i.e., i is chosen in Step (i)).

(iii) Compute the acceptance probability $\omega_1(\alpha_{m,t}; y_t)\omega_2(\alpha_{m,t}; \alpha_{j,t+1|T})$ for fixed j (i.e., j is chosen in Step (i)).
   (a) If $\alpha_{m,t}$ is accepted with probability $\omega_1(\alpha_{m,t}; y_t)\omega_2(\alpha_{m,t}; \alpha_{j,t+1|T})$, take $\alpha_{m,t}$ as a random draw of α_t from $P(\alpha_t|Y_T)$, which is denoted by $\alpha_{j,t|T}$.
   (b) If $\alpha_{m,t}$ is not accepted with probability $\omega_1(\alpha_{m,t}; y_t)\omega_2(\alpha_{m,t}; \alpha_{j,t+1|T})$, go back to Step (ii) and generate another random draw of η_t.

(iv) Repeat Steps (ii) and (iii) until accepted.

(v) Repeat Steps (i) – (iv) n times, because we need to generate n random draws of α_t from $P(\alpha_t|Y_T)$, i.e., $\alpha_{j,t|T}$ for $j = 1, \cdots, n$.

Note that in Step (iii) a uniform random number (say, U) has to be generated to check whether $\alpha_{m,t}$ is accepted or not. That is, $\alpha_{m,t}$ is accepted if $U < \omega_1(\alpha_{m,t}; y_t)\omega_2(\alpha_{m,t}; \alpha_{j,t+1|T})$ and it is rejected otherwise.

Thus, the random draws $\alpha_{i,t|T}$ for $i = 1, \cdots, n$ and $t = 1, \cdots, n$ are generated by the backward recursive algorithm. Let $a_{t|T}$ and $\Sigma_{t|T}$ be the smoothing mean and variance of the state-variable α_t.

$$a_{t|T} = \int \alpha_t P(\alpha_t|\alpha_T) \mathrm{d}\alpha_T$$
$$\approx \frac{1}{n} \sum_{i=1}^{n} \alpha_{i,t|T},$$

$$\Sigma_{t|T} = \int (\alpha_t - a_{t|T})(\alpha_t - a_{t|T})' P(\alpha_t|\alpha_T) \mathrm{d}\alpha_T$$
$$\approx \frac{1}{n} \sum_{i=1}^{n} (\alpha_{i,t|T} - a_{t|T})(\alpha_{i,t|T} - a_{t|T})',$$

which simply corresponds to the arithmetic mean and variance.

Summarizing the prediction, filtering and smoothing algorithms using rejection sampling, we have the followings:

- For the L-step ahead prediction density,

$$P(\alpha_{t+L}|Y_t) = \frac{1}{n} \sum_{i=1}^{n} P(\alpha_{t+L}|\alpha_{i,t+L-1|t}),$$

- For the filtering density,

$$P(\alpha_t|Y_t) \propto \frac{1}{n} \sum_{i=1}^{n} \omega_1(\alpha_t; y_t) P(\alpha_t|\alpha_{i,t-1|t-1}),$$

- For the smoothing density,

$$P(\alpha_t|Y_T) \propto \frac{1}{n} \sum_{i=1}^{n} \sum_{j=1}^{n} \frac{1}{q_{j,t}} \omega_1(\alpha_t; y_t) \omega_2(\alpha_t; \alpha_{j,t+1|T}) P(\alpha_t|\alpha_{i,t-1|t-1}),$$

where $\omega_1(\alpha_t; y_t)$, $\omega_2(\alpha_t; \alpha_{j,t+1|T})$ and $q_{j,t}$ are explicitly obtained from the density functions $P(y_t|\alpha_t)$, $P(\alpha_{j,t+1|T}|\alpha_t)$ and $P(\alpha_{j,t+1|T}|Y_t)$, respectively. $\alpha_{i,t-1|t-1}$ for $i = 1, \cdots, n$ denote the random draws generated from the filtering density at time $t-1$, i.e., $P(\alpha_{t-1}|Y_{t-1})$ and $\alpha_{j,t+1|T}$ for $j = 1, \cdots, n$ represent the random draws from the smoothing density at time $t+1$, i.e., $P(\alpha_{t+1}|Y_T)$. For each algorithm (i) – (iii), random draws are generated as follows.

For (i), suppose that $\alpha_{i,t+L-1|t}$ for $i = 1, \cdots, n$ are available. Pick up one out of the n random draws $\alpha_{i,t+L-1|t}$, $i = 1, \cdots, n$, randomly with probability $1/n$ and generate a random draw of α_{t+L} from $P(\alpha_{t+L}|\alpha_{i,t+L-1|t})$. Repeat the procedure n times and we have n random draws of α_{t+L}, which are denoted by $\alpha_{j,t+L|t}$, $j = 1, \cdots, n$. Thus, $\alpha_{j,t+L|t}$, $j = 1, \cdots, n$, are generated from $\alpha_{i,t+L-1|t}$, $i = 1, \cdots, n$, for $L = 1, 2, \cdots$. Note that the initial random draws of the prediction algorithm (i.e., $\alpha_{i,t|t}$, $i = 1, \cdots, n$) are obtained from the filtering algorithm.

For (ii), suppose that $\alpha_{i,t-1|t-1}$ for $i = 1, \cdots, n$ are available. Pick up one out of the n random draws $\alpha_{i,t-1|t-1}$, $i = 1, \cdots, n$, randomly with probability $1/n$, generate a random draw of α_t from $P(\alpha_t|\alpha_{i,t-1|t-1})$, and accept it with probability $\omega_1(\alpha_t; y_t)$. Repeat the procedure n times and we have n random draws of α_t, which are denoted by $\alpha_{j,t|t}$, $j = 1, \cdots, n$. Thus, $\alpha_{j,t|t}$, $j = 1, \cdots, n$, are generated from $\alpha_{i,t-1|t-1}$, $i = 1, \cdots, n$, for $t = 1, \cdots, T$. Note that $\alpha_{i,0|0}$, $i = 1, \cdots, n$, are obtained from the initial density function $P(\alpha_0|Y_0) = P(\alpha_0)$.

For (iii), suppose that $\alpha_{j,t+1|T}$ for $j = 1, \cdots, n$ are available. Pick up one out of the n random draws $\alpha_{i,t-1|t-1}$, $i = 1, \cdots, n$, randomly with probability $1/n$ and one out of the n random draws $\alpha_{j,t+1|T}$, $j = 1, \cdots, n$, with probability $1/q_{j,t}$, generate a random draw of α_t from $P(\alpha_t|\alpha_{i,t-1|t-1})$, and accept it with probability $\omega_1(\alpha_t; y_t)\omega_2(\alpha_t; \alpha_{j,t+1|T})$. Repeat the procedure n times and we have n random draws of α_t, which are denoted by $\alpha_{m,t|T}$, $m = 1, \cdots, n$. Thus, $\alpha_{m,t|T}$, $m = 1, \cdots, n$, are generated from $\alpha_{j,t+1|T}$, $j = 1, \cdots, n$, for $t = T-1, T-2, \cdots, 1$. Note that $\alpha_{i,t-1|t-1}$, $i = 1, \cdots, n$ and $t = 1, \cdots, T$, are obtained from the filtering algorithm.

Thus, for the rejection sampling procedure, first, pick up $\alpha_{i,t-1|t-1}$ and $\alpha_{j,t+1|T}$, respectively. The distribution which we want to sample is dominated by the $P(\alpha_t|\alpha_{i,t-1|t-1})$ density, which is derived from the transition equation. Therefore, we may use rejection sampling to obtain a random observation from $P(\alpha_t|Y_t)$ or $P(\alpha_t|Y_T)$. That is, we choose i with probability $1/n$ and j with probability $1/q_{j,t}$, sample a $P(\alpha_t|\alpha_{i,t-1|t-1})$ random variable and accept it with probability $\omega_1(\alpha_t; y_t)$ or $\omega_1(\alpha_t; y_t)\omega_2(\alpha_t; \alpha_{j,t+1|T})$. We can recursively generate the random numbers for prediction, filtering and smoothing.

Note as follows. In equation (7.21), we require an extra computational burden to compute $q_{j,t}$. Therefore, to reduce a computational disadvantage, we approximately use $q_{j,t} = n$ for all j. Then, the posterior density $P(\alpha_t|Y_T)$ shown in equation (7.21) is approximated by:

$$P(\alpha_t|Y_T) \propto \frac{1}{n^2}\sum_{i=1}^{n}\sum_{j=1}^{n}\omega_1(\alpha_t; y_t)\omega_2(\alpha_t; \alpha_{j,t+1|T})P(\alpha_t|\alpha_{i,t-1|t-1}).$$

Thus, we generate the random numbers of α_t, given $\alpha_{i,t-1|t-1}$ and $\alpha_{j,t+1|T}$, where i and j are chosen with probability $1/n$ and probability $1/n$, respectively.

Until now, we have shown a general solution to the smoothing algorithm. We sometimes have the case which does not need rejection sampling. Rejection sampling is often criticized from the computational point of view. If we do not have to use rejection sampling, more efficient estimator can be obtained in the sense of computational burden. We show some examples to implement the rejection sampling procedure.

Example 1 (Linear Model). We consider a simple linear and normal case. Both the measurement and transition equations are linear and the error terms are assumed to be normal. The state-space model in this case is represented as:

$$\begin{aligned}
&\text{(Measurement equation)} \qquad y_t = Z_t\alpha_t + \epsilon_t,\\
&\text{(Transition equation)} \qquad \alpha_t = T_t\alpha_{t-1} + \eta_t,
\end{aligned} \tag{7.22}$$

$$\begin{pmatrix} \epsilon_t \\ \eta_t \end{pmatrix} \sim N\left(\begin{pmatrix} 0 \\ 0 \end{pmatrix}, \begin{pmatrix} H_t & 0 \\ 0 & Q_t \end{pmatrix} \right),$$

where Z_t, T_t, H_t and Q_t are assumed to be known.

In general, from the Bayes' formula, we can obtain the posterior density $P(\alpha_t|y_t,\alpha_{t-1},\alpha_{t+1})$ from the density functions $P(y_t|\alpha_t)$, $P(\alpha_t|\alpha_{t-1})$ and $P(\alpha_{t+1}|\alpha_t)$ as follows:

$$P(\alpha_t|y_t,\alpha_{t-1},\alpha_{t+1}) = \frac{P(y_t|\alpha_t)P(\alpha_t|\alpha_{t-1})P(\alpha_{t+1}|\alpha_t)}{\displaystyle\int P(y_t|\alpha_t)P(\alpha_t|\alpha_{t-1})P(\alpha_{t+1}|\alpha_t)\mathrm{d}\alpha_t},$$

which implies that we have the following proportional relationship:

$$P(y_t|\alpha_t)P(\alpha_t|\alpha_{t-1})P(\alpha_{t+1}|\alpha_t) \propto P(\alpha_t|y_t,\alpha_{t-1},\alpha_{t+1}).$$

Furthermore, in the case of the system (7.22), $P(\alpha_t|y_t,\alpha_{t-1},\alpha_{t+1})$ is represented by:

$$P(\alpha_t|y_t,\alpha_{t-1},\alpha_{t+1}) = \Phi(\alpha_t - \Sigma_t^*\mu_t^*, \Sigma_t^*),$$

where $\Phi(\alpha_t - \Sigma_t^*\mu_t^*, \Sigma_t^*)$ denotes the normal density with mean $\Sigma_t^*\mu_t^*$ and variance Σ_t^*. Let us define μ_t^* and Σ_t^* as:

$$\begin{aligned}
\mu_t^* &= Q_t^{-1}T_t\alpha_{t-1} + Z_t'H_t^{-1}y_t + \alpha_{t+1}Q_{t+1}^{-1}T_{t+1},\\
\Sigma_t^{*-1} &= Z_t'H_t^{-1}Z_t + Q_t^{-1} + T_{t+1}'Q_{t+1}^{-1}T_{t+1}.
\end{aligned}$$

See Carlin, Polson and Stoffer (1992) for the above manipulation. Therefore, the smoothing density $P(\alpha_t|Y_T)$ is rewritten as:

$$\begin{aligned}
&P(\alpha_t|Y_T)\\
&= P(\alpha_t|Y_t)\int \frac{P(\alpha_{t+1}|Y_T)P(\alpha_{t+1}|\alpha_t)}{P(\alpha_{t+1}|Y_t)}\mathrm{d}\alpha_{t+1}\\
&= \int \frac{P(\alpha_t|Y_t)P(\alpha_{t+1}|Y_T)P(\alpha_{t+1}|\alpha_t)}{P(\alpha_{t+1}|Y_t)}\mathrm{d}\alpha_{t+1}\\
&\propto \int \frac{P(y_t|\alpha_t)P(\alpha_t|Y_{t-1})P(\alpha_{t+1}|\alpha_t)P(\alpha_{t+1}|Y_T)}{P(\alpha_{t+1}|Y_t)}\mathrm{d}\alpha_{t+1}\\
&= \int\!\!\int \frac{1}{P(\alpha_{t+1}|Y_t)} P(y_t|\alpha_t)\,P(\alpha_t|\alpha_{t-1})P(\alpha_{t+1}|\alpha_t)\\
&\qquad\qquad \times P(\alpha_{t-1}|Y_{t-1})P(\alpha_{t+1}|Y_T)\mathrm{d}\alpha_{t-1}\mathrm{d}\alpha_{t+1}
\end{aligned}$$

$$\propto \iint \frac{1}{P(\alpha_{t+1}|Y_t)} \Phi(\alpha_t - \Sigma_t^* \mu_t^*, \Sigma_t^*) P(\alpha_{t-1}|Y_{t-1}) \times P(\alpha_{t+1}|Y_T) \mathrm{d}\alpha_{t-1} \mathrm{d}\alpha_{t+1}$$

$$= \iint \frac{1}{\int P(\alpha_{t+1}|\alpha_t) P(\alpha_t|Y_t) \mathrm{d}\alpha_t} \Phi(\alpha_t - \Sigma_t^* \mu_t^*, \Sigma_t^*) P(\alpha_{t-1}|Y_{t-1}) \times P(\alpha_{t+1}|Y_T) \mathrm{d}\alpha_{t-1} \mathrm{d}\alpha_{t+1}$$

$$\approx \iint \frac{1}{\frac{1}{n}\sum_{m=1}^{n} P(\alpha_{t+1}|\alpha_{m,t|t})} \Phi(\alpha_t - \Sigma_t^* \mu_t^*, \Sigma_t^*) P(\alpha_{t-1}|Y_{t-1}) \times P(\alpha_{t+1}|Y_T) \mathrm{d}\alpha_{t-1} \mathrm{d}\alpha_{t+1}$$

$$\approx \frac{1}{n} \sum_{i=1}^{n} \sum_{j=1}^{n} \frac{1}{\sum_{m=1}^{n} P(\alpha_{j,t+1|T}|\alpha_{m,t|t})} \Phi(\alpha_t - \Sigma_t^* \mu_{ij,t}^*, \Sigma_t^*)$$

$$\propto \frac{1}{n} \sum_{i=1}^{n} \sum_{j=1}^{n} \frac{1}{q_{j,t}} \Phi(\alpha_t - \Sigma_t^* \mu_{ij,t}^*, \Sigma_t^*),$$

where $\mu_{ij,t}^*$ is given by:

$$\mu_{ij,t}^* = Q_t^{-1} T_t \alpha_{i,t-1|t-1} + Z_t' H_t^{-1} y_t + \alpha_{j,t+1|T} Q_{t+1}^{-1} T_{t+1}.$$

Accordingly, for the rejection sampling smoothing algorithm in this case, from $P(y_t|\alpha_t)$, $P(\alpha_t|\alpha_{t-1})$ and $P(\alpha_{t+1}|\alpha_t)$, the posterior density $P(\alpha_t|Y_T)$ is represented as average of the normal densities with mean $\Sigma_t^* \mu_{ij,t}^*$ and variance Σ_t^*, i.e.,

$$\frac{1}{n} \sum_{i=1}^{n} \sum_{j=1}^{n} \frac{1}{q_{j,t}} \Phi(\alpha_t - \Sigma_t^* \mu_{ij,t}^*, \Sigma_t^*),$$

where $q_{j,t}$ satisfies the following two conditions:

(i) $$q_{j,t} \propto P(\alpha_{j,t+1|T}|Y_t) = \int P(\alpha_{j,t+1|T}|\alpha_t) P(\alpha_t|Y_t) \mathrm{d}\alpha_t = \frac{1}{n} \sum_{m=1}^{n} P(\alpha_{j,t+1|T}|\alpha_{m,t|t}),$$

(ii) $$\sum_{j=1}^{n} \frac{1}{q_{j,t}} = 1.$$

$\alpha_{i,t-1|t-1}$ and $\alpha_{j,t+1|T}$ are picked up with probability $1/n$ and probability $1/q_{j,t}$, respectively. Thus, the random number of α_t given Y_T is recursively obtained.

Example 2 (Nonlinear Model). In the case where the transition equation is nonlinear and the measurement equation is linear, we have the following state-space model:

$$
\begin{aligned}
&\text{(Measurement equation)} \qquad && y_t = Z_t \alpha_t + \epsilon_t, \\
&\text{(Transition equation)} \qquad && \alpha_t = g_{1,t}(\alpha_{t-1}) + \eta_t,
\end{aligned} \tag{7.23}
$$

$$
\begin{pmatrix} \epsilon_t \\ \eta_t \end{pmatrix} \sim N\left(\begin{pmatrix} 0 \\ 0 \end{pmatrix}, \begin{pmatrix} H_t & 0 \\ 0 & Q_t \end{pmatrix} \right),
$$

where Z_t, $g_{1,t}(\cdot)$, H_t and Q_t are assumed to be known.

First, note as follows. From $P(y_t|\alpha_t)$ and $P(\alpha_t|\alpha_{t-1})$, the posterior density $P(\alpha_t|y_t, \alpha_{t-1})$ is obtained, i.e.,

$$
P(\alpha_t|y_t, \alpha_{t-1}) = \frac{P(y_t|\alpha_t)P(\alpha_t|\alpha_{t-1})}{\displaystyle\int P(y_t|\alpha_t)P(\alpha_t|\alpha_{t-1})\mathrm{d}\alpha_t},
$$

and accordingly, we have the following proportional relationship:

$$
P(y_t|\alpha_t)P(\alpha_t|\alpha_{t-1}) \propto P(\alpha_t|y_t, \alpha_{t-1}).
$$

Taking into account linearity and normality assumptions, the posterior density $P(\alpha_t|y_t, \alpha_{t-1})$ is computed as the following normal density:

$$
P(\alpha_t|y_t, \alpha_{t-1}) = \Phi(\alpha_t - \Sigma_t \mu_t, \Sigma_t),
$$

where μ_t and Σ_t are given by:

$$
\begin{aligned}
\mu_t &= Q_t^{-1} g_{1,t}(\alpha_{t-1}) + Z_t' H_t^{-1} h_{1,t}(y_t), \\
\Sigma_t^{-1} &= Z_t' H_t^{-1} Z_t + Q_t^{-1}.
\end{aligned}
$$

Utilizing the above equation, the smoothing algorithm (7.10) are rewritten as:

$$
\begin{aligned}
&P(\alpha_t|Y_T) \\
&= P(\alpha_t|Y_t) \int \frac{P(\alpha_{t+1}|Y_T)P(\alpha_{t+1}|\alpha_t)}{P(\alpha_{t+1}|Y_t)} \mathrm{d}\alpha_{t+1} \\
&= \iint \frac{1}{P(\alpha_{t+1}|Y_t)} P(y_t|\alpha_t)\, P(\alpha_t|\alpha_{t-1}) P(\alpha_{t+1}|\alpha_t) \\
&\qquad\qquad \times P(\alpha_{t-1}|Y_{t-1}) P(\alpha_{t+1}|Y_T) \mathrm{d}\alpha_{t-1} \mathrm{d}\alpha_{t+1} \\
&\propto \iint \frac{1}{P(\alpha_{t+1}|Y_t)} \Phi(\alpha_t - \Sigma_t\mu_t, \Sigma_t) P(\alpha_{t+1}|\alpha_t) \\
&\qquad\qquad \times P(\alpha_{t-1}|Y_{t-1}) P(\alpha_{t+1}|Y_T) \mathrm{d}\alpha_{t-1} \mathrm{d}\alpha_{t+1} \\
&\approx \iint \frac{1}{\displaystyle\sum_{m=1}^{n} P(\alpha_{t+1}|\alpha_{m,t|t})} \Phi(\alpha_t - \Sigma_t\mu_t, \Sigma_t) P(\alpha_{t+1}|\alpha_t) \\
&\qquad\qquad \times P(\alpha_{t-1}|Y_{t-1}) P(\alpha_{t+1}|Y_T) \mathrm{d}\alpha_{t-1} \mathrm{d}\alpha_{t+1}
\end{aligned}
$$

$$\approx \frac{1}{n}\sum_{i=1}^{n}\sum_{j=1}^{n} \frac{1}{\sum_{m=1}^{n} P(\alpha_{j,t+1|T}|\alpha_{m,t|t})} \Phi(\alpha_t - \Sigma_t\mu_{i,t}, \Sigma_t) P(\alpha_{j,t+1|T}|\alpha_t)$$

$$\propto \frac{1}{n}\sum_{i=1}^{n}\sum_{j=1}^{n} \frac{1}{q_{j,t}} \Phi(\alpha_t - \Sigma_t\mu_{i,t}, \Sigma_t)\omega_2(\alpha_t;\alpha_{j,t+1|T}),$$

where $\mu_{i,t}$ denotes:

$$\mu_{i,t} = Q_t^{-1} g_{1,t}(\alpha_{i,t-1|t-1}) + Z_t' H_t^{-1} h_{1,t}(y_t),$$

and $\omega_2(\alpha_t;\alpha_{j,t+1|T})$ is called the acceptance probability, which is taken as:

$$\omega_2(\alpha_t;\alpha_{j,t+1|T})$$
$$= \exp\left(-\frac{1}{2}\big(\alpha_{j,t+1|T} - g_{1,t+1}(\alpha_t)\big)' Q_{t+1}^{-1}\big(\alpha_{j,t+1|T} - g_{1,t+1}(\alpha_t)\big)\right).$$

Note that the acceptance probability $\omega_2(\alpha_t;\alpha_{j,t+1|T})$ is equal to the exponential part of the normal density.

For the smoothing algorithm with the system (7.23), from the three distribution functions $P(y_t|\alpha_t)$, $P(\alpha_t|\alpha_{t-1})$ and $P(\alpha_{t+1}|\alpha_t)$, the posterior density $P(\alpha_t|Y_T)$ shown in equation (7.21) cannot be obtained explicitly. Therefore, the random numbers from the posterior density $P(\alpha_t|Y_T)$ are generated utilizing rejection sampling. We pick up one out of the n random draws $\alpha_{i,t-1|t-1}$, $i = 1, \cdots, n$, with probability $1/n$ and one out of the n random draws $\alpha_{j,t+1|T}$, $j = 1, \cdots, n$, with probability $1/q_{j,t}$. Then, we generate a random draw of α_t from $\Phi(\alpha_t - \Sigma_t\mu_{i,t}, \Sigma_t)$ and accept it with probability $\omega_2(\alpha_t;\alpha_{j,t+1|T})$.

Example 3 (Nonlinear Model). In the case where the transition equation is linear and the measurement equation is nonlinear, we have the following state-space model:

$$\begin{aligned} &\text{(Measurement equation)} \qquad && y_t = h_{1,t}(\alpha_t) + \epsilon_t, \\ &\text{(Transition equation)} && \alpha_t = T_t\alpha_{t-1} + \eta_t, \end{aligned} \tag{7.24}$$

$$\begin{pmatrix} \epsilon_t \\ \eta_t \end{pmatrix} \sim N\left(\begin{pmatrix} 0 \\ 0 \end{pmatrix}, \begin{pmatrix} H_t & 0 \\ 0 & Q_t \end{pmatrix} \right),$$

where $h_{1,t}$, $T_t(\cdot)$, H_t and Q_t are assumed to be known.

From $P(\alpha_t|\alpha_{t-1})$ and $P(\alpha_{t+1}|\alpha_t)$, the posterior density $P(\alpha_t|\alpha_{t-1}, \alpha_{t+1})$ is represented as:

$$P(\alpha_t|\alpha_{t-1},\alpha_{t+1}) = \frac{P(\alpha_t|\alpha_{t-1})P(\alpha_{t+1}|\alpha_t)}{\int P(\alpha_t|\alpha_{t-1})P(\alpha_{t+1}|\alpha_t)\mathrm{d}\alpha_t},$$

and accordingly, we have the following proportional relationship:

$$P(\alpha_t|\alpha_{t-1})P(\alpha_{t+1}|\alpha_t) \propto P(\alpha_t|\alpha_{t-1},\alpha_{t+1}).$$

In the case of the state-space model given by (7.24), the posterior density $P(\alpha_t|\alpha_{t-1},\alpha_{t+1})$ is obtained as:

$$P(\alpha_t|\alpha_{t-1},\alpha_{t+1}) = \Phi(\alpha_t - \Sigma_t\mu_t^*, \Sigma_t^*),$$

where μ_t and Σ_t are given by:

$$\mu_t^* = Q_t^{-1}T_t\alpha_{t-1} + \alpha_{t+1}Q_{t+1}^{-1}T_{t+1},$$
$$\Sigma_t^{*-1} = Q_t^{-1} + T_{t+1}'Q_{t+1}^{-1}T_{t+1}.$$

Utilizing the above equation, the smoothing algorithm (7.10) are rewritten as:

$$\begin{aligned}
&P(\alpha_t|Y_T)\\
&= P(\alpha_t|Y_t)\int \frac{P(\alpha_{t+1}|Y_T)P(\alpha_{t+1}|\alpha_t)}{P(\alpha_{t+1}|Y_t)}\mathrm{d}\alpha_{t+1}\\
&= \iint \frac{1}{P(\alpha_{t+1}|Y_t)} P(y_t|\alpha_t)P(\alpha_t|\alpha_{t-1})P(\alpha_{t+1}|\alpha_t)\\
&\qquad\qquad \times P(\alpha_{t-1}|Y_{t-1})P(\alpha_{t+1}|Y_T)\mathrm{d}\alpha_{t-1}\mathrm{d}\alpha_{t+1}\\
&\propto \iint \frac{1}{P(\alpha_{t+1}|Y_t)} P(y_t|\alpha_t)\Phi(\alpha_t - \Sigma_t\mu_t^*, \Sigma_t^*)\\
&\qquad\qquad \times P(\alpha_{t-1}|Y_{t-1})P(\alpha_{t+1}|Y_T)\mathrm{d}\alpha_{t-1}\mathrm{d}\alpha_{t+1}\\
&\approx \iint \frac{1}{\sum_{m=1}^{n} P(\alpha_{t+1}|\alpha_{m,t|t})} P(y_t|\alpha_t)\Phi(\alpha_t - \Sigma_t\mu_t^*, \Sigma_t^*)\\
&\qquad\qquad \times P(\alpha_{t-1}|Y_{t-1})P(\alpha_{t+1}|Y_T)\mathrm{d}\alpha_{t-1}\mathrm{d}\alpha_{t+1}\\
&\approx \frac{1}{n}\sum_{i=1}^{n}\sum_{j=1}^{n} \frac{1}{\sum_{m=1}^{n} P(\alpha_{j,t+1|T}|\alpha_{m,t|t})} P(y_t|\alpha_t)\Phi(\alpha_t - \Sigma_t\mu_{ij,t}^*, \Sigma_t^*)\\
&\propto \frac{1}{n}\sum_{i=1}^{n}\sum_{j=1}^{n} \frac{1}{q_{j,t}}\omega_1(\alpha_t;y_t)\Phi(\alpha_t - \Sigma_t\mu_{ij,t}^*, \Sigma_t^*),
\end{aligned}$$

where $\mu_{ij,t}^*$ denotes:

$$\mu_{ij,t} = Q_t^{-1}T_t\alpha_{i,t-1|t-1} + \alpha_{j,t+1|T}Q_{t+1}^{-1}T_{t+1},$$

and $\omega_1(\alpha_t;y_t)$ is called the acceptance probability, which is taken as:

$$\omega_1(\alpha_t;y_t) = \exp\left(-\frac{1}{2}\big(y_t - h_{1,t}(\alpha_t)\big)'H_t^{-1}\big(y_t - h_{1,t}(\alpha_t)\big)\right).$$

Thus, for the smoothing algorithm in this case, from $P(y_t|\alpha_t)$, $P(\alpha_t|\alpha_{t-1})$ and $P(\alpha_{t+1}|\alpha_t)$, the posterior density $P(\alpha_t|y_t,\alpha_{t-1},\alpha_{t+1})$ cannot be obtained explicitly. Therefore, the random numbers from the posterior density (7.21) are generated utilizing rejection sampling. We generate a random draw from the normal density with mean $\Sigma_t^* \mu_t^*$ and variance Σ_t^*, i.e., $\Phi(\alpha_t - \Sigma_t^* \mu_t^*, \Sigma_t^*)$. Picking up one out of the n random draws $\alpha_{i,t-1|t-1}$, $i = 1, \cdots, n$, with probability $1/n$ and one out of the n random draws $\alpha_{j,t+1|T}$, $j = 1, \cdots, n$, with probability $1/q_{j,t}$, a random number of α_t given Y_T is obtained from $\Phi(\alpha_t - \Sigma_t^* \mu_{ij,t}^*, \Sigma_t^*)$. Then, we accept the normal random draw with probability $\omega_1(\alpha_t; y_t)$.

Example 4 (Nonlinear Model). In the case where both the transition equation and the measurement equation are nonlinear, we have the following state-space model:

$$
\begin{aligned}
&\text{(Measurement equation)} \qquad && y_t = h_{1,t}(\alpha_t) + \epsilon_t, \\
&\text{(Transition equation)} \qquad && \alpha_t = g_{1,t}(\alpha_{t-1}) + \eta_t,
\end{aligned}
\tag{7.25}
$$

$$
\begin{pmatrix} \epsilon_t \\ \eta_t \end{pmatrix} \sim N\left(\begin{pmatrix} 0 \\ 0 \end{pmatrix}, \begin{pmatrix} H_t & 0 \\ 0 & Q_t \end{pmatrix} \right),
$$

where $h_{1,t}$, $g_{1,t}(\cdot)$, H_t and Q_t are assumed to be known. We can directly apply equation (7.10) to the state-space model given by (7.25). That is, the random numbers from the smoothing density are generated utilizing rejection sampling. Pick up one out of the n random draws $\alpha_{i,t-1|t-1}$, $i = 1, \cdots, n$, with probability $1/n$ and one out of the n random draws $\alpha_{j,t+1|T}$, $j = 1, \cdots, n$, with probability $1/q_{j,t}$, generate a random draw from $N\big(g_{1,t}(\alpha_{i,t-1|t-1}), Q_t\big)$, and accept it with probability $\omega_1(\alpha_t; y_t)\omega_2(\alpha_t; \alpha_{j,t+1|T})$, where $\omega_1(\alpha_t; y_t)$ and $\omega_2(\alpha_t; \alpha_{j,t+1|T})$ are given by:

$$
\omega_1(\alpha_t; y_t) = \exp\left(-\frac{1}{2} \big(y_t - h_{1,t}(\alpha_t)\big)' H_t^{-1} \big(y_t - h_{1,t}(\alpha_t)\big) \right),
$$

$$
\begin{aligned}
&\omega_2(\alpha_t; \alpha_{j,t+1|T}) \\
&= \exp\left(-\frac{1}{2} \big(\alpha_{j,t+1|T} - g_{1,t+1}(\alpha_t)\big)' Q_{t+1}^{-1} \big(\alpha_{j,t+1|T} - g_{1,t+1}(\alpha_t)\big) \right).
\end{aligned}
$$

Both $\omega_1(\alpha_t; y_t)$ and $\omega_2(\alpha_t; \alpha_{j,t+1|T})$ indicate the exponential part of normal distributions.

7.4 Summary

Thus, in this chapter, the filtering algorithm introduced in Chapter 4 has been extended to the prediction and smoothing algorithms for the numerical integration procedure (Sections 7.2.1 and 7.3.1), the importance sampling procedure (Sections 7.2.2 and 7.3.2), the density-based Monte-Carlo procedure (Sections 7.2.3 and 7.3.3), and the rejection sampling procedure (Sections 7.2.4 and 7.3.4). In the state-space model, using the current information, prediction implies estimating the state-variable at future period, filtering is to estimate the state-variable at current period, and smoothing implies obtaining the state-variable at past period. Thus, the smoothing estimate uses the extra information, compared with prediction and filtering. In addition, the one-step ahead prediction estimate is required to obtain the filtering estimate and in turn both the one-step ahead prediction estimate and the filtering estimate depend on the smoothing estimate. According to the linear recursive algorithms, from equations (2.9) – (2.15), $a_{t|t}$ and $\Sigma_{t|t}$ utilize $a_{t|t-1}$ and $\Sigma_{t|t-1}$, and moreover, from equations (7.11) – (7.13), $a_{t|T}$ and $\Sigma_{t|T}$ are functions of $a_{t+1|t}$, $\Sigma_{t+1|t}$, $a_{t|t}$ and $\Sigma_{t|t}$. As for the density-based recursive algorithms, from equations (2.16) and (2.17), $P(\alpha_t|Y_t)$ utilizes $P(\alpha_t|Y_{t-1})$, and moreover, from equation (7.10), $P(\alpha_t|Y_T)$ is a function of $P(\alpha_{t+1}|Y_t)$ and $P(\alpha_t|Y_t)$. Thus, smoothing has more information than prediction and filtering. Therefore, the smoothing estimate is more efficient than the prediction and filtering estimates in the sense of mean square error. However, the smoothing algorithm has the disadvantages that it is more complicated than the filtering algorithm and that it takes more computational time.

8. Summary and Concluding Remarks

This book is summarized as follows.

- In Chapters 3 and 4, several nonlinear and nonnormal filters are introduced and developed, i.e.,
  - the extended Kalman filter (EKF) in Section 3.2.1,
  - the second-order nonlinear filter (SNF) in Section 3.2.2,
  - the Monte-Carlo simulation filter (MSF) in Section 3.2.3,
  - the single-stage iteration filter (SIF) in Section 3.4,
  - the Gaussian sum filter (GSF) in Section 4.2,
  - the numerical integration filter (NIF) in Section 4.3,
  - the importance sampling filter (ISF) in Section 4.4,
  - the density-based Monte-Carlo filter (DMF) in Section 4.5 and
  - the rejection sampling filter (RSF) in Section 4.6.

  Thus, each nonlinear filter is described.
- It is analyzed what has to be approximated when applying the nonlinear measurement and transition equations approximated by the Taylor series expansions to the conventional linear recursive Kalman filter algorithm, which is discussed in Section 3.2.
- For the nonlinear filters based on the Taylor series expansions, it is shown in Section 3.3 that under a certain functional form of either the measurement equation or the transition equation there is no correlation between the error terms (i.e., residuals) in the linearized system represented by equations (3.7) and (3.8). Recall that one of the approximations in linearization of the nonlinear measurement and transition equations is to assume that the correlated error terms (i.e., residuals) are uncorrelated.
- A re-interpretation is given to the single-stage iteration filter (SIF) in Section 3.4.
- Comparing the nonlinear filters by the Monte-Carlo experiments, it is shown in Chapter 5 that the numerical integration filter (NIF) in Section 4.3, the importance sampling filter (ISF) in Section 4.4 and the rejection sampling filter (RSF) in Section 4.6 are better than the other estimators while the extended Kalman filter (EKF) in Section 3.2 and the Gaussian sum filter (GSF) in Section 4.2 are clearly biased.

- An estimation of permanent and transitory consumption is taken as an application to the nonlinear filters (Chapter 6). A ratio of permanent consumption relative to total consumption is estimated for each country, i.e., Japan, U.S., U.K., Spain, France and Italy. For almost all the countries, consumption depends on income, and therefore in this sense it is shown that the permanent income hypothesis does not hold in reality.
- In Chapter 7, the nonlinear and nonnormal filters discussed in Chapter 4 (i.e., the numerical integration filter (NIF), the importance sampling filter (ISF), the density-based Monte-Carlo filter (DMF) and the rejection sampling filter (RSF)) are extended to the L-step ahead prediction and the fixed-interval smoothing estimators.

These are mentioned in more detail as follows.

We have the extended Kalman filter (EKF), the second-order nonlinear filter (SNF) and the other higher-order nonlinear filters as the most heuristic traditional nonlinear filters. In Chapter 3, the three nonlinear filters are derived from the same theoretical framework (i.e., the extended Kalman filter (EKF), the second-order nonlinear filter (SNF) and the Monte-Carlo simulation filter (MSF)), where the nonlinear measurement and transition equations are approximated by the Taylor series expansions and the approximated state-space model is directly applied to the standard Kalman filter algorithm. Moreover, we have analyzed in Section 3.2 what is approximated when the traditional nonlinear filters are used in practice. There, we have seen that four approximations are imposed. For the first approximation, the expectations in the algorithm have to be computed by utilizing the Taylor series expansions. In approximating the nonlinear functions to evaluate the expectations, we have the approximation that the non-zero mean error terms are assumed to be zero mean errors, because an expectation of a nonlinear function of a random variable is not generally equivalent to a nonlinear function of an expectation of a random variable, i.e., $\mathrm{E}(g(x)) \neq g(\mathrm{E}(x))$ for a nonlinear function $g(\cdot)$ and a random variable x unless $g(x)$ is linear in x. Second, the state-vectors are assumed to be uncorrelated with the error terms for both the measurement and transition equations. Third, the error terms (i.e., residuals) in the approximated system are approximated to be uncorrelated with each other even if they are correlated. The fourth approximation is that the error terms (i.e., residuals) obtained by transforming the nonlinear measurement and transition equations are assumed to be normal. This approximation results in normality of the state-variables even if they are not normal.

According to the extended Kalman filter (EKF) and the second-order nonlinear filter (SNF), we have to expand the nonlinear measurement and transition equations with respect to the state-variables α_t and α_{t-1} and the error terms ϵ_t and η_t. we have introduced the filtering algorithm based on the Monte-Carlo stochastic simulations (i.e., the Monte-Carlo simulation filter (MSF)), where the random numbers are generated for the state-variables (i.e.,

α_t and α_{t-1}) and the error terms (i.e., ϵ_t and η_t) to evaluate the expectations more precisely. See Section 3.2.3 for the Monte-Carlo simulation filter (MSF).

Also, it is shown that, for the Taylor series expansion approach, there is no correlation between the error terms (i.e., residuals) in the case where either $h_t(\alpha_t, \epsilon_t)$ or $g_t(\alpha_{t-1}, \eta_t)$ can be written in a linear function with respect to the state-variable or the error term. Two theorems are obtained depending on a functional form of $h_t(\alpha_t, \epsilon_t)$ and $g_t(\alpha_{t-1}, \eta_t)$. This implies that we can obtain a better approximation if we take such a functional form for the measurement equation or the transition equation. See Section 3.3.

Moreover, the single-stage iteration filter (SIF) is derived from the mixed estimation method, i.e., the Goldberger-Theil estimator, which differs from an interpretation made in Wishner, Tabaczynski and Athans (1969). According to the mixed estimation method, the iterative procedure appears because of nonlinearity of the measurement equation, not that of the transition equation. There, the conventional nonlinear generalized least squares method is used for derivation. Thus, the single-stage iteration filter (SIF) should be distinguished from the other three nonlinear filters (i.e., EKF, SNF and MSF) which utilize the Taylor series expansions. See Section 3.4.

The disadvantage of the first three algorithms introduced in Chapter 3 (i.e., the extended Kalman filter (EKF), the second-order nonlinear filter (SNF) and the Monte-Carlo simulation filter (MSF)) is that we have to assume approximately normal distributions for the error terms (i.e., residuals), because normality assumption is required in the derivation procedures to obtain the linear recursive conventional Kalman filter algorithm. Otherwise, we cannot obtain the filtering estimates of the state-variables. This normality approximation of the error terms (i.e., residuals) results in the normal state-vector. From such an approximation, it is possible that the extended Kalman filter (EKF) and the second-order nonlinear filter (SNF) indicate the biased estimators. For the Monte-Carlo simulation filter (MSF), however, there is a smaller possibility of bias, because we generate the random numbers for the errors and the state-variables to evaluate the expectations of nonlinear functions more correctly. In addition, Kalman filter models based on the assumption of normal distributions are known to be non-robust (see Meinhold and Singpurwalla (1989)). This implies that when a discrepancy arises between the prior density and the observed data, the posterior distribution becomes unrealistic. Therefore, density approximation might be more appropriate than approximation of nonlinear functions.

On the other hand, for the single-stage iteration filter (SIF), we do not have to impose the normality assumption for the error terms because this filter is derived from the mixed estimation approach where we do not have to assume any distribution for the errors. In the derivation process, however, the transition equation has to be approximated by the first-order Taylor series expansion, the second-order Taylor series expansion or the Monte-Carlo stochastic simulations. The bias of the single-stage iteration filter (SIF) arises from approximation of the transition equation. Therefore, if the transition equation is approximated by the first- or the second-order Taylor series expansion, the filtering estimates obtained from the single-stage iteration filter (SIF) are also biased, but less biased than the extended Kalman filter (EKF) and the second-order nonlinear filter (SNF), because for the single-stage iteration filter (SIF) the conventional optimization procedure such as the Newton-Raphson method is taken when the filtering estimate of the state-vector is updated (see equation (3.45)).

When we utilize the Taylor series expansions for approximation of the nonlinear functions, finally, note that the general comments are given as follows. Approximation by the Taylor series expansions is broadly used in the case of nonlinear estimation. When approximating nonlinear functions by the Taylor series expansions, however, we need to recognize that the structure of the original function is completely different from that of the approximated function. Especially, the approximated error terms consist of two parts, i.e., the stochastic part and the deterministic part. According to the nonlinear filters based on the Taylor series expansions, we have to assume the approximated error terms to be normal, which implies that we must assume that both the stochastic and deterministic parts are normally distributed. Clearly, it is not appropriate to assume the deterministic part to be normal.

In Chapter 3, we have discussed about the traditional nonlinear filters, which are based on the Taylor series expansions. However, it is evident that the traditional nonlinear filters give us the biased estimators. Another approach to reduce the biased filtering estimates comes from approximating the probability density functions, i.e., $P(\alpha_t|Y_{t-1})$ and $P(\alpha_t|Y_t)$, which is developed in Chapter 4. The filtering algorithms derived from the underlying probability densities such as $P_\eta(\eta_t)$ and $P_\epsilon(\epsilon_t)$ (i.e., the probability density functions of the error terms ϵ_t and η_t) are more fundamental than those from the Taylor series expansions, and the density-based filtering estimates might be more precise in the sense that they give us asymptotically unbiased and efficient estimators, since we do not need to impose the assumption that the state-variable α_t is normally distributed. The conventional linear recursive Kalman filter algorithm is given by the first- and the second-moments of the underlying distributions, i.e., mean and variance, provided that the measurement and transition equations are linear and that normality is assumed for the error terms. In the nonlinear case, the estimated state-variables are biased when we linearize the nonlinear functions and apply them to the linear case. Hence, approximation of the densities, rather than that of the nonlinear functions, is essential to the nonlinear filtering problem. In Chapter 4, five nonlinear filters were introduced and developed, based on the density approximation, i.e., the Gaussian sum filter (GSF), the numerical integration filter (NIF), the importance sampling filter (ISF), the density-based Monte-Carlo filter (DMF) and the rejection sampling filter (RSF).

For the Gaussian sum filter (GSF), however, it is known that the filtering estimates might be less biased but still biased (see Anderson and Moore (1979)), because for each segment the nonlinear measurement and transition equations are linearized and the extended Kalman filters (EKF) are implemented. Therefore, the Gaussian sum filter (GSF) is the weighted sum of the extended Kalman filters (EKF). As mentioned above, the filtering estimates from the extended Kalman filter (EKF) are biased. The weighted sum of biased estimates are also biased. Therefore, since the Gaussian sum filter (GSF) approximates the densities $P(\alpha_t|Y_{t-1})$ and $P(\alpha_t|Y_t)$ as the weighted sum of normal distributions, the filtering estimates obtained from the Gaussian sum filter (GSF) are less biased than the extended Kalman filter (EKF) and the second-order nonlinear filter (SNF), but still biased. According to the Monte-Carlo studies in Chapter 5, it is observed from Tables 5.1, 5.2, 5.10, 5.11, 5.22 and 5.23 that the Gaussian sum filter (GSF) approaches the extended Kalman filter (EKF) as time t is large. The Gaussian sum filter (GSF) requires n initial values, and therefore it is equivalent to the weighted average of n extended Kalman filters with different initial values. For one of the reasons, Appendix A5.1 shows that the filtering estimates do not depend on the initial values for large t, which implies that the filtering estimates with the biased initial nodes approach the filtering estimates with the true initial nodes.

According to the numerical integration filter (NIF) proposed by Kitagawa (1987) and Kramer and Sorenson (1988), each density function of $P(\alpha_t|\alpha_{t-1})$, $P(y_t|\alpha_t)$, $P(\alpha_t|Y_{t-1})$ and $P(\alpha_t|Y_t)$ is approximated by numerical integration. The numerical integration filter (NIF) is discussed in Section 4.3. Moreover, practically, it might be appropriate that the nodes for numerical integration are constructed utilizing the extended Kalman filter (EKF) estimates. In the case where the distribution of the state-vector shifts over time, a range of the nodes possibly becomes unrealistic and therefore the estimated state-vectors are possibly biased as time t is large. The extended Kalman filter (EKF) is utilized to avoid this situation, which is applied to the Monte-Carlo experiments in Chapter 5 and the empirical studies in Chapter 6. Finally, note that the numerical integration filter (NIF) proposed in Kitagawa (1987) and Kramer and Sorenson (1988) has the disadvantages:

- Location of nodes has to be set by a researcher.
- We have to derive the densities $P(\alpha_t|\alpha_{t-1})$ and $P(y_t|\alpha_t)$ by hand.
- Computational time increases more than proportionally as the dimension of the state variable is high.

Next, utilizing the Bayesian density approximation (for example, see Geweke (1988, 1989a, 1989b)), i.e., the Monte-Carlo integration method with importance sampling, a recursive filtering algorithm of the weight functions, not the density functions, is introduced in Section 4.4, which is called the importance sampling filter (ISF). There, the weight functions are represented by the ratio of two density functions. One of the two densities is $P(\alpha_t|Y_{t-1})$ or $P(\alpha_t|Y_t)$, which is the conditional distribution of α_t given information Y_{t-1} or Y_t. Another density, called the importance density, is an appropriately chosen distribution function of α_t, denoted by $P_\alpha(\alpha_t)$. We generate the random numbers of α_t from $P_\alpha(\alpha_t)$, which is typically normal but not necessarily. According to the importance sampling procedure based on the weight functions, the filtering algorithm does not require numerical integration and we do not need to sort the nodes in order of size while the nodes have to be sorted in the algorithm of the numerical integration procedure. Therefore, the importance sampling filter (ISF) needs less computation than the numerical integration estimator, especially in the higher-dimensional cases of the state-vector. The problems of the importance sampling filter (ISF) developed by Tanizaki (1991, 1993), Tanizaki and Mariano (1994) and Mariano and Tanizaki (1995) are:

- The importance density $P_\alpha(\alpha_t)$ has to be appropriately chosen by a researcher.
- We need to derive the densities $P(\alpha_t|\alpha_{t-1})$ and $P(y_t|\alpha_t)$ by hand, which is similar to the numerical integration filter (NIF).

The density-based Monte-Carlo filter (DMF) is shown in Section 4.5, which improves the problems of the numerical integration filter (NIF) and the importance sampling filter (ISF). That is, we have to compute $P(y_t|\alpha_t)$ by hand but not $P(\alpha_t|\alpha_{t-1})$, and we do not need *ad hoc* assumptions such as choice of the nodes for the numerical integration procedure, choice of the importance density for the Monte-Carlo integration with importance sampling and choice of the density function of nuisance parameters for the Monte-Carlo integration with Gibbs sampling. The measurement equation is utilized for deriving the density $P(y_t|\alpha_t)$ while the transition equation is used to generate the random numbers of the state variable α_t. In order to implement numerical integration, the nodes have to be chosen by a researcher. Also, Monte-Carlo integration with importance sampling requires the importance density, which is appropriately taken by a researcher. Thus, the features of the density-based Monte-Carlo filter (DMF) are that the random numbers of the state-variable α_t (i.e., $\alpha_{i,t}$, $i = 1, \cdots, n$) are generated from the transition equation (3.2) for all time t, and that the filtering algorithm requires the functional form of the density function of y_t given α_t, i.e., $P(y_t|\alpha_t)$, which has to be computed by hand. The density-based Monte-Carlo filter (DMF) does not yield the recursive algorithm, which is different from the other nonlinear filters introduced in this book. There, we do not need *ad hoc* assumptions such as choice of the nodes for the numerical integration filter (NIF) or that of the importance density for the importance sampling filter (ISF). However, the density-based Monte-Carlo filter (DMF) proposed by Tanizaki and Mariano (1995a, 1995b) also has some problems:

– If the range of the state variable is restricted, the density-based Monte-Carlo filter (DMF) performs quite better. Otherwise, the density-based Monte-Carlo filter (DMF) does not work.

Density approximation by numerical integration (i.e., NIF) includes only the computation errors, while density approximation by Monte-Carlo integration (i.e., ISF and DMF) includes not only the computation errors but also the simulation errors. See Appendix A5.3 for the computation errors and the simulation errors. In any case, for all time t, we need to perform the adjustment such that each integration included in the filtering algorithm is equal to one, i.e.,

$$\int P(\alpha_t|Y_s)\mathrm{d}\alpha_t = 1,$$

for $s = t-1, t$, which is represented as follows:

- For the numerical integration filter (NIF),

$$\sum_{i=1}^{n} P(\alpha_{i,t}|Y_s)(\alpha_{i,t} - \alpha_{i-1,t}) = 1,$$

  for $s = t-1, t$.
- For the importance sampling filter (ISF),

$$\frac{1}{n}\sum_{i=1}^{n} \omega_{i,t|s} = 1$$

  for $s = t-1, t$,
- For the density-based Monte-Carlo filter (DMF),

$$\sum_{i=1}^{n} \omega_{i,t} = 1.$$

Unless the above adjustment is restricted for each time t, precision of the filtering estimates becomes poor as time t increases.

As an alternative simple procedure, Tanizaki and Mariano (1995b) proposed the rejection sampling filter (RSF), where the filtering estimates are evaluated by random draws only. For the random number generation, rejection sampling may be adopted, which is a method of the random number generation from any distribution. See Appendices A4.6 and A5.4 for rejection sampling. The rejection sampling filter does not need to derive the functional form of $P(y_t|\alpha_t)$. (however, as shown in Section 4.6, it is used to obtain the acceptance probability $\omega_1(\alpha_t; y_t)$.) Thus, an attempt is made in Section 4.6 to generate the random draws directly from the filtering density, without approximating any density function. That is, we consider directly generating the random draws of α_t from the filtering density $P(\alpha_t|Y_t)$, where we do not need choice of the nodes, that of the importance density and the Monte-Carlo approximation of the filtering density. As a general solution to this procedure, rejection sampling is utilized, which is called the rejection sampling filter (RSF) in this book (even though rejection sampling is not utilized depending on the state-space model; see Examples 1 and 2 in Section 4.6). As mentioned above, rejection sampling is one of the random number generation procedures and it enables us to generate random draws from any distribution function in a Bayesian framework. In a filtering framework, rejection sampling is applied in the following way. The transition equation (3.2) is used to generate random draws of the state-variable while the measurement equation (3.1) is utilized to derive the acceptance probability $\omega_1(\alpha_t; y_t)$. The rejection sampling filter proposed in Tanizaki (1995b), Tanizaki and Mariano (1995b) and Mariano and Tanizaki (1996) has the following disadvantages:

- When the acceptance probability $\omega_1(\alpha_t; y_t)$ is small enough, we cannot generate random draws. Therefore, sometimes, it does not work in practice.

Certainly, for the Gaussian sum filter (GSF), the numerical integration filter (NIF), the importance sampling filter (ISF), the density-based Monte-Carlo filter (DMF) and the rejection sampling filter (RSF), precision of the filtering estimates might be improved to some extent as number of the nodes or the random draws increases (see Chapter 5). However, an increase in the nodes or the random draws greatly increases the computational burden, and an implementation can therefore be sometimes inapplicable in practice. Clearly, approximating the underlying densities is better than linearizing the nonlinear functions because approximation of the nonlinear functions leads to the biased estimates but approximating the densities using the nodes or the random numbers is asymptotically consistent and efficient. In order for the nonlinear filters based on the density approximation to obtain more precise filtering estimates, however, we have an increase in computational time because of an increase in the number of nodes or random draws. There is a trade-off between the Taylor series expansion approaches (Chapter 3) and the density-based approximation approaches (Chapter 4).

In Chapter 5, for all of the nonlinear filters introduced in this book, the filtering estimates are compared with the true state-variables which are simulated by the Monte-Carlo experiments in Sections 5.2. According to the bias (BIAS) and the root mean square error (RMSE) criteria between the artificially simulated state-variables and the filtering estimates, we have obtained the following results:

- The numerical integration filter (NIF), the importance sampling filter (ISF) and the rejection sampling filter (RSF) are better than any other estimators while the extended Kalman filter (EKF) and the Gaussian sum filter (GSF) are biased.

- The filtering estimates by the numerical integration filter (NIF), the importance sampling filter (ISF) and the rejection sampling filter (RSF) are close to the true state-variables which are artificially simulated, compared with the extended Kalman filter (EKF), the second-order nonlinear filter (SNF), the Monte-Carlo simulation filter (MSF), the single-stage iteration filter (SIF), the Gaussian sum filter (GSF).
- The rejection sampling filter (RSF) has a superior performance over the other nonlinear filters. In addition, RSF has the features of simplicity of computer programming and no *ad hoc* assumptions.

The Monte-Carlo experiments have shown the result that the rejection sampling filter (RSF) is very close to the numerical integration filter (NIF) and the importance sampling filter (ISF).

As an application, in Chapter 6, we take an example of estimating permanent consumption, taking into account nonlinearity of the Euler equation, variable interest rate and transitory consumption. Hall (1978) proposed a new approach to the permanent income hypothesis. The essence is as follows. If consumption is based on the permanent income hypothesis, by which kind of model can a sequence of consumption be written? According to the approach proposed by Hall (1978), the permanent income hypothesis can be tested without estimating the consumption function as conventionally done.

Per capita permanent consumption are estimated in Chapter 6, using the nonlinear filtering techniques. Total consumption (i.e., c_t) is represented by a sum of permanent consumption (i.e., c_t^p), transitory consumption (i.e., c_t^T) and the other factor independent of the permanent income hypothesis (i.e., c_t^*), which is given by the identity equation, i.e., $c_t = c_t^p + c_t^* + c_t^T$. We consider that permanent consumption is based on the Euler equation derived by the utility maximization problem of the representative agent, i.e., $\mathrm{E}_{t-1}\left(\dfrac{\beta R_{t-1} u'(c_t^p)}{u'(c_{t-1}^p)}\right) = 1$, where $u'(\cdot)$ denotes the first derivative of the utility function of the representative agent. Transitory consumption is regarded as a random shock, i.e., $c_t^T = \epsilon_t$. Hence, the measurement equation is represented as the identity equation in this case, while the transition equation is given by the Euler equation.

In this example, the unknown parameters β, α, γ_1, γ_2, σ_ϵ and σ_η have to be estimated, and at the same time we need to compute the filtering estimates, performing the nonlinear filters. β denotes the discount factor and α represents a measure of the relative risk aversion. The approximated likelihood functions given by equations (2.30), (4.17), (4.26), (4.36), (4.42) and (4.44) are maximized with respect to the unknown parameters for estimation. The bias of the parameter estimates arises by using the approximated likelihood function such as equation (2.30). Note that the likelihood function to be maximized in EKF, SNF, MSF and SIF is equation (2.30), where

$F_{t|t-1}$ and $y_{t|t-1}$ in equation (2.30) are approximately computed in each filtering algorithm (see Chapter 3). For the filtering estimation methods which correspond to the likelihood functions (4.26), (4.36), (4.42) and (4.44), the estimates of unknown parameters might be asymptotically unbiased, because the nonlinear measurement and transition equations are not linearized in the likelihood functions (4.26), (4.36), (4.42) and (4.44).

Maximization of the likelihood functions are performed by a simple grid search. There, one of the unknown parameters β, α, γ, σ_ϵ and σ_η is changed by 0.1, 0.001 or 0.00001, given the other four parameters, to maximize the likelihood function. For each parameter, the same procedure is repeated until all the parameters are the same values as the last iteration, where the 0.001% convergence criterion is utilized.

It is known that the life cycle permanent income hypothesis holds only in the short run and that there would be a income factor or a trend factor in the long run. Therefore, we have assumed in Chapter 6 that consumption consists of three parts: permanent consumption (i.e., c_t^p), transitory consumption (i.e., $c_t^T \equiv \epsilon_t$) and the other factor such as income (i.e., $c_t^* = \gamma_1 \widehat{Y}_t + \gamma_2 Y_{t-1}$). It is clearly concluded from the value of the log-likelihood function that that we must take into account permanent consumption, transitory consumption and the income factor into the life cycle permanent income hypothesis. In this sense, the permanent income hypothesis does not hold in reality even if we take into account transitory consumption, nonlinearity of the Euler equation and the variable gross rate of return on savings, because consumption depends on income as well as lagged consumption.

Until Chapter 6, we deal with the filtering problem only. In Chapter 7, we consider both prediction (i.e., L-step ahead prediction) and smoothing problems. Prediction, filtering and smoothing imply the following estimation methods:

- Prediction is helpful to estimate the future state-variable at the present time.
- Filtering is used to estimate the present state-variable at the present time.
- Smoothing is an estimation of the past state-variable at the present time.

The above three estimators are differentiated by information set. The relationship among prediction, filtering and smoothing is as follows.

- The one-step ahead prediction estimate is utilized to obtain the filtering estimate.
- We need the filtering estimate to derive the smoothing estimate.

The prediction and smoothing algorithms are derived in the same fashion as the nonlinear filters discussed in Chapter 4. There, we have discussed the prediction and smoothing algorithms by the numerical integration procedure, the importance sampling procedure, the density-based Monte-Carlo method and the rejection sampling approach. For prediction, generally we have considered L-step ahead prediction algorithms. We have the three kinds of smoothing, i.e., the fixed-point smoothing, the fixed-lag smoothing and the fixed-interval smoothing. Of the three, we have considered the fixed-interval smoothing, which might be the most useful tool in a field of economics. Thus, the density-based nonlinear filters are easily extended to the prediction and smoothing algorithms in Chapter 7.

References

Alspach, D.L. and Sorenson, H.W., 1972, " Nonlinear Bayesian Estimation Using Gaussian Sum Approximations, " *IEEE Transactions on Automatic Control*, Vol.AC-17, No.4, pp.439 – 448.

Anderson, B.D.O. and Moore, J.B., 1979, *Optimal Filtering*, Prentice-Hall, New York.

Aoki, M., 1987, *State Space Modeling of Time Series*, Springer-Verlag.

Belsley, D.A., 1973, " On the determination of Systematic Parameter Variation in the Linear Regression Model, " *Annals of Economic and Social Measurement*, Vol.2, pp.487 – 494.

Belsley, D.A. and Kuh, E., 1973, " Time-Varying Parameter Structures: An Overview, " *Annals of Economic and Social Measurement*, Vol.2, No.4, pp.375 – 379.

Boswell, M.T., Gore, S.D., Patil, G.P. and Taillie, C., 1993, " The Art of Computer Generation of Random Variables, " in *Handbook of Statistics*, Vol.9, edited by Rao, C.R., pp.661 – 721, North-Holland.

Branson, W.H., 1979, *Macroeconomic Theory and Policy* (second edition), Haper & Row, Publishers, Inc.

Brockwell, P.A. and Davis, R.A., 1987, *Time Series Theory and Models*, Springer-Verlag.

Brown, B.W. and Mariano, R.S., 1984, " Residual-based Procedures for Prediction and Estimation in a Nonlinear Simultaneous System, " *Econometrica*, Vol.52, No.2, pp.321 – 343.

Brown, B.W. and Mariano, R.S., 1989, " Measures of Deterministic Prediction Bias in Nonlinear Models, " *International Economic Review*, Vol.30, No.3, pp.667 – 684.

Burmeister, E. and Wall, K.D., 1982, " Kalman Filtering Estimation of Unobserved Rational Expectations with an Application to the German Hyperinflation, " *Journal of Econometrics*, Vol.20, pp.255 – 284.

Burridge, P. and Wallis, K.F., 1988, " Prediction Theory for Autoregressive Moving Average Processes, " *Econometric Reviews*, Vol.7, No.1, pp.65 – 95.

Campbell, J.Y., 1987, " Does Saving Anticipate Declining Labor Income? An Alternative Test of the Permanent Income, " *Econometrica*, Vol.55, No.6, pp.1249 – 1273.

Campbell, J.Y. and Deaton, A., 1989, " Why is Consumption So Smooth?, " *Review of Economic Studies*, Vol.56, pp.357 – 374.

Campbell, J.Y. and Mankiw, N.G., 1987, " Are Output Fluctuations Transitory?, " *Quarterly Journal of Economics*, Vol.102, pp.857 – 880.

Campbell, J.Y. and Mankiw, N.G., 1990, " Permanent Income, Current Income, and Consumption, " *Journal of Business and Economics*, Vol.8, No.3, pp.265 – 279.

Carlin, B.P. and Polson, N.G., 1991, " Inference for Nonconjugate Bayesian Models Using the Gibbs Sampler, " *Canadian Journal of Statistics*, Vol.19, pp.399–405.

Carlin, B.P., Polson, N.G. and Stoffer, D.S., 1992, " A Monte Carlo Approach to Nonnormal and Nonlinear State Space Modeling, " *Journal of the American Statistical Association*, Vol.87, pp.493 – 500.

Chow, G.C., 1983, *Econometrics*, McGraw-Hill Book Company.

Conrad, W. and Corrado, C., 1979, " Application of the Kalman Filter to Revisions in Monthly Retail Sales Estimates, " *Journal of Economic Dynamic and Control*, Vol.1, pp.177 – 198.

Cooley, T.F., 1977, " Generalized Least Squares Applied to Time Varying Parameter Models: A Comment, " *Annals of Economic and Social Measurement*, Vol.6, No.3, pp.313 – 314.

Cooley, T.F. and Prescott, E.C., 1976, " Estimation in the presence of stochastic parameter variation, " *Econometrica*, Vol.44, pp.167 – 183.

Cooley, T.F., Rosenberg, B. and Wall, K.D., 1977, " A Note on Optimal Smoothing for Time Varying Coefficient Problems, " *Annals of Economic and Social Measurement*, Vol.6, No.4, pp.453 – 456.

Cooper, J.P., 1973, " Time-Varying Regression Coefficients: A Mixed Estimation Approach and Operational Limitations of the General Markov Structure, " *Annals of Economic and Social Measurement*, Vol.2, No.4, pp.525 – 530.

Deaton, A., 1987, " Life Cycle Models of Consumption: Is the Evidence Consistent with the Theory?, " in *Advances in Econometrics*, Fifth World Congress, Vol.II, edited by Bewley, T.F., pp.121 – 146, Cambridge University Press.

Dempster, A.P., Laird, N.M. and Rubin, D.B., 1977, " Maximum Likelihood from Incomplete Data via the EM Algorithm, " *Journal of the Royal Statistical Society*, Ser.B, Vol.39, pp.1 – 38 (with discussion) .

Diderrich, G.T., 1985, " The Kalman Filter from the Perspective of Goldberger-Theil Estimators, " *The American Statistician*, Vol.39, pp.193 – 198.

Diebold, F.X. and Nerlove, M., 1989, " Unit Roots in Economic Time Series: A Selective Survey, " in *Advances in Econometrics*, Vol.8, pp.3 – 69, JAI Press.

Diebold, F.X. and Rudebusch, G.D., 1991, " Is Consumption Too Smooth? Long Memory and the Deaton Paradox, " *The Review of Economics and Statistics*, pp.1 – 9.

Dziechciarz, J., 1989, " Changing and Random Coefficient Models. A Survey, " in *Statistical Analysis and Forecasting of Economic Structural Change*, edited by Hackl, P., Springer-Verlag.

Engle, R.F. and Watson, M.W., 1987, " The Kalman Filter: Applications to Forecasting and Rational expectations Models, " in *Advances in Econometrics, Fifth World Congress, Vol.I*, Cambridge University Press.

Flavin, M.A., 1981, " The Adjustment of Consumption to Changing Expectations about Future Income, " *Journal of Political Economy*, Vol.89, No.5, pp.974 – 1009.

Friedman, M., 1957, *A Theory of the Consumption Function*, Princeton University Press.

Garbade, K., 1976, " Two Methods of Examining the Stability of Regression Coefficients, " *Journal of the American Statistical Association*, Vol.72, pp.54 – 63.

Gardner, G., Harvey, A.C. and Phillips, G.D.A., (1980), " An Algorithm for Maximum Likelihood Estimation Autoregressive-Moving Average Models by means of Kalman Filtering, " *Applied Statistics*, Vol.29, No.3, pp.311 – 322.

Gelb, A., 1974, *Applied Optimal Estimation*, MIT Press.

Gelfand, A.E., Hills, S.E., Racine-Poon, H.A., and Smith, A.F.M., 1990, " Illustration of Bayesian Inference in Normal Data Models Using Gibbs Sampling, " *Journal of the American Statistical Association*, Vol.85, No.412, pp.972 – 985.

Gelfand, A.E. and Smith, A.F.M., 1990, " Sampling-Based Approaches to Calculating Marginal Densities, " *Journal of the American Statistical Association*, Vol.85, No.410, pp.398 – 409.

Geman, S. and Geman, D., 1984, " Stochastic Relaxation, Gibbs Distributions, and the Bayesian Restoration of Images, " *IEEE Transactions on Pattern Analysis and Machine Intelligence*, Vol.Pami-6, No.6, pp.721 – 741.

Geweke, J., 1988, " Antithentic Acceleration of Monte Carlo Integration in Bayesian Inference, " *Journal of Econometrics*, Vol.38, pp.73 – 90.

Geweke, J., 1989a, " Modeling with Normal Polynomial Expansions, " in *Economic Complexity: Chaos, Sunspots, Bubbles and Nonlinearity*, edited by Barnett, W.A., Geweke, J. and Shell, K., Cambridge University Press.

Geweke, J., 1989b, " Bayesian Inference in Econometric Models Using Monte Carlo Integration, " *Econometrica*, Vol.57, pp.1317 – 1339.

Hall, R.E., 1978, " Stochastic Implications of the Life Cycle-Permanent Income Hypothesis: Theory and Evidence, " *Journal of Political Economy*, Vol.86, No.6, pp.971 – 987.

Hall, R.E. 1990, *The Rational Consumer*, The MIT Press.

Hall, R.E. and Mishkin, F.S., 1982, " The Sensitivity of Consumption to Transitory Income: Estimates from Panel Data on Households, " *Econometrica*, Vol.50, No.2, pp.461 – 481.

Hamori, S., 1992a, " Test of C-CAPM for Japan: 1980-1988, " *Economics Letters*, Vol.38, No.1, pp.67 – 72.

Hamori, S., 1992b, " On the Structural Stability of Preference Parameters Obtained from Japanese Financial Market Data, " *Economics Letters*, Vol.40, No.4, pp.459 – 464.

Hamori, S., 1993, " Test of the International Equity Integration of Japan, " *Economics Letters*, Vol.42, No.1, pp.41 – 76.

Hamori, S. and Kitasaka, S., 1996, " A Numerical Analysis of the Monetary Aspects of the Japanese Economy: The Cash-in-Advance Approach, " forthcoming in *Applied Financial Economics*.

Hamori, S., Kitasaka, S. and Tanizaki, H., 1996, " On a Test for Structural Stability of Euler Conditions Parameters Estimated via Generalized Method of Moments Estimators: Small Sample Properties, " *Econometric Reviews*, Vol.15, No.1.

Hannan, E.J. and Deistler, M., 1988, *The Statistical Theory of Linear System*, John Wiley & Sons.

Harvey, A.C., 1981, *Time Series Models*, Philip Allen Publishers Limited, Oxford.

Harvey, A.C., 1987, " Applications of the Kalman Filter in Econometrics, " in *Advances in Econometrics, Fifth World Congress, Vol.I*, Cambridge University Press.

Harvey, A.C., 1989, *Forecasting, Structural Time Series Models and the Kalman Filter*, Cambridge University Press.

Hayashi, F., 1985a, " The Effects of Liquidity Constraints on Consumption: A Cross-sectional Study, " *Quarterly Journal of Economics*, Vol.100, pp.183 – 206.

Hayashi, F., 1985b, " The Permanent Income Hypothesis and Consumption Durability: Analysis Based on Japanese Panel Data, " *Quarterly Journal of Economics*, Vol.100, pp.1083 – 1113.

Howrey, E.P., 1978, " The Use of Preliminary Data in Econometric Forecasting, " *The Review of Economics and Statistics*, Vol.60, pp.193 – 200.

Howrey, E.P., 1984, " Data Revision, Reconstruction, and Prediction: An Application to Inventory Investment, " *The Review of Economics and Statistics*, Vol.66, pp.386 – 393.

Jazwinski, A.H., 1970, *Stochastic Processes and Filtering Theory*, Academic Press, New York.

Johnston, J., 1984, *Econometric Methods* (third edition), McGraw-Hill Book Company.

Kalman, R.E., 1960, " A New Approach to Linear Filtering and Prediction Problems, " *Journal of Basic Engineering, Transactions ASME*, Ser.D, Vol.82, pp.35 – 45.

Kalman, R.E. and Bucy, R.S., 1961, " New Results in Linear Filtering and Prediction Theory, " *Journal of Basic Engineering, Transactions ASME*, Ser.D, Vol.83, pp.95 – 108.

Kitagawa, G., 1987, " Non-Gaussian State-Space Modeling of Nonstationary Time Series, " *Journal of the American Statistical Association*, Vol.82, pp.1032 – 1063 (with discussion).

Kirchen, A., 1988, *Schäzung zeitveränderlicher Strukturparameter in ökonometrischen Prognosemodellen*, Frankfurt/Main: Athenäum.

Kocherlakota, N.R., 1990, " On the 'Discount' Factor in Growth Economies, " *Journal of Monetary Economics*,Vol.25, pp.43 – 48.

Kokoska, S. and Nevison, C., 1989, *Statistical Tables and Formulae* (Springer Texts in Statistics), Springer-Verlag.

Kramer, S.C. and Sorenson, H.W., 1988, " Recursive Bayesian Estimation Using Piece-wise Constant Approximations, " *Automatica*, Vol.24, No.6, pp.789 – 801.

Knuth, D.E., 1981, *The Art of Computer Programming*, Vol.2 (Seminumerical Algorithms, 2nd ed.), Addison-Wesley.

Laumas, G.S. and Mehra, Y.P., 1976, " The Stability of the Demand for Money Function: The Evidence from Quarterly Data, " *The Review of Economics and Statistics*, Vol.58, pp.464 – 468.

Mankiw, N.G., 1981, " The Permanent Income Hypothesis and The Real Interest Rate, " *Economics Letters*, Vol.7, pp.307 – 311.

Mankiw, N.G. and Shapiro, M.D., 1985, " Trends, Random Walks, and Tests of the Permanent Income Hypothesis, " *Journal of Monetary Economics*, Vol.16, pp.165 – 174.

Mariano, R.S. and Brown, B.W., 1983, " Asymptotic Behavior of Predictors in a Nonlinear Simultaneous System, " *International Economic Review*, Vol.24, No.3, pp.523 – 536.

Mariano, R.S. and Brown, B.W., 1989, " Stochastic Simulation, Prediction and Validation of Nonlinear Models, " in *Economics in Theory and Practice: An Eclectic Approach*, edited by Klein, L.R. and Marquez, J., pp.17 – 36, Kluwer Academic Publishers.

Mariano, R.S. and Tanizaki, H., 1995, " Prediction of Final Data with Use of Preliminary and/or Revised Data, " *Journal of Forecasting*, Vol.14, No.4, pp.351 – 380.

Mariano, R.S. and Tanizaki, H., 1996, " Simulation-Based Inference in Nonlinear State-Space Models: Application to Testing the Permanent Income Hypothesis, " forthcoming in *Simulation-Based Inference in Econometrics: Methods and Applications*, edited by Mariano, R.S., Weeks, M. and Schuermann, T., Cambridge University Press.

McFadden, D., 1989, " A Method of Simulated Moments for Estimation of Discrete Response Models without Numerical Integration, " *Econometrica*, Vol.57, pp.995 – 1026.

McNelis, P.D. and Neftci, S.N., 1983, " Policy-dependent Parameters in the Presence of Optimal Learning: An Application of Kalman Filtering to the Fair and Sargent Supply-side Equations, " *The Review of Economics and Statistics*, Vol.65, pp.296 – 306.

Meinhold, R.J. and Singpurwalla, N.D., 1989, " Robustification of Kalman Filter Models, " *Journal of the American Statistical Association*, Vol.84, pp.479 – 486.

Muellbauer, J., 1983, " Surprises in the Consumption Function, " *Economic Journal*, Supplement, pp.34 – 50.

Nicholls, D.F. and Pagan, A.R., 1985, " Varying Coefficient Regression, " in *Handbook of Statistics*, Vol.5, edited by Hannan, E.J., Krishnaiah, P.R. and Rao, M.M., Elsevier Science Publishers B.V., pp.413 – 449.

O'Hagan, A., 1994, *Kendall's Advanced Theory of Statistics*, Vol.2B (Bayesian Inference), Edward Arnold.

Pagan, A.R., 1975, " A Note on the Extraction of Components from Time Series, " *Econometrica*, Vol.43, pp.163 – 168.

Pagan, A.R., 1980, " Some Identification and Estimation Results for Regression Models with Stochastically Varying, " *Journal of Econometrics*, Vol.13, pp.343 – 363.

Rosenberg, B., 1977, " Estimation Error Covariance in Regression with Sequentially Varying Parameters, " *Annals of Economic and Social Measurement*, Vol.6, No.4, pp.457 – 462.

Rund, P.A., 1991, " Extensions of Estimation Methods Using the EM Algorithm, " *Journal of Econometrics*, Vol.49, pp.305 – 341.

Sant, D.T., 1977, " Generalized Least Squares Applied to Time Varying Parameter Models, " *Annals of Economic and Measurement*, Vol.6, No.3, pp.301 – 311.

Sarris, A.H., 1973, " A Bayesian Approach to Estimation of Time Varying Regression Coefficients, " *Annals of Economic and Social Measurement*, Vol.2, No.4, pp.501 – 523.

Shao, J., 1989, " Monte Carlo Approximations in Bayesian Decision Theory, " *Journal of the American Statistical Association*, Vol.84, No.407, pp.727 – 732.

Shumway, R.H. and Stoffer, D.S., 1982, " An Approach to Time Series Smoothing and Forecasting Using the EM Algorithm, " *Journal of Time Series Analysis*, Vol.3, PP.253 – 264.

Sorenson, H.W. and Alspach, D.L., 1971, " Recursive Bayesian Estimation Using Gaussian Sums, " *Automatica*, Vol.7, pp.465 – 479.

Tanizaki, H., 1987, " Theory and Application of Kalman Filter Model, " Master Thesis, Kobe University (in Japanese).

Tanizaki, H., 1989, " The Kalman Filter Model under the Assumption of the First-Order Autoregressive Process in the Disturbance Terms, " *Economics Letters*, Vol.31, No.2, pp.145 - 149.

Tanizaki, H., 1991, " Nonlinear Filters: Estimation and Applications, " Ph.D. Dissertation, University of Pennsylvania.

Tanizaki, H., 1993a, *Nonlinear Filters: Estimation and Applications* (Lecture Notes in Economics and Mathematical Systems, No.400), Springer-Verlag.

Tanizaki, H., 1993b, " Kalman Filter Model with Qualitative Dependent Variable, " *The Review of Economics and Statistics*, Vol.75, No.4, pp.747 - 752.

Tanizaki, H., 1993c, *Application of State-Space Model to Economics: Estimation of U.S. and Japanese Macro-Econometric Model by Time-Varying Parameter Model* (Research Monograph in Department of Economics, Kobe-Gakuin University, No.9), Nippon-Hyoron-Sya (in Japanese).

Tanizaki, H., 1995a, " Asymptotically Exact Confidence Intervals of CUSUM and CUSUMSQ Tests: A Numerical Derivation Using Simulation Technique, " *Communications in Statistics, Simulation and Computation*, Vol.24, No.4, pp.1019 - 1036.

Tanizaki, H., 1995b, " Non-Linear and Non-Normal Filter Based on Monte-Carlo Technique, " Unpublished Manuscript.

Tanizaki, H. and Mariano, R.S., 1994, " Prediction, Filtering and Smoothing in Nonlinear and Nonnormal Cases Using Monte-Carlo Integration, " *Journal of Applied Econometrics*, Vol.9, No.2, pp.163 - 179 (in *Econometric Inference Using Simulation Techniques*, Chap.12, edited by van DijK, H.K., Manfort, A. and Brown, B.W., John Wiley & Sons, pp.245 - 261, 1995).

Tanizaki, H. and Mariano, R.S., 1995a, " A Comparison of Three Non-Linear and Non-Normal Filters, " Unpublished Manuscript.

Tanizaki, H. and Mariano, R.S., 1995b, " Nonlinear and Nonnormal State-Space Modeling with Monte-Carlo Stochastic Simulations, " forthcoming in *Journal of Econometrics* (presented at The Seventh World Congress of the Econometric Society, Keio University, 1995).

Tanizaki, H. and Mariano, R.S., 1996, " Nonlinear Filters Based on Taylor Series Expansions, " forthcoming in *Communications in Statistics, Theory and Methods*, Vol.25, No.6.

Tanner, M.A. and Wong, W.H., 1987, " The Calculation of Posterior Distributions by Data Augmentation, " *Journal of the American Statistical Association*, Vol.82, No.398, pp.528 - 550 (with discussion).

Watanabe, N., 1985, " Notes on the Kalman Filter with Estimated Parameters, " *Journal of Time Series*, Vol.6, No.4, pp.269 - 278.

Watson, M.W. and Engle, R.F., 1983, " Alternative Algorithms for the Estimation of Dynamic Factor, MIMIC and Varying Coefficient Regression Models, " *Journal of Econometrics*, Vol.23, pp.385 - 400.

West, K.D., 1988, " The Insensitivity of Consumption to News about Income, " *Journal of Monetary Economics*, Vol.21, pp.17 – 33.

Wishner, R.P., Tabaczynski, J.A. and Athans, M., 1969, " A Comparison of Three Non-Linear Filters, " *Automatica*, Vol.5, pp.487 – 496.

Zeger, S.L. and Karim, M.R., 1991, " Generalized Linear Models with Random Effects: A Gibbs Sampling Approach, " *Journal of the American Statistical Association*, Vol.86, No.413, pp.79 – 86.

Index

Springer-Verlag and the Environment

We at Springer-Verlag firmly believe that an international science publisher has a special obligation to the environment, and our corporate policies consistently reflect this conviction.

We also expect our business partners – paper mills, printers, packaging manufacturers, etc. – to commit themselves to using environmentally friendly materials and production processes.

The paper in this book is made from low- or no-chlorine pulp and is acid free, in conformance with international standards for paper permanency.

Printing: Weihert-Druck GmbH, Darmstadt
Binding: Buchbinderei Schäffer, Grünstadt